Proceedings in Life Sciences

Compartments in Algal Cells and Their Interaction

Edited by

W. Wiessner D. G. Robinson R. C. Starr

With 141 Figures

Springer-Verlag
Berlin Heidelberg New York Tokyo 1984

Prof. Dr. W. WIESSNER, Pflanzenphysiologisches Institut der
Universität Göttingen, Abt. für Experimentelle Physiologie, Untere Karspüle 2,
3400 Göttingen, FRG

Prof. D. G. ROBINSON, Pflanzenphysiologisches Institut der
Universität Göttingen, Abt. Cytologie

Dr. R. C. STARR, Ashbel Smith Professor of Botany,
The University of Texas at Austin, Austin, TX 78712, USA

ISBN-13:978-3-642-69688-6 e-ISBN-13:978-3-642-69686-2
DOI: 10.1007/978-3-642-69686-2

Library of Congress Cataloging in Publication Data. Main entry under title: Compartments in algal cells and their interaction. (Proceedings in life sciences) 1. Allgae-Cytology-Addresses, essays, lectures. 2. Cell compartmentation - Addresses, essays, lectures. 3. Cell interaction - Addresses, essays, lectures. I. Wiessner, Wolfgang, 1933- . II. Robinson, D. (David), 1947- . III. Starr, R. C. (Richard C.), 1924- . IV. Series. QK565.C66 1984 589.3'87 84-1407

2131/3130-543210

Preface

The successful functioning of eukaryotic cells is dependent upon a highly organized intracellular cooperation and interaction between the numerous compartments which are present in them. In recent years unicellular algae have become increasingly recognized as favoured objects for studies on structural, biochemical and developmental aspects of compartmentation. Prompted by these facts, Professor A. Pirson, former Directior of the Plant Physiology Institute in Göttingen, suggested organizing a symposium entitled *Compartments in Algal Cells and Their Interaction.*

This was then organized under the auspices of the Akademie der Wissenschaften in Göttingen with the financial support of the Stiftung Volkswagenwerk. The meeting took place in Göttingen in September of 1983 and was attended by a number of internationally renowned colleagues. The restricted size of the meeting, together with the convivial atmosphere, led to an optimal exchange of views between the participants. This book presents the proceedings of this meeting and contains all the papers presented. We feel, however, that it contains information of use to a much larger circle of scientists than to phycologists alone.

We should like to take this opportunity to express our thanks to all those involved in the convention, support and in day-to-day operation of the meeting. We also thank the contributors for their quick and effective cooperation and, last but not least, the editorial staff of Springer Verlag, who enabled this book to appear only 9 months after the meeting took place, thus ensuring its up-to-date nature.

Göttingen

Robinson, Starr, and Wiessner

April 1984

Contents

Contributors

You will find the addresses at the beginning of the respective contributions

The Cytological Viewpoint of Functional Compartmentation

E. SCHNEPF[1]

Introduction

Compartmentation means the subdivision of a cell or of a tissue in different reaction spaces with restricted or controlled exchange of substances (Ziegler et al. 1981). From a cytological point of view a compartment is delimited against another compartment by a biomembrane; the cytoplasm is demarcated from the environment by the plasmalemma. Correspondingly each endomembrane separates two inner compartments. Simultaneously, it connects them metabolically, allowing certain molecules and ions to pass. The principle of separation and connection allows many fundamental metabolic processes, e.g., ATP synthesis in the respiratory chain and in photophosphorylation.

In the very beginning of ultrastructural research the various compartments were believed to be "plasmatic" as it is manifested by the terms "Golgi plasm", "reticulum plasm" etc. (see, e.g., Ruska 1960). Nearly 20 years ago, after investigations in *Geosiphon pyriforme,* I suggested a new conception of cell compartmentation (Schnepf 1964). I would like to re-present these ideas – some of them seem to be rather trivial now – in the form of 4 theses, and to give four additional statements which follow from these theses (cf., e.g., Sitte 1980). Then we have to examine critically: what has stood the test of 20 years of further research and what has to be rejected or, at least, modified? Finally, I shall mention briefly two special examples of functional compartmentation.

Rules of Cell Compartmentation

Theses:

a) A compartment is surrounded and completely enclosed by a biomembrane. This thesis is another expression of the cytological definition of a compartment. It implies that karyoplasm and cytoplasm form a common "protoplasmatic" or, briefly, "plasmatic" compartment, the nucleocytoplasmatic matrix, because they are in open connection through the pores of the nuclear envelope.

1 Zellenlehre, Universität Heidelberg, Im Neuenheimer Feld 230, D-6900 Heidelberg 1, FRG

Compartments in Algal Cells and Their Interaction
ed. by W. Wiessner, D. Robinson, and R.C. Starr

b) Each biomembrane separates a plasmatic compartment either from the nonplasmatic environment of the cell or from a likewise nonplasmatic inner compartment. A "plasmatic" compartment has the following characteristics (cf. Sitte 1980). (1) It is the site of nucleic acid and protein synthesis; DNA and ribosomes are situated here. (2) It is the site of ATP production and ATP consumption. Though most ATP is synthesized at a membrane, it is the plasmatic face of the membrane where the membrane-bound ATPase has its reactive center. (3) It is the compartment in which the elements of the cytoskeleton, microtubules and microfilaments, are situated. (4) It is the site where α-glucans and proteins generally occur. Nonplasmatic compartments rather contain β-glucans and glycoproteids. Usually, the C-terminus of integral membrane proteins is situated at the plasmatic, the N-terminus at the nonplasmatic side of a membrane. Membrane-bound proton pumps transport protons from plasmatic into nonplasmatic compartments. Phosphorolysis and hydrolysis are characteristic catabolic processes of plasmatic and nonplasmatic compartments respectively. Correspondingly, the eukaryotic cell has two or three plasmatic compartments: The nucleocytoplasmatic matrix, the mitochondrial matrix and the matrix of the plastids. All other compartments are nonplasmatic ones.

c) Compartments of the same kind can fuse, plasmatic and nonplasmatic compartments do not fuse. The main nonplasmatic compartments of eukaryotic cells are represented by the ER, the Golgi apparatus, vacuoles, lysosomes, microbodies etc. They evolve from or fuse with each other. Generally they are not connected permanently but by intermediate vesicles which bud off from one compartment and fuse with another one. These compartments may be, indirectly rather than directly, connected with the nonplasmatic environment of the cell, as seen during exocytosis and endocytosis. The plasmatic compartments of a cell do not merge, but the nucleocytoplasmatic matrices of different cells do so, e.g., during fusion of gametes. Even such different plants as *Helianthus* and *Vicia* can form open plasmodesmata between the protoplasts of the two partners in grafts (Kollmann, personal communication).

d) Membranes (and, likewise, compartments) do not arise de novo but form from existing membranes (and compartments, respectively). Omnis membrana e membrana. The classical illustration of this thesis is the membrane flow system. Membranes, hence, represent elements of inheritance without nucleic acids.

Implications of the Theses

a) Membranes have a sidedness. Delimiting and connecting at any given time a plasmatic from (and with) a nonplasmatic compartment, a membrane is functionally and structurally asymmetrical, a rather trivial statement now. The terminology of the freeze-fracture technique: p- (plasmatic) face and e- (external = nonplasmatic) face for the membrane halves which result from the splitting through the membranes has its roots in these ideas.

b) Membranes do not branch. If a membrane would branch it could not everywhere separate a plasmatic from a nonplasmatic compartment.

c) Membranes do not have free ends. Each compartment is closed; otherwise it would not be a compartment. Because membranes and compartments do not arise de novo, open ends should not exist in vivo.

d) When a cell (or an electron micrograph of a cell) is pierced (or crossed by a line), an even number of membranes is passed. This is a consequence of the thesis, that a membrane always separates a plasmatic from a nonplasmatic compartment. It has some practical meaning, facilitating the interpretation of electron micrographs.

Critical Examination of the Theses

It is not necessary to emphasize here that these theses are valid only for normal cells in a good state. During degeneration processes, myelin figures may arise. The membranes of these myelin figures do not represent biomembranes – functionally active membranes – in the narrow sense which is used here. Similarly, vacuoles, phagosomes and lysosomes may contain cell-own or alien membraneous structures or debris. They have been, but are no longer, active biomembranes.

Nearly every electron micrograph of a thin section of a well-preserved eukaryotic cell confirms these theses. Let me present only a few observations (cf. Sitte 1980). Secretory products which are formed within an nonplasmatic compartment, e.g., within the Golgi apparatus, leave the cell via exocytosis. In contrast, particles like milk fat globules which are formed within the cytoplasmatic matrix are pinched off and become surrounded by a part of the plasmalemma. In the same way certain animal viruses leave the cell, enclosed by a portion of the host plasmalemma (Simons et al. 1982).

When a cell is mechanically hurt or, even more impressive, when a droplet of endoplasm of a *Physarum* plasmodium is isolated mechanically, the initially "naked" surface becomes covered by a new plasmalemma within a few seconds, not by de novo formation of free pieces of membranes which eventually join but, as shown by Wohlfahrth-Bottermann and Stockem (1970) in *Physarum,* by the peripheral alignment and subsequent fusion of mucilage vesicles.

Nevertheless, when a cell is pierced by a thin glass capillary, filled with a solution of electrolytes, for instance in electrophysiological experiments, the plasmalemma is locally interrupted. The "nonplasmatic" lumen of the capillary comes into open contact with the plasmatic matrix. Obviously, the edge of the remaining plasmalemma is in intimate contact with the wall of the capillary so that the cell remains sealed on the whole. Unambiguously, this rather artificial situation does not correspond with the rules of compartmentation (Fig. 1A).

Another example for "open" membranes is more ambiguous. Robinson et al. (1976) described that, during cytokinesis, the plasmalemma of the chlorococcalean alga *Eremosphaera* is newly formed by the fusion of "open membranes". Is *Eremosphaera* cell division unique among eukaryotes? A new study should reexamine this special case.

Some caution is also necessary with respect to the compartmentation of metabolites. In *Dictyostelium,* there is besides the "regular" glycogen pool within the plasmatic matrix another, separate one within special vacuoles (Müller and Hohl 1975).

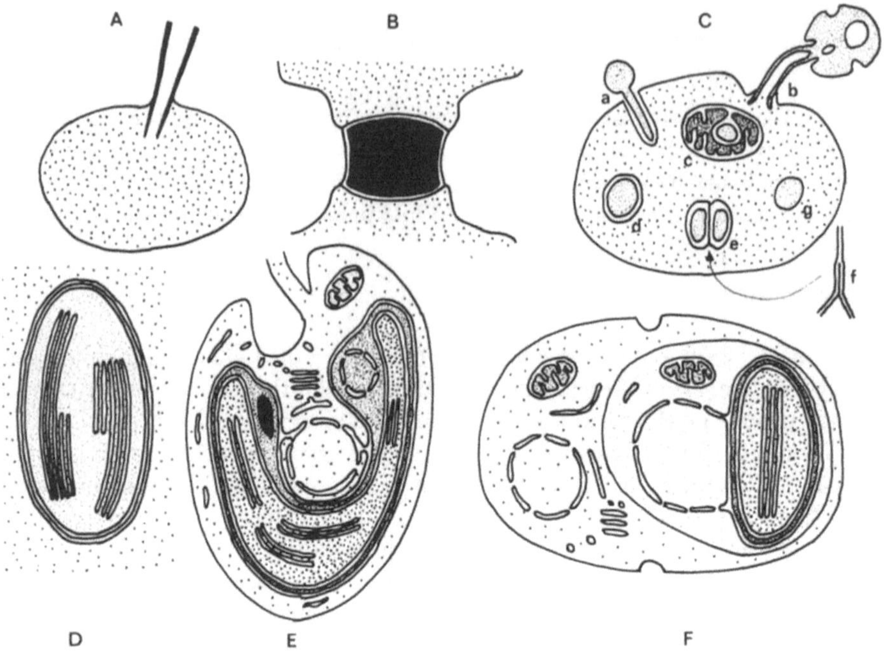

Fig. 1A-F. Examples for usual and unusual cell compartmentation, schematically. Plasmatic phases are *stippled,* nonplasmatic not. Cell walls are not drawn. For details see text; **A** No membrane between the plasmatic compartment of a cell and the nonplasmatic lumen of a cell-piercing glass capillary; **B** Pit connection in the red alga *Gracilaria.* The plug *(black)* is surrounded by the plasmalemma, which is continuous between the cells, and the "cap membrane". Membrane bifurcation? **C** A cell with various parasites and endocytobiots. *a* invagination of the host plasmalemma by a fungal haustorium; *b* opening of the host plasmalemma by the feeding tube of the parasitic dinoflagellate, *Paulsenella*; *c* a bacterium within the nonplasmatic space of a mitochondrial crista; *d* endocytobiot within a phagosome or its derivative; *e* bacteria-containing vacuoles with partly fused membranes: membrane bifurcation; *f* detail from (*e*), demonstrating the course of the membrane layers; the middle portion shows an "inner compound membrane"; *g* endocytobiot not surrounded by a host membrane: two plasmatic compartments separated by one membrane only; **D** Chloroplast ensheathed by three membranes, as, e.g., after being fed from an opisthobranch or in Euglenophyceae and in most Dinophyceae; **E** Compartmentation in *Cryptomonas.* The flagellate (*stippling* of the nucleocytoplasmatic matrix *fine,* starch grain *black*) with a chloroplast (*stippling* of the matrix *coarse* and *dense*); **F** Compartmentation in *Peridinium balticum.* The dinoflagellate (*stippling* of the nucleocytoplasmatic matrix *coarse* and *loose*) – houses but separated from it by one membrane only – a chrysophycean-like endocytobiot (*stippling* of the nucleocytoplasmatic matrix *fine* and *loose*) which has included the putative remnant of a eukaryotic endocytobiot (*stippling fine* and *dense*) with a chloroplast (*stippling* of the matrix *coarse* and *dense*). The mitochondrial matrices are *stippled finely* and *densely*

Glycogen as an α-glucan, however, should not occur in a nonplasmatic compartment. Physiological data on glycogen mobilization likewise indicate the existence of two pools here: one is degraded by amylase, the other by phosphorylase. A possible explanation of the occurrence of glycogen in vacuoles is provided by the observation that the vacuoles in question contain also membraneous debris; they seem to represent autolysomes. If so, glycogen would not be formed here but secondarily deposited.

Repeatedly there are data on the occurrence of microtubules with nonplasmatic compartments (e.g., Hoffman and Olson 1980). However, not each small tubule is a microtubule. An unambiguous identification of the material in question is frequently lacking.

Are there exceptions from the rule that every membrane separates a plasmatic from a nonplasmatic compartment?

Gram-negative bacteria have the so-called "outer membrane" besides the plasmalemma. If this outer membrane is a biomembrane in the narrow sense used here and not a membrane-like cell wall layer, the rules of compartmentation are infringed here. Various data indicate its membraneous character.

However, it does not house, e.g., the systems for active transport and is consequently not a biomembrane in the narrow sense used here, but corresponds functionally rather with a cell wall (for survey see, e.g., DiRienzo et al. 1978). A membrane should be defined by its function rather than by its architecture!

In eukaryotes, a "double plasma membrane" is also occasionally observed, e.g., by Lane and Harrison (1979) in the gut microvilli of the insect, *Rhodnius.* The "outer plasmalemma" has the appearance of a membrane. The space between the two membranes, however, is invaded by lanthanum, indicating that it is extraplasmatic and that the outer membrane perhaps rather represents a special kind of "peritrophic membrane" (which is no biomembrane in the sense used here). Its precise nature remains to be ascertained.

A likewise not well understood situation is observed at the pit connections of some red algae. There is general agreement that the plasma membranes of two cells are continuous through the pit connection. The central part of the pit connection is plugged with a structure of various appearance. There are red algae in which the plug is clearly intracellular, e.g., in the parasitic red alga *Holmsella australis.* In the host of *Holmsella, Gracilaria furcellata* (Wetherbee and Quirk 1982), as in several other red algae (Pueschel 1977), the plug is separated from the cytoplasm by a membrane-like structure, the "cap membrane". Due to the junction between the plasmalemma and the cap membrane the plug seems to be entirely membrane-bounded and essentially nonplasmatic here (Fig. 1B). Between cells of the parasite, *Holmsella,* and the host, *Gracilaria,* there are also pit connections. Even here the plasma membranes are continuous. A cap membrane is produced only on the host surface, generally not on the parasite surface, though occasionally the parasite side is enclosed by a cap membrane as well. There remains the question whether the cap membrane is, indeed, a biomembrane or a membrane-like structure. The differences in occurrence indicate that it is not a true biomembrane. On the other hand, there are freeze-fracture electron micrographs showing that a fracture plane following the "normal" plasmalemma continues into the cap membrane. They suggest a real continuity between the plasmalemma and the cap membrane. This junction would imply a bifurcation of the membranes (Pueschel 1977).

Generally, the plasma membranes of host and parasite remain existing at their interface when the parasite partly is situated extracellularly, be one or both partners ensheathed additionally by a cell wall (e.g., between *Scenedesmus* and the fungus *Chytridium*, Schnepf et al. 1971b) or be the partners without walls at the interface (e.g., between *Coscinodiscus* and the oomycete, *Lagenisma*, Schnepf et al. 1978) (Fig. 1Ca). Between *Scenedesmus* and the parasitic *Aphelidium*, walls are lacking, the plasma membranes are closely adjacent and form locally a pentalaminar so-called outer compound membrane, in which the outer layers of the trilaminar membranes appear to be united. Consequently, parts of the host plasmalemma can be dissolved without opening the host cytoplasm (Schnepf et al. 1971a): examples, where a membrane has a free edge.

A somewhat similar situation is realized in the diatom *Streptotheca* when parasitized by the dinoflagellate *Paulsenella* (Fig. 1Cb). *Paulsenella* attaches to the host cell and drives a 12-15 μm long, cytoplasmic feeding tube between epi- and hypovalva into the interior of the cell. Distally, the feeding tube forms a funnel-like opening which becomes closely appressed onto the host cytoplasm. This is sucked up within 30-40 min (Drebes and Schnepf 1982), completely in narrow *Streptotheca* cells but only partially in broad ones (Schnepf and Drebes, in preparation). Our electron microscopical investigations on sucking *Paulsenella* cells (Schnepf and Drebes, in preparation) have shown that the feeding tube is delimited on its outside by the plasmalemma which, at the apical opening, continues into the inner membrane of the feeding tube channel. Finally it is, at least discontinuously, in connection with the membrane of the developing food vacuole. The host plasmalemma is interrupted at the opening of the feeding tube, where the host cytoplasm becomes sucked up. This local break is not lethal because the free edge of the remaining host plasmalemma firmly adheres to the parasite plasmalemma at the tip of the feeding tube. In this way, similar as at an injected glass capillary, the host cytoplasm is sealed from leaking out, profitable not only for the parasite, but also for a broad host cell. When the feeding process is finished, the feeding tube retracts. If only a portion of the cytoplasm without the nucleus is lost, the membrane opening is closed; the cell survives. As yet, it has not been possible to trace the membrane layers at the edge of the host plasmalemma exactly enough to state unambiguously whether there is a true membrane bifurcation or whether the edge ends in an outer compound membrane, as observed in *Aphelidium*. The abnormal situation during the feeding of a parasite results in deviations from the rules of compartmentation and allows the occurrence of opened membranes.

Completely endoplasmatic alien organisms live under other conditions (Powell 1982). The destruction of the enveloping host membranes does not cause an outflow of the host cytoplasm. Nevertheless the plasma membranes of the host and of the endobiot are retained frequently also here: Symbiotic green algae in ciliates are situated within a membrane-bounded vacuole (e.g., Reisser 1981) which is derived from the phagosome, as do symbiotic *Rhizobium* cells in roots of Leguminosae (Fig. 1Cd).

When *Rhizobium* cells are crowded in root nodule cells, neighboring host membrane envelopes come into close contact (Fig. 1Ce) and form inner compound membranes. Occasionally, even a single symmetrical tripartite membrane seems to be formed (Fig. 1Cf). The expanse of fused membrane forks at its periphery, i.e., is confluent with both of the envelopes from which it is derived (Gunning 1970). The rules of compartmentation seem to be infringed here.

This is also true in associations in which the endoplasmatic organism is not enclosed by a special host membrane. Compatible with the rules of compartmentation is, however, that a *Ricksettsia* lives within the mitochondrial cristae of the myxomycete, *Didymium* (Duval 1972) without a special host membrane, because this space is a nonplasmatic compartment (Fig. 1Cc). Incompatible is a situation where the cell lies without host membrane within the nucleocytoplasmatic matrix (Fig. 1Cg), as observed in mycoplasmas (e.g., Esau et al. 1976), in normal bacteria (e.g., Hausmann 1972) and in intracellularly growing fungi: *Catenaria* in *Allomyces* (Powell 1982), *Sorosphaera* in *Veronica* (Braselton and Miller 1975, see also for further examples with one or two membranes in endobiotic parasite-host interfaces).

When chlorococcalean algae are injected into amoebae, they lie initially, for 3 h, free within the host cytoplasm. Later they appear to become associated with discontinuous pieces of "nontrilaminar membranes" as the authors, Jeon and Lorch (1982), call them. These pieces are suggested to fuse and to form de novo individual vacuole membranes. Later, the algae are found in lysosomes. If these "nontrilaminar membranes" indeed are biomembranes, this kind of membrane formation would be a rather spectacular exception of the rules of compartmentation.

Contradictory are the data on the number of membranes around chloroplasts of *Codium* and of other algae after their acquisition by Opisthobranchia. These chloroplasts remain functionally intact within the slugs for a long time. Some workers (e.g., Hawes 1979) describe that they are retained inside phagosomes (Fig. 1D), while other believe that the phagosome membrane disappears soon after ingestion (e.g., Brandley 1981). A chloroplast surrounded by two as well as by three membranes would be compatible with the rules of compartmentation, because we have excepted the situation within phagosomes and lysosomes explicitly. The loss of one membrane would restore the "regular" conditions.

However, there is clear evidence that chloroplasts bounded by three enveloping membranes do exist (Fig. 1D). They are found in all Euglenophyceae and in most Dinophyceae (an exception seems to be *Prorocentrum*). Gibbs (1981a) summarized the examples and concludes from the lack of ribosomes on the outermost membrane and the absence of connections with other membranes that the third membrane surrounding the chloroplasts of *Euglena* is derived from the plasmalemma of a eukaryotic endocytobiot, the cytoplasm of which is largely reduced. The "phagosome membrane" would no longer exist then. Though this question is perhaps not answered finally, the presence of three membranes and, therewith, a clear contradiction to the rules of compartmentation cannot be denied: two plasmatic compartments are separated by three membranes.

In cryptomonads (Fig. 1E), the chloroplast is surrounded by two membranes. In addition, there is an outer envelope consisting of two membranes. Its lumen is continuous with the lumen of the nuclear envelope and separates a periplastidal space from the nucleocytoplasmatic matrix of the cell. The periplastidal space contains ribosomes, starch grains, and a nucleus-like structure, the nucleomorph, which is surrounded by an envelope with nuclear pore-like interruptions (Gibbs 1981b). It is generally accepted that the chloroplast plus the periplastidal space with the nucleomorph represent an ancestral eukaryotic symbiot.

The chloroplasts of Chloromonadophyceae, Prymnesiophyceae, Chrysophyceae, Xanthophyceae, Eustigmatophyceae, Bacillariophyceae and Phaeophyceae likewise are surrounded by an additional outer envelope, consisting of two membranes. The periplastidal space, however, does not contain a nucleomorph. Gibbs (1981a,b) favors the idea that also here a eukaryotic cell, the cytoplasm of which is more reduced than in Cryptophyceae and has nearly disappeared, is the ancestor of the chloroplast. The occurrence of ribosomes along the outer membrane of the outer envelope and its continuity with the nuclear envelope has, however, also be taken as an argument for the assumption that the outer envelope is derived from the cell's own ER. In any case, the "normal" compartmentation is retained in these algae.

Whereas the compartmentation of these algae corresponds with the general rules, that of *Peridinium balticum* and a few other dinoflagellates does not. Here only one membrane separates two plasmatic compartments (Fig. 1F). *Peridinium balticum* has a "normal" dinokaryon and the normal dinophycean cytoplasm, which does not contain chloroplasts, but a second, truely eukaryotic nucleus which has its own cytoplasm with own mitochondria and chloroplasts. The latter allowed to state that the endosymbiotic organism is a chrysophyte the chloroplast of which is, as shown above, perhaps also derived from a eukaryot: a series of endocytobioses. Both cytoplasms are separated by a *single* membrane (Thomas and Cox 1973).

Summarizing the results of this critical review we can state that the rules of cell compartmentation are valid in most cases, especially in normal cells. However, there are also a few exceptions, mainly with respect to the separation of plasmatic and nonplasmatic compartments. These exceptions are of high interest for the conceptions of the evolution of the cell and are generally restricted to what is or what is believed to be an association between different cells. Free ends of membranes may occur very occasionally; generally they are observed only in outer compound membranes which are formed in pathological situations. Cell compartmentation is governed by rules, not by laws.

Special Aspects of Functional Compartmentation

Compartmentation and Morphogenesis

Compartmentation is not only the condition of various metabolic processes but also, though often neglected, directly involved in morphogenetic processes. An especially striking example is the formation of the silicious scales of *Synura.* They arise within special vesicles which serve as a template for their morphogenesis. The modelling of the scale vesicle, for its part, is the result of a complicated cooperation between the scale vesicle itself, the underlying outer envelope of the chloroplast, which forms the scale vesicle by local expansions and protrusions, and cytoskeletal elements like microtubules and microfilaments (Schnepf and Deichgräber 1969).

Compartments and Ion Transport

One functional difference between a plasmatic and a nonplasmatic compartment results from the sidedness of the separating membranes in ion transport. Generally, protons are pumped into the nonplasmatic compartment: into the thylakoids during photosynthesis, into the mitochondrial cristae during respiration, into the cell wall during growth, etc. The proton gradients are used for ATP generation but also, e.g., to drive other transport processes. The ion fluxes can be disturbed by ion ionophores.

We used monensin to interrupt the vesicle flow from the Golgi apparatus in various cells. Monensin is believed to be an ionophore which binds Na^+, K^+ and H^+, and mediates a one-for-one cation exchange. It causes the swelling of compartments of the Golgi apparatus in animal and plant cells (Tartakoff 1983).

We observed recently (Schnepf 1983), that *Funaria* caulonema cells, treated with monensin, cease to grow, in part due to the perturbation of the Golgi apparatus. Additionally, the mitochondrial cristae swell slightly as do, quite drastically, the thylakoids of the chloroplast. The thylakoid swelling depends on light. Its degree is considerably reduced when the ionophore is applied in darkness. The other compartments become inflated invariably under these conditions. Our explanation corresponds with the suggestion of Tartakoff (1983): Driven by light, the thylakoid membranes pump protons into the thylakoids which exchange for Na^+ or K^+ resulting in a passive water influx. A gradient in the degree of thylakoid swelling, which is low in the apical region of the tip cell and high in other regions, reflects a gradient in the photosynthetic capacity of the chloroplasts and/or a gradient in the amount of cations available for exchange (Schnepf, in preparation). The swelling of the other compartments likewise seems to indicate that their membranes have active proton pumps. Consequently, we can conclude that the swelling of the compartments of the Golgi apparatus indicates that they are about to reduce their internal pH by proton pumping.

We tried to confirm our observations with other plants including various algae (Schnepf and Herth, in preparation) but there the thylakoids swelled less or not at all after monensin treatment. However, the broad range ionophore X537A causes the same light dependent thylakoid swellings not only in chloroplasts of *Funaria* protonema cells but also, e.g., of *Poterioochromonas* (Schnepf and Herth, in preparation). As monensin, X537A gives rise to inflations of compartments of the Golgi apparatus also.

References

Brandley BK (1981) New Phytol 89:679–686
Braselton JP, Miller CE (1975) Arch Microbiol 104:97–99
DiRienzo JM, Nakamura K, Inouye M (1978) Ann Rev Biochem 47:481–532
Drebes G, Schnepf E (1982) Helgol Meeresunters 35:501–515
Duval JC (1972) J Microsc 13:31–46
Esau K, Magyarosy AC, Breazeale V (1976) Protoplasma 90:189–203
Gibbs SP (1981) Ann N Y Acad Sci 361:193–208

Gibbs SP (1981b) Int Rev Cytol 72:49–99
Gunning BES (1970) J Cell Sci 7:307–317
Hausmann K (1972) Z Naturforsch 27b:1572
Hawes CR (1979) New Phytol 83:445–450
Hoffman LH, Olson GE (1980) Exp Cell Res 127:1–14
Jeon KW, Lorch IJ (1982) Exp Cell Res 141:351–356
Lane NJ, Harrison JB (1979) J Cell Sci 39:355–372
Müller U, Hohl HR (1975) Protoplasma 85:199–207
Powell MJ (1982) Bot Gaz 143:176–187
Pueschel CM (1977) Protoplasma 9:15–30
Reisser W (1981) Protoplasma 105:273–284
Robinson DG, Sachs H, Mayer F (1976) Planta (Berl) 129:75–82
Ruska H (1960) Sitzber Ges Beförd ges Naturwiss Marburg 82:3–38
Schnepf E (1964) Arch Mikrobiol 49:112–131
Schnepf E (1983) Naturwissenschaften 70:260
Schnepf E, Deichgräber G (1969) Protoplasma 68:85–106
Schnepf E, Deichgräber G, Drebes G (1978) Arch Microbiol 116:133–139
Schnepf E, Deichgräber G, Hegewald E, Soeder C-J (1971a) Arch Mikrobiol 75:230–245
Schnepf E, Hegewald E, Soeder C-J (1971b) Arch Mikrobiol 75:209–229
Simons K, Garoff H, Helenius A (1982) Sci Am 246/2:46–54
Sitte P (1980) In: Nover L, Lynen F, Mothes K (eds) Cell compartmentation and metabolic channeling. Fischer, Jena pp 17–32
Tartakoff AM (1983) Cell 32:1026–1028
Tomas RN, Cox ER (1973) J Phycol 9:304–323
Wetherbee R, Quirk WM (1982) Protoplasma 110:153–165
Wohlfarth-Bottermann KE, Stockem W (1970) Wilhelm Roux' Archiv 164:321–340
Ziegler H, Lin H-PP, Schnabl H (1981) Ber Dtsch Bot Ges 94:193–202

Organization and Capacity of the Chloroplast Genome in Algae

E. STUTZ, P. E. MONTANDON, E. ROUX, B. RUTTI and B. SCHLUNEGGER[1]

Introduction

Only two kinds of algae have been studied on a molecular level in greater detail, namely, *Chlamydomonas reinhardii* and *Euglena gracilis.* In all other cases the chloroplast DNA is as yet poorly characterized. An intensive study of other algal chloroplast DNA (cpDNA), however, would be of great interest in view of the very wide spectrum of anatomical and functional properties of algal chloroplasts that should have its parallel in a wide spectrum of different kinds of chloroplast genomes.

In Table 1 we list physicochemical data of chloroplast DNA's from several algae. The list is not exhaustive, but this limited collection already indicates that algal cpDNA's can be very different in terms of overall genetic complexity and G + C content. The most complex algal cpDNA so far reported is that from *Acetabularia cliftonii* (150 x 10^8), the smallest is that of *Codium fragile* (0.56 x 10^8). It would be interesting to know whether a correspondingly large difference in genetic information exists between the two plastidial genomes. Similarly large differences exist in the average G + C content of cpDNA's, e.g., 25% for *Euglena gracilis* and 47% for *Acetabularia cliftonii.* Such differences suggest important structural differences in the organization of the two genomes. All chloroplast genomes, probably with the exception of *Acetabularia,* occur in circular form. Furthermore, it is accepted that chloroplasts, retaining the capacity to proliferate, must contain at least one copy of the genome. However, functional chloroplasts seem to exist in *Acetabularia acetabulum* (Woodcock and Bogorad 1970) which lack DNA, probably as a result of asymmetric DNA partitioning during chloroplast division (Lüttke and Bonotto 1981).

We include in this list *Polytoma obtusum* which is a colorless algae and a relative to *Chlamydomonas,* containing leucoplasts. It has an extremely low G + C content, probably due to extensive deletions, but most remarkable, this genome has retained the genes for rRNA (Siu et al. 1975).

Cyanophora paradoxa, a flagellate of uncertain taxonomic position is capable of growing photoautotrophically and its cyanelles have many features in common with chloroplasts. The cyanelle DNA is about 115 megadaltons, contains two inverted repeats with rDNA sequences, i.e., its a chloroplast type genome comparable to those found in higher plants (Mucke et al. 1980).

1 Laboratoire de Biochimie de l'Institut de Botanique, Université de Neuchâtel, Suisse

Compartments in Algal Cells and Their Interaction
ed. by W. Wiessner, D. Robinson, and R.C. Starr

Table 1. Physicochemical properties of algal plastid DNA

Taxonomic position and species		Buoyant density g/cm3 (CsCl)	dG + dC %	Size EM (μ m)	RE (Kbp)	Ploidy	References
Division	Class/Order/Species						
Euglenophytes	*Euglenales*						
	Euglena gracilis	1.685	24-26	40-44	125-143	59-72	Edelman et al. (1964); Ray and Hanawalt (1964); Stutz and Vandrey (1971); Manning and Richards (1972); Gray and Hallick (1978); Rawson et al. (1978); Crouse et al. (1978)
Chrysophytes	*Xanthophyceae*						
	Olisthodiscus luteus	1.691	31	46	150	13-44	Aldrichs and Cattolico (1981); Ersland et al. (1981); Cattolico (1978); Lüttke (1980); Herrmann et al. (1980)
	Vaucheria sessilis			37			
	Chrysophyceae						
	Ochromonas danica	1.690	30				Charles (1977)
Phaeophytes	*Phaeophyceae*						
	Pylaiella littoralis	1.693	34				Dalmon and Loiseaux (1981)
	Sphacellaria sp.	1.695	36				Dalmon and Loiseaux (1981)
Rhodophytes	*Rhodophyceae*						
	Porphyridium cruentum	1.694	35				Charles (1977)
Chlorophytes	*Chlorophyceae*						
	Chlamydomonas reinhardii	1.695	36	62	196	26-52	Sager and Ishida (1963); Bastia et al. (1971); Howell and Walker (1976); Behn and Herrmann (1977); Rochaix (1978)

Chlorophytes	*Polytoma obtusum* *(leucoplast)*		17				Siu et al. (1975)
	Chlorella ellipsoidea	1.695	36	56	175		Yamada et al. (1982)
	Chlorella pyrenoidosa	1.687	28			20	Bayen and Rode (1973); Dalmon Bayen (1975)
	Scenedesmus obliquus	1.694	35				Pryke et al. (1979)
	Bryopsidophyceae *Codium fragile*	1.596	37	*27*	85		Hedberg et al. (1981)
	Acetabularia cliftonii	*1.706*	47	Linear up to *400* and mini-circles		3	Green et al. (1977); Padmanabhan and Green (1978)
	Acetabularia acetabulum	1.703	44				Mazza et al. (1979, 1980)
	Conjugatophyceae *Spirogyra* sp.	1.690	30				Takaya and Sasaki (1973)
	Cyanophora paradoxa	1.692	33		127-138		Mucke et al. (1980); Bohnert et al. (1982); Löffelhardt et al. (1983)

Taxonomy according to Round (1965).

The Chloroplast DNA's of *Euglena gracilis* and *Chlamydomonas reinhardii*:

Structural Features

Chloroplast DNA's from *E. gracilis* and *C. reinhardii* are extensively studied both with respect to structure and function (for recent reviews see, respectively, Hallick 1983; Rochaix 1981). A comparative analysis of the two genomes is interesting for several reasons and especially in view of the large difference in taxonomic position of the two unicellular algae: *Chlamydomonas reinhardii* belongs to the division of chlorophytes, order Volvocales, while *Euglena gracilis* belongs to the division of euglenophytes, order Euglenales (Round 1966). Euglenophytes are not in a direct line with higher plants rather they represent an evolutionary dead end.

In Fig. 1 we show the EcoRI restriction sites map of both genomes in a linearized form. The two maps are approximately aligned with respect to the locus coding for the large subunit protein of the 1,5-bisphosphate-carboxylase (LS). The results of a comparative structureal analysis may be summarized as follows: (a) The two EcoRI restriction sites maps have no common fragment pattern in any region of the genome, as expected for two only distantly related genomes. (b) The *C. reinhardii* genome is about 46 kbp longer than that of *E. gracilis* and carries two inverted repeats of about 19 kbp in size which are 48 kbp apart (small single copy region). This structural feature is typical for many higher plant chloroplast genomes. (c) In *C. reinhardii*

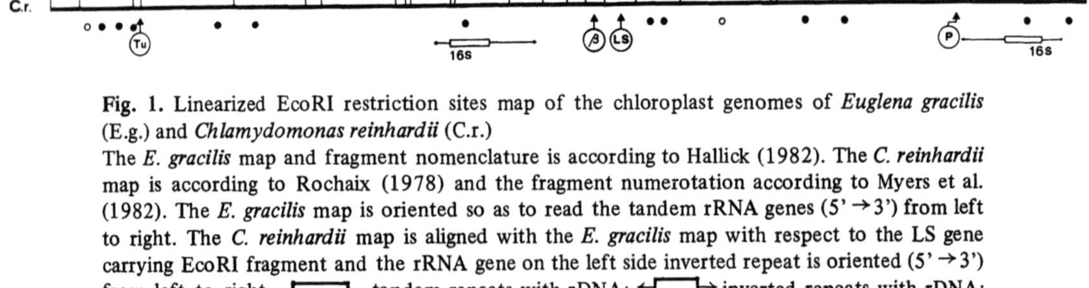

Fig. 1. Linearized EcoRI restriction sites map of the chloroplast genomes of *Euglena gracilis* (E.g.) and *Chlamydomonas reinhardii* (C.r.)
The *E. gracilis* map and fragment nomenclature is according to Hallick (1982). The *C. reinhardii* map is according to Rochaix (1978) and the fragment numerotation according to Myers et al. (1982). The *E. gracilis* map is oriented so as to read the tandem rRNA genes (5' → 3') from left to right. The *C. reinhardii* map is aligned with the *E. gracilis* map with respect to the LS gene carrying EcoRI fragment and the rRNA gene on the left side inverted repeat is oriented (5' → 3') from left to right. -▭- tandem repeats with rDNA; ←▭→ inverted repeats with rDNA; ● fragments strongly hybridizing with tRNA; ○ fragments weakly hybridizing with tRNA; (LS) gene for the large subunit protein of the ribulose 1,5-biphosphate carboxylase; (P) gene for the 32,000 MW thylakoid protein of the photosystem II reaction center; (Tu) ~ gene for elongation factor EFTu; (ß) gene for the ß-subunit of the coupling factor CF1; (Z) site of variable length. (After Stutz 1983)

the rDNA region is part of two distant inverted repeats but in *E. gracilis* the rDNA is part of three direct continuous repeats of about 6.2 kbp. (d) The tRNA genes are scattered in both cases throughout the genome. Several clusters of tRNA genes have been mapped in the case of *E. gracilis.* (e) The relative position of the few proteins genes mapped so far with certainty on both genomes is similar. Both genomes seem to contain small repetitive sequences which are interspersed throughout the genome in case of *C. reinhardii* (Rochaix 1972). For *E. gracilis* cpDNA short inverted repeats were physically mapped in the vicinity of the rDNA region (Koller and Delius 1982a). The EcoRI map shown in Fig. 1 is that of *Euglena gracilis,* strain Z. A similar structural analysis was made with *Euglena gracilis,* strain *bacillaris* (Helling et al. 1979) which gave essentially the same fragment pattern, except for differences in the rDNA regions.

The Ribosomal DNA Region

In Fig. 2 we show structural details of the chloroplast rDNA region of *E. gracilis* and *C. reinhardii.* It is known that in both cases the rRNA genes are arranged on the same strand and in the order 5' 16S RNA-spacer-large ribosomal subunit RNA's.

The rRNA operons of *E. gracilis* are part of three continuous repeats (Gray and Hallick 1978; Jenni and Stutz 1978; Rawson et al. 1978) and the genes for stable

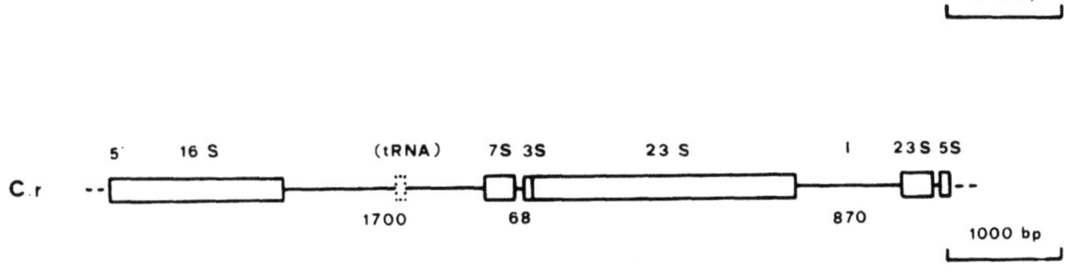

Fig. 2. Detailed structural features of the chloroplast rDNA region of *Euglena gracilis* (E.g.) and *Chlamydomonas reinhardii* (C.r.) (for references see text)
The rRNA genes are oriented (5' → 3') from left to right. For *E. gracilis* rDNA the 16S - 23S intergenic spacer is amplified showing the arrangement of the tRNA genes for isoleucine and alanine. The -100 to -400 region 5' to the next 16S RNA gene is also amplified showing the approximate position of the pseudo-tRNA genes for isoleucine and tryptophane. In the 16S - 7S intergenic spacer of the *C. reinhardii* rDNA we placed a tRNA gene to indicate that this region hybridized with 4S RNA, however no specific results are yet available to the best of our knowledge. *Numbers below the bars* indicate the number of nucleotides in spacers and intron (I). (Stutz 1983)

rRNAs are encoded within a region of about 5 kbp leaving a stretch of about 1.2 kbp between the 3' end of the 5S gene of one operon and the 5' end of the 16S gene of the next repeat (Orozco et al. 1980a). From hybridization (Keller et al. 1980) and nucleotide sequencing experiments (Graf et al. 1980; Orozco et al. 1980b) it has been established that the 16S - 23S spacer contains the tRNA genes for isoleucine and alanine. It is noteworthy that three out of seven rRNA operons of *E. coli* contain the same kind of tRNA genes (Nomura and Morgan 1977), which gives further support to the hypothesis that chloroplasts are descendants of prokaryotic cells.

The entire 16S rDNA was sequenced including flanking regions (Graf et al. 1982). The 5' and 3' termini of the 16S gene were determined by comparison with the results from T1-ribonuclease analysis of the 16S rRNA (Zablen et al. 1975). The structural gene comprises 1491 nucleotides and its sequence is 72% homologous to that of *E. coli* (Brosius et al. 1978) and 80% to that of maize chloroplasts (Schwarz and Kössel 1980). Two deletions of 9 and 23 nucleotides are identical in size and position with deletions found in 16S rDNA of maize and tobacco (Tohdoh and Sugiura 1982) chloroplasts, and what seems to be a characteristic structural feature of chloroplast 16S rRNA. A crucial difference in the sequence occurs at position 1487, where an A is found instead of a C what changes the Shine-Dalgarno sequence, required for mRNA interaction (Shine and Dalgarno 1974).

The 1.2 kbp region of the repeat was partially sequenced (Orozco et al. 1980b). It was shown that a DNA segment between position -100 to -400 5' to the 16S rRNA gene in the leader part shows great homology with the intergenic 16S - 23S spacer and contains a cluster of pseudo-tRNA genes. These authors and others (Miyata et al. 1982) conclude that the leader region with its pseudogene cluster is the result of a (imperfect ?) gene duplication event, and subsequent rapid mutations during evolution.

In *E. gracilis*, strain Z, the three repeats are probably identical in length and sequence. Minor differences do exist, e.g., in the spacer regions and the 5S DNA region (Karabin et al. 1983). A somewhat different situation is found in *E. gracilis*, strain *bacillaris*, where the rDNA repeats are not of identical length due to small deletions in the leader part (El-Geweley et al. 1981). In another strain, *E. gracilis*, strain Z.S., there exists only one operon (Wurtz and Buetow 1981) while in still another *bacillaris* strain five rDNA repeats seem to exist (B. Koller, personal communications). The conclusion, so far, would be that within the species *E. gracilis* different strains exist with variable numbers of rDNA repeats.

Another pecularity concerning the rDNA region was observed in *E. gracilis*, Z, where in addition to the three rDNA repeats with a complete set of RNA genes, an extra 16S rRNA gene was mapped (Jenni and Stutz 1979). This extra 16S gene (or pseudogene ?) and adjacent regions were sequenced. The structural part of this extra 16S gene is 98% homologous to the "normal" 16S gene. Sequence homology stops abruptly 16 positions after the 3' end of the structural part and no genes nor pseudogenes for tRNAs and/or 23S RNA are found within the next 573 positions. 5' to the structural part of the extra 16S gene homology to the leader sequences in the regular operons continues for at least 85 positions (sequencing data) and probably more, as shown by reannealing studies (Koller and Delius 1982b). It seems that this extra 16S gene is not actively transcribed.

The organization of the rDNA segment in *C. reinhardii* is depicted in Fig. 2. Two unique features can be recognized which are not seen in *E. gracilis,* nor in chloroplast rDNA of other plants (Rochaix and Malnoe 1978). (a) The 23S rRNA gene is intercepted by a 870 bp intron (Allet and Rochaix 1979). (b) The 23S rRNA gene is preceded by a 7S and a 3S rRNA gene, which are both associated with the large ribosomal subunit. The sequence of the 7S gene shows considerable sequence homology with the 5' terminus of the *E. coli* 23S gene, the conclusion being that the 5' end of the large subunit rRNA gene of *C. reinhardii* has been split into two small genes (Rochaix and Darlix 1982). *C. reinhardii* like *E. gracilis* cpDNA lacks a 4.5S rRNA gene, seen in flowering plants and, e.g., in the fern *Dryopteris acuminata* (Takaiwa et al. 1982).

The 5' terminus of the 23S rRNA chloroplast gene of *E. gracilis* was sequenced (Graf et al. 1980; Orozco et al. 1980b) and considerable sequence homology was found with the corresponding parts of the *E. coli* (Brosius et al. 1980) and *Zea mays* (Edwards and Kössel 1981) 23S RNA genes. Furthermore Hallick et al. (1981) sequenced the 3' end of the *E. gracilis* chloroplast 23S rRNA gene and adjacent regions and found that it is about 60% homologous with the bacterial 23S rRNA gene and the 4.5S chloroplast RNA gene of several higher plants, e.g., tobacco (Takaiwa and Sugiura 1980). The tentative conclusion would be that both the algal and bacterial 3' end of the 23S RNA have the same position and function in the large ribosomal subunit as the 4.5S RNA, this especially since these RNA regions can be folded into a common secondary structure.

The Transfer DNA Region

In Fig. 1 we have marked for both algal genomes those areas which hybridize with tRNA's. Malnoe and Rochaix (1978) mapped 12 to 14 regions on the circular chloroplast genome of *C. reinhardii* which hybridized with 4S RNA. Further details are lacking.

From hybridization experiments it was known that *E. gracilis* cpDNA contains tRNA genes (Schwartzbach et al. 1976; McCrea and Hershberger 1976; Gruol and Haselkorn 1976). According to the newest hybridization data (Kuntz et al. 1982) a minimum of 10 EcoRI fragments, namely EcoRI·A,B,G,H,I,J or J' (these two co-migrate), K or L (these two co-migrate), P,Q,T,U or V (these three co-migrate) contain genes for tRNA. More specific data are listed in Table 2. The chloroplast genome of the bacillaris strain shows a similar distribution of the tRNA genes, but detailed sequencing data are lacking (El-Geweley et al. 1982).

DNA Regions Coding for Protein Genes

As seen in Fig. 1 several protein-coding genes have been mapped in particular genes for the large subunit protein of the ribulose 1,5-bisphosphate carboxylase (LS), for the 32,000 MW thylakoid protein (P) and the elongation factor EF-Tu (Tu); according to a newer nomenclature these genes are named *rbc* L, *psb* A or *tuf* A gene, respectively (tentative rules established at Porto-Portese, NATO-FEBS Meeting, 1982).

Table 2. Location of specific tRNA genes on EcoRI fragments of the *E. gracilis*, Z, chloroplast genome

EcoRI fragment		
A [S]	Phe	Nickoloff and Hallick 1982
A [H]	Phe, Gly	Kuntz et al. 1982
B [H]	Leu_1	Kuntz et al. 1982
G [S]	Val,Asn, Arg,Leu	Orozco and Hallick 1982b
G [H]	Asn	Kuntz et al. 1982
V – H [S]	Tyr,His,Met,Trp,Glu,Gly	Hollingsworth and Hallick 1982
I [H]	Leu_2	Kuntz et al. 1982
J or J' [H]	Lys	Kuntz et al. 1982
P_1, P_2, P_3 [H] [S]	Ile, Ala	Keller et al. 1980 Orozco et al. 1980b Graf et al. 1980
Q [H]	Thr	Kuntz et al. 1982
Q [S]	Thr,Gly,Met,Ser,Gln	Karabin and Hallick 1983
T, U or V	Tyr	Kuntz et al. 1982

[H] : Identified by hybridization with specific tRNA's.
[S] : Identified by DNA sequencing.

Dron et al. (1982) totally sequenced the *Chlamydomonas rbc* L gene, which is 1425 N long and is highly homologous to those of spinach (Zurawski et al. 1981) and maize (McIntosh et al. 1980). The *Chlamydomonas rbc* L gene does not contain introns and this contrary to the *Euglena rbc* L gene which must contain introns (Stiegler et al. 1980a). So far about 50% of the gene which extends over 6 kb have been sequenced and the gene may contain up to nine introns (R.B. Hallick, personal communication). Similarly, the *psb* A gene which was mapped on the EcoI fragment (Keller et al. 1982) contains introns (M. Keller, personal communication) as does the *psb* A gene from *Chlamydomonas* (see Goldschmidt-Clermont, this vol.).

Passavant et al. (1983) reported that the *tuf* A gene is located on the Eco-N fragment of *Euglena gracilis* chloroplast DNA. Rutti et al. (1981) had previously shown that this DNA fragment is actively transcribed and codes for a major stromal protein. Finally, the entire Eco·N transcribed region was sequenced by Montadon and Stutz (1983) and it was shown that the *tuf* A gene contains two introns of 103 and 110 positions. Sequences around the splicing sites do not match the consensus sequences established for eukaryotic genes, nor do the introns contain open reading frames. It is, however, notworthy that contrary to higher plant chloroplast protein-coding genes, the chloroplast genes of algae can be split. We may add that Watson and Surzycki (1982) have also mapped the *tuf* A gene on the *Chlamydomonas* chloroplast genome, and sequencing work is in progress.

Selective Transcription and Regulatory Sites in *E. gracilis* Chloroplast DNA

At all developmental stages examined so far, chloroplast rRNA's are the most abundant chloroplast transcripts (Chelm et al. 1979; Rawson et al. 1981). This is also true for in vitro transcription experiments using a transcriptionally active chromosome (Rushlow et al. 1980). This means that the chloroplast-specific DNA-dependent RNA polymerase preferentially recognizes specific sequences in the rDNA region.

The structural requirements (sequences) for this preferential rRNA synthesis are presently unknown. As already mentioned, 5' to the 16S RNA gene pseudo tRNA genes have been identified in the strain Z (Orozco et al. 1980b) as well as in the *bacillaris* strain (Helling 1982, personal communication). Their presence may be crucial for the control of transcription and the secondary/or tertiary structure of the co-transcripts may be important for post-transcriptional processing signals as found in bacterial systems (Young and Steitz 1978). Regulatory sequences (5'TATRATR) were also identified 5' to several of the sequenced tRNA gene clusters (Orozco et al. 1982b). Furthermore, it could be that the tRNA genes in the proximity of protein genes scattered throughout the genome have dual function, i.e., in addition to act as structural genes they may assure regulatory functions controlling the transcription of the correlated protein gene (Hudson et al. 1981).

Length Heterogeneity of Chloroplast Genomes

As a first approximation, we may consider the multiple circular DNA molecules in a chloroplast to be identical in length and nucleotide sequence. Restriction enzyme analyses of all chloroplast genomes so far tested support this assumption. For the *E. gracilis* chloroplast genome, however, a site specific size variation was located in EcoRI·B (Jenni et al. 1981). This region (Z-region, Fig. 1) can vary in length, due to deletions/insertions, by about 600 to 800 bp. It was shown by further mapping experiments and by electron microscopic studies (Schlunegger et al. 1982) that this variable zone is framed by two small inverted repeats. The entire segment was sequenced and a repeated motive of 54 bases could be identified (Schlunegger, to be published). Furthermore, it could be shown by EM analysis that the origin of DNA replication is close to that Z-region, suggesting that the observed size variation is functionally related to the replication event (Koller and Delius 1982).

Heizmann et al. 1981, have analyzed the plastid DNA of bleached mutants of *E. gracilis*. Some of these plastidial DNA's contain a defective chloroplast genome preferentially retaining the rDNA region (ϕ^--mutants). They argue that the origin of replication may be near the rDNA region and that this site is preserved in the defective DNA as is the case in spontaneous ρ^--mutants of yeast.

In summary we can say thal algal chloroplast genomes display much more variation in size, composition and genetic complexity than angiosperm chloroplast genomes. Many special structural features have already been described and of particular importance will be the study of gene dispersal, gene regulation and expression including algal species from other classes like, e.g., the phaeophyceae or rhodophyceae which so far have been little analyzed on a molecular level.

Acknowledgments. We are grateful to Christiane Bachmann for clerical help. Our research referenced in this paper is supported by Fonds National Suisse de la Recherche Scientifique.

References

Aldrich J, Cattolico RA (1981) Plant Physiol (Bethesda) 68:641–647
Allet B, Rochaix JD (1979) Cell 18:55–60
Bayen M, Rode A (1973) Eur J Biochem 39:413–420
Bastia D, Chiang KS, Swift H, Siersma P (1971) Proc Nat Acad Sci USA 68:1157–1161
Behn W, Hermann RG (1977) Molec Gen Genet 157:25–30
Bohnert HJ, Crouse EJ, Pouyet J, Mucke H, Löffelhardt W (1982) Eur J Biochem 126:381–288
Brawerman G, Eisenstadt JM (1964) Biochim Biophys Acta 91:477–485
Brosius J, Dull TJ, Noller HF (1980) Proc Natl Acad Sci USA 77:201–204
Brosius J, Palmer ML, Kennedy PJ, Noller HF (1978) Proc Natl Acad Sci USA 75:4801–4805
Cattolico RA (1978) Plant Physiol 62:558–562
Charles D (1977) Plant Sci Lett 8:35–44
Chelm BK, Hallick RB, Gray PW (1970) Proc Natl Acad Sci USA 76:2258–2262
Crouse EJ, Schmitt JM, Bohnert HJ, Gordon K, Driesel AJ, Herrmann RG (1978) In: Akoyunoglou G, Argyroudi-Akoyunoglou JH (eds) Chloroplast development. Elsevier/North Holland, Amsterdam, pp 565–572
Dalmon J, Bayen M (1975) Arch Mikrobiol 103:57–61
Dalmon J, Loiseaux S (1981) Plant Sci Lett 21:241–248
Dron M, Rahire M, Rochaix JD (1982) J Mol Biol 162:775–794
Edelman M, Cowan CA, Epstein HT, Schiff JA (1964) Proc Natl Acad Sci USA 52:1214–1219
Edwards K, Kössel H (1981) Nucleic Acids Res 9:2853–2869
El-Geweley MR, Lomax MI, Lau ET, Helling RB, Farmerie W, Barnett WE (1981) Mol Gen Genet 181:296–305
El-Geweley MR, Helling RB, Farmerie W, Barnett WE (1982) Gene (Amst) 17:337–339
Ersland DR, Aldrich J, Cattolico RA (1981) Plant Physiol 68:1468–1473
Graf L, Kössel H, Stutz E (1980) Nature (Lond) 286:908–910
Graf L, Roux E, Stutz E, Kössel H (1982) Nucleic Acids Res 10:6369–6381
Gray PW, Hallick RB (1978) Biochemistry 17:284–289
Green BR, Muir BL, Padmanabhan U (1977) In: Woodcock CFL (ed) Progress in Acetabularia research. Acad Press, London New York, pp 107–122
Gruol DJ, Haselkorn R (1976) Biochim Biophys Acta 447:82–95
Hallick RB (1983) In: Buetow DE (ed) The Biology of Euglena, vol IV. Academic Press, New York
Hedberg MF, Huang YS, Hommersand MH (1981) Science (Wash DC) 213:445–447
Heizmann P, Doly J, Hussein Y. Nicolas P, Nigon V, Bernardi G (1981) Biochim Biophys Acta 653:412–415
Helling RB, El-Geweley MR, Lomax MI, Baumgartner JE, Schwarzbach SD, Barnett WE (1979) Mol Gen Genet 174:1–10
Herrmann RG, Palta K, Kowallik KV (1980) Planta (Berl) 148:319–327
Hollingworth MJ, Hallick RB (1982) J Biol Chem 257:12795–12799
Howell SH, Walker LL (1976) Biochim Biophys Acta 418:249–256
Hudson L, Rossi J, Landy A (1981) Nature (Lond) 294:422–427
Jenni B, Fasnacht M, Stutz E (1981) FEBS Lett 125:175–179
Jenni B, Stutz E (1978) Eur J Biochem 88:127–134
Jenni B, Stutz E (1979) FEBS Lett 102:95–99
Karabin GD, Hallick RB (1983) J Biol Chem 258:5512–5518
Karabin GD, Narita JO, Dodd JR, Hallick RB (1983) J Biol Chem 258:14790–14796
Keller M, Burkard G, Bohnert HJ, Mubumbila M, Gordon K, Steinmetz A, Heiger D, Crouse EJ, Weil JH (1980) Biochem Biophys Res Commun 95:47–54

Keller M, Rutti B, Stutz E (1982) FEBS Lett 149:133–137
Koller B, Delius H (1982a) FEBS Lett 139:86–92
Koller B, Delius H (1982b) FEBS Lett 140:198–202
Kuntz M, Keller M, Crouse EJ, Burkard G, Weil JH (1982) Current Genetics 6:63–70
Löffelhardt W, Mucke H, Crouse EJ, Bohnert HJ (1983) Current Gen 7:139–144
Lüttke A (1980) Plant Sci Lett 18:191–199
Lüttke A, Bonotto S (1981) Planta (Berl) 153:269–275
Malnoe P, Rochaix JD (1978) Mol Gen Genet 166:269–275
Manning JE, Richards OC (1972) Biochim Biophys Acta 259:285–296
Mazza A, Bonotto S, Felluga B, Casale A, Sassone-Corsi P (1979) In: Bonotto S, Kefeli V, Puiseux-Dao S (eds) Developmental biology of Acetabularia, Elsevier/North-Holland Biomedical Press, pp 115–130
Mazza A, Casale A, Sassone-Corsi P, Bonotto S (1980) Biochem Biophys Res Comm 93:668–674
McCrea JM, Hershberger Cl (1976) Nucleic Acids Res 3:2005–2017
McIntosh L, Poulsen C, Bogorad L (1980) Nature (Lond) 288:556–560
Miyata T, Kikuno R, Ohshima Y (1982) Nucleic Acids Res 10:1771–1780
Montandon PE, Stutz E (1983) Nucleic Acids Res 11:5877–5892
Mucke H, Löffelhardt W, Bohnert HJ (1980) FEBS Lett 111:347–352
Myers AM, Grant DM, Rabert DK, Harris EH, Boynton JE, Gillham NW (1982) Plasmid 7: 133–151
Nickoloff JA, Hallick RB (1982) Nucleic Acids Res 10:8191–8210
Nomura M, Morgan EA (1977) Annu Rev Genet 11:297–347
Orozco EM Jr, Gray PW, Hallick RB (1980a) J Biol Chem 255:10991–10996
Orozco EM Jr, Hallick RB (1982a) J Biol Chem 257:3258–3264
Orozco EM Jr, Hallick RB (1982b) J Biol Chem 257:3265–3275
Orozco EM JR, Rushlow KE, Dodd JR, Hallick RB (1980b) J Biol Chem 255:10997–11003
Padmanabhan U, Green BR (1978) Biochim Biophys Acta 521:67–73
Passavant CW, Stiegler GL, Hallick RB (1983) J Biol Chem 258:693–695
Pryke JA, Cranney MK, Jones DS (1979) Plant Sci Lett 16:125–128
Rawson JRY, Boerma CL, Andrews WH, Wilkerson CG (1981) Biochemistry 20:2639–2644
Rawson JRY, Kushner SR, Vapnek D, Alton NK, Boerma CL (1978) Gene (Amst) 3:191–209
Ray DS, Hanawalt PC (1964) J Mol Biol 9:812–824
Rochaix JD (1972) Nature (Lond) 238:76–78
Rochaix JD (1978) J Mol Biol 126:597–617
Rochaix JD (1981) Experientia (Basel) 37:324–332
Rochaix JD, Darlix JL (1982) J Mol Biol 159:383–395
Rochaix JD, Malnoe P (1978) Cell 15:661–670
Round FE (1966) In: The Biology of the Algae. Edward Arnold, London
Rushlow KE, Orozco EM, Lipper C, Hallick RB (1980) J Biol Chem 255:3780–3792
Rutti B, Keller M, Ortiz W, Stutz E (1981) FEBS Lett 134:15–19
Sager R, Ishida MR (1963) Proc Natl Acad Sci USA 50:725–730
Schlunegger B, Fasnacht M, Stutz E, Koller B, Delius H (1982) Biochim Biophys Acta 739: 114–121
Schwartzbach SD, Hecker LI, Barnett WE (1976) Proc Natl Acad Sci USA 73:1984–1988
Schwarz ZS, Kössel H (1980) Nature (Lond) 283:739–742
Shine J, Dalgarno L (1974) Proc Natl Acad Sci 71:1342–1346
Siu CH, Chiang KS, Swift H (1975) J Mol Biol 98:369–391
Stiegler GL, Matthews HM, Bingham SE, Hallick RB (1980a) Nucleic Acids Res 10:3427–3444
Stutz E (1970) FEBS Lett 8:25–28
Stutz E (1983) Bot Helv Acta, in press
Stutz E, Vandrey JP (1971) FEBS Lett 17:277–280
Takaiwa F, Kusuda M, Sugiura M (1982) Nucleic Acids Res 10:2257–2260
Takaiwa F, Sugiura M (1980) Mol Gen Genet 180:1–4
Takaya K, Sasaki K (173) Plant Cell Physiol 14:237–248
Tohdoh N, Sugiura M (1982) Gene (Amst) 17:213–218

Watson JC, Surzycki SJ (1982) Proc Natl Acad Sci USA 79:2264–2267
Wells R, Sager R (1971) J Mol Biol 58:611–622
Woodcock CLF, Bogorad L (1970) J Cell Biol 44:361–375
Wurtz EA, Buetow DE (1981) Current Genetics 3:181–187
Yamada T (1982) Plant Physiol 70:92–96
Young RA, Steitz JA (1978) Proc Natl Acad Sci USA 75:3593–3597
Zablen LB, Kissil MS, Woese CR, Buetow DE (1975) Proc Natl Acad Sci USA 72:2418–2422
Zurawski G, Perrot B, Bottomley W, Whitfeld PR (1981) Nucleic Acids Res 9:3251–3270

Structure and Expression of Chloroplast and Nuclear Genes in *Chlamydomonas reinhardii*

M. GOLDSCHMIDT-CLERMONT[1], M. DRON[2], J. M. ERICKSON[1], J.-D. ROCHAIX[1], M. SCHNEIDER[1], R. SPREITZER[1] and J.-M. VALLET[1]

Introduction

Chlamydomonas reinhardii, a facultative photoautotroph, can also be grown heterotrophically on acetate in the dark. Thus, mutants impaired in photosynthetic functions can be recovered as acetate requirers and their specific defects analyzed. While a part of the genetic information necessary for chloroplast function is found in the nucleus and displays classical Mendelian inheritance, another part is found in the chloroplast and undergoes non-Mendelian, uniparental transmission. In the occasional biparental zygotes, chloroplast gene markers can be observed to recombine. This unique property of *Chlamydomonas* has led to the derivation of a formal genetic map of the chloroplast genome (Sager 1972; Gilham 1978). An extensive nuclear genetic map also exists (Harris 1982).

The genetic systems of these two compartments cooperate in the biogenesis of the functional chloroplast. Our efforts have been directed at gaining insight into the structure, function and coordinate regulation of these genomes. Such an analysis involves two steps: the first is to define structural features like the organization of the chloroplast DNA or the nucleotide sequence of particular genes. The second step is to relate structure to function: in this respect, combining a biochemical and molecular analysis with the classical genetic approach is powerful and promising. We have begun to correlate the genetic and physical maps of the chloroplast, and to analyze the effect of specific mutations on the interactions of the chloroplast and nucleocytoplasmic compartments. The ability to introduce genes into the organism by transformation is another useful tool in relating structure and function. The significance of particular structural features of genes can thus be assessed by altering them in vitro and testing their function in vivo. We are developing transformation vectors that are capable of replicating autonomously both in *Chlamydomonas* and in *E. coli.*

Structure of the Chloroplast Genome

The chloroplast DNA of *C. reinhardii* consists of 190 kbp (190,000 base pair) circular molecules that are present in multiple copies in the single large chloroplast of

1 Department of Molecular Biology, University of Geneva, 30 quai Ernest-Ansermet, CH-1211 Geneva 4, Switzerland

2 Present address: Université de Paris-Sud, Orsay, France

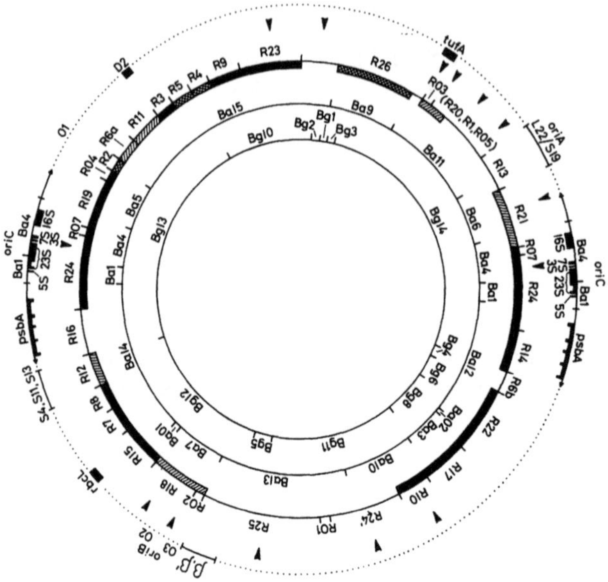

Fig. 1. Physical map of the chloroplast genome of *C. reinhardii*. The *three inner circles* represent, starting from the center, the BglII, BamH1 and EcoRI restriction maps (Rochaix 1978). The steady state levels of transcripts from the different EcoRI fragments are indicated as high (▬), medium (▨) or low (▩). *Dark wedges* show the positions of the 4S RNA genes (Malnoe and Rochaix 1979). In the *outer dotted circle, thin lines bounded by arrowheads* represent the inverted repeat with the rRNA and *psbA* genes (Rochaix and Malnoe 1978; Erickson et al. 1983a). *Black rectangles* indicate the positions of the genes for the large subunit of ribulose bisphosphate carboxylase/oxygenase (*rbcL*) and for the 32,000 thylakoid membrane polypeptide, D2 (Malnoe et al. 1979). Based on homology to their *E. coli* counterparts, the tentative locations of genes for elongation factor EF-Tu (*tufA*), ribosomal proteins (S4, S11, S13, S29, L22) and RNA polymerase subunits (ββ') are also indicated (Watson and Surzycki 1982, 1983). The location of sequences that allow autonomous replication of plasmids in yeast (01, 02, 03) and in *C. reinhardii* (ori A, B, C) are shown (Rochaix et al. 1983b)

each cell. A variety of biochemical techniques have been used to characterize this DNA molecule. A restriction map was established (Fig. 1) and a collection of recombinant clones covering the entire genome was prepared (Rochaix 1978). This work provided the basis for locating genes on the physical map (Fig. 1, outer circle) by nucleic acid hybridization with homologous probes (tRNA, rRNA) (Malnoe and Rochaix 1979; Rochaix and Malnoe 1978) or heterologous probes from *E. coli* (*tufA*, Watson and Surzycki 1982) or from higher plants (*psbA*, Erickson et al. 1983a). Genes on recombinant plasmids have also been identified by coupled transcription and translation in vitro in an *E. coli* system and by hybrid arrested in vitro translation of the mRNA (e.g., *rbcL*, Malnoe et al. 1979).

We have studied some chloroplast genes in greater detail and determined their nucleotide sequences. This analysis has revealed interesting features of the structure and function of the chloroplast genome. Transcription signals are related to those of prokaryotes: strong homology can be seen between the essential sequences of *E coli* promoters and terminators and the corresponding regions of *C. reinhardii* genes (Dron et al. 1982). In the initiation of prokaryotic translation, the 3' end of 16S ribosomal RNA interacts with a short sequence just upstream of the starting codon (Shine and Dalgarno 1974). A similar interaction can be inferred when the 3' end of chloroplast 16S rRNA is compared with sequences just upstream of the initiator codon of *Chlamydomonas* chloroplast genes (Dron et al. 1982). This resemblance of chloroplast and prokaryotic transcription and translation signals has also been noted in other plants (reviewed by Whitfeld and Bottomley 1983). We have found that, unlike in prokaryotes, some *C. reinhardii* chloroplast genes are interrupted by introns. This is the case for the 23S rRNA genes, which contain an 888bp intron (Rochaix and Malnoe 1978) and the *psbA* gene (Erickson et al. 1983a). Four introns interrupt the sequence of this gene which codes for the rapidly turned over 32,000 MW protein of photosystem II (Hoffman-Falk et al. 1982). The *psbA* and *rbcL* genes of *Euglena gracilis* also have introns (Stutz et al., this vol., Stiegler et al. 1982), but none have so far been found in the corresponding genes from higher plants. Conversely, the algal chloroplast tRNA genes are not known to have introns while some tRNA genes of higher plants are interrupted (reviewed by Whitfeld and Bottomley 1983; Rochaix 1983a).

Relating Structure and Function

Our sequencing studies have also provided a basis for correlating the genetic and physical maps of the *C. reinhardii* chloroplast. We have found a change in the nucleotide sequence of the large subunit (LS) of ribulose bisphosphate carboxylase/oxygenase (RuBPCO) (Dron et al. 1983), in a uniparental mutant with defective RuBPCO (Spreitzer and Mets 1980). This point mutation results in an amino acid substitution of glycine by aspartic acid near one of the presumptive active sites. We have also recently sequenced the *psbA* gene in a uniparental mutant of *C. reinhardii* that is resistant to the herbicides DCMU and atrazine: it differs from the wild type at a single base pair, resulting in the substitution of serine by alanine in the predicted sequence of the 32,000 MW protein of photosystem II (Erickson et al. 1983b). In addition to providing another site for the correlation of the genetic and physical maps, this finding has other important implications: it provides us with a well defined selectable marker to attempt chloroplast transformation (see below) and establishes the molecular basis for herbicide resistance in the mutant.

The nuclear genome encodes numerous polypeptides that are imported into the chloroplast and assembled with organelle encoded polypeptides into large multimolecular complexes. A simple system to study this interaction is the chloroplast enzyme ribulose bisphosphate carboxylase/oxygenase (RuBPCO) which consists of eight identical large subunits (LS) coded in the chloroplast and eight small subunits (SS) coded in the nucleus. The SS is translated as a precursor in the cytoplasm and imported, processed and assembled in the chloroplast (reviewed by Ellis 1981). We have recently isolated SS genes from a recombinant lambda phage library of *C. reinhardii* nuclear DNA by hybridization to a mixture of synthetic oligonucleotides whose sequence were inferred from the known protein sequence of the amino terminus of the SS (Schmidt et al. 1979). Our preliminary analysis indicates that there is a small family of SS genes in *C. reinhardii.* This raises the possibility that different copies of the gene may be expressed differentially during the life cycle or under different environmental conditions. The cloned SS genes are now being analyzed in detail. Together with the well-characterized LS gene, they provide hybridization probes to investigate the coordinate expression of the genes under different growth conditions. The availability of uniparental LS mutants (Spreitzer and Mets 1980; Spreitzer and Ogren 1983) is central to this approach: they allow us to analyze the effect of the inactivation or the absence of RuBPCO on the expression of the genes. Our preliminary results indicate that the levels of accumulation of LS mRNA and of SS mRNA (from one of the members of the gene family) are not tightly coupled to the level of activity of RuBPCO.

The ability to introduce DNA by transformation is an important tool for the isolation of genes and for the study of their expression and regulation. We have developped a transformation system and shown that a foreign gene (from yeast) can be introduced into *Chlamydomonas,* integrated in the genome and expressed so that it complements a nuclear mutation (Rochaix and van Dillewijn 1982). A further step has been the isolation of DNA sequences that allow the autonomous replication of plasmid molecules in *Chlamydomonas.* Vectors have thus been constructed that can replicate both in *E. coli* and in *C. reinhardii.* We are also trying to transform the chloroplast genome, using the herbicide resistant *psbA* gene as a selectable marker.

Concluding Remarks

The structural analysis of the chloroplast genes of *C. reinhardii* and their comparison with those of other algae, of higher plants and of prokaryotes have given many insights into the structure and evolution of the chloroplast genome. This work also establishes a basis for investigating the function of this organelle and its cooperation with the nucleo-cytoplasmic compartment. *Chlamydomonas reinhardii* will continue to be an organism of choice for this analysis because it provides a unique opportunity to combine classical genetic studies with molecular approaches.

Acknowledgments. We are grateful to A. Chollais (Biogen SA, Geneva) for providing us with synthetic oligonucleotides. This work was supported by grant 3.258.082 from the Swiss National Foundation.

References

Dron M, Rahire M, Rochaix J-D (1982) J Mol Biol 182:775–793
Dron M, Rahire M, Rochaix J-D, Mets L (1983) Plasmid 9:321–324
Ellis RJ (1981) Annu Rev Plant Physiol 32:111–137
Erickson JM, Schneider M, Vallet JM, Dron M, Bennoun P, Rochaix JD (1983a) In: Sybesma C (ed) Proceedings of the 6th International Congress on Photosynthesis. M Nijhoff and W Junk, The Netherlands (in press)
Erickson JM, Rahire M, Bennoun P, Delepelaire P, Diner B, Rochaix J-D (1983b) In preparation
Gilham NW (1978) Organelle Heredity. Raven Press, New York, pp 347–438

Harris EH (1982) In: O'Brien SJ (ed) Genetic maps. National Cancer Institute, Frederick, Maryland, pp 168–174
Hoffmann-Falk H, Mattoo AK, Marder JB, Edelman M, Ellis RJ (1982) J Biol Chem 257:4583–4587
Malnoe P, Rochaix J-D (1979) Mol Gen Genet 166:269–275
Malnoe P, Rochaix J-D, Chua N-H, Spahr P-F (1979) J Mol Biol 133:417–434
Rochaix J-D (1978) J Mol Biol 126:597–617
Rochaix J-D (1983a) In: Reanney DC, Chambon P (eds) International review of cytology. Special volume: Genome evolution in Prokaryotes and Eukaryotes (in press)
Rochaix J-D, Dron M, Schneider M, Vallet J-M, Erickson JM (1983b) In: Ahmad F, Downey K, Schultz J, Voellmy RW (eds) 15th Miami Winter Symposium. Advances in gene technology: Molecular genetics of plants and animals (in press)
Rochaix J-D, Malnoe P (1978) Cell 15:661–670
Rochaix J-D, van Dillewijn J (1982) Nature (Lond) 296:70–72
Sager R (1972) Cytoplasmic genes and organelles. Academic Press, New York
Schmidt GW, Devillers-Thiery A, Desruisseaux H, Bloebel G, Chua NH (1979) J Cell Biol 83: 615–622
Shine J, Dalgarno L (1974) Proc Nat Acad Sci USA 71:1342–1346
Spreitzer RJ, Mets L (1980) Nature (Lond) 285:114–115
Spreitzer RJ, Ogren WL (1983) Proc Nat Acad Sci USA (in press)
Stiegler GL, Matthews HM, Bingham SE, Hallick RB (1982) Nucl Acids Res 10:3427–3444
Watson JC, Surzycki SJ (1982) Proc Nat Acad Sci USA 79:2264–2267
Watson JC, Surzycki SJ (183) Current Genetics 7:201–210
Whitfeld PR, Bottomley W (1983) Ann Rev Plant Physiol 34:279–310

Interactions Between Compartments in *Acetabularia* During Gene Expression

H. G. SCHWEIGER, H. BANNWARTH, S. BERGER, E. DE GROOT, G. NEUHAUS and G. NEUHAUS-URL[1]

Eukaryotic cells are provided with more than one gene expression system. In plant cells, and that is true also for algal cells, we deal with three systems which are subjected to a pronounced compartmentation. The first is the eukaryote-characteristic, nucleo-cytosol compartment. This in itself is divided by the nuclear envelope which separates the nucleus and the cytosol. In addition to the nucleo-cytosol compartment plant cells dispose of the chloroplasts and also of mitochondria. In the following some experiments will be discussed which illustrate the interactions between the cytoplasm and the nucleus.

These experiments were performed on the unicellular and uninucleate marine green alga *Acetabularia* and almost exclusively on the species *Acetabularia mediterranea* (Schweiger and Berger 1979). The fully developed cell at the end of the vegetative phase consists of the rhizoid, the stalk and the cap. The stalk carries a number of whorls. The fact that cells of *Acetabularia* may reach a length of more than 50 mm makes *Acetabularia* a favourable organism for cell biological experiments. Techniques which can be easily used with *Acetabularia* include cell surgery by which the cell may be fragmented resulting in nucleate as well as anucleate fragments (Hämmerling 1932). Another easily performed technique is the isolation of the cell nucleus by hand and its implantation into an anucleate or nucleate fragment of *Acetabularia* (Hämmerling 1955).

Cells usually survive such operations. The different types of experiment might answer a number of questions which in particular underline the role of the different compartments and their interactions.

The influence of cytoplasm on the cell nucleus has been demonstrated in a rather simple experiment (Hämmerling 1958; Berger and Schweiger 1975) (Fig. 1). The experiment was based on the observation that the ultrastructure of the cell nucleus significantly depends on the age of the cell or its developmental stage respectively. The nucleus of a young cell, i.e., a cell in the stage of early cap formation, has a number of nucleoli which in their ultrastructure exhibit symptoms of high activity. Furthermore, the nuclear enevelope is covered by a thin cytoplasmic layer and surrounded by highly vacuolated cytoplasm. On the other hand, in an old cell which is characterized by a fully developed cap the nucleus is surrounded by a pronounced cytoplasmic layer without any vacuoles. The cell nucleus of an old cell has only one nucleolus which is sphere-like and vacuolized. From these observations it follows

1 Max-Planck-Institut für Zellbiologie, Rosenhof, 6802 Ladenburg b. Heidelberg, FRG

Compartments in Algal Cells and Their Interaction
ed. by W. Wiessner, D. Robinson, and R.C. Starr

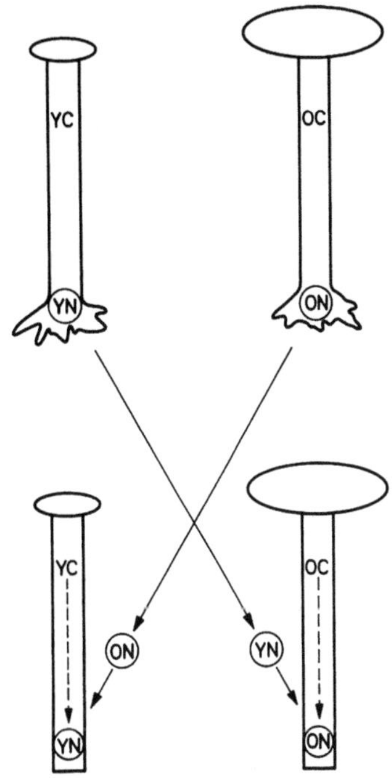

Fig. 1. Schematic drawing of the influence of the cytoplasm on the cell nucleus. An old nucleus (*ON*) implanted into young cytoplasm (*YC*) within a short time shows the symptoms of rejuvenation while a young nucleus (*YN*) implanted into old cytoplasm (*OC*) shows the symptoms of aging

that old and young cells can be easily distinguished on the basis of the ultrastructure of the nucleus.

The experiment which demonstrated the influence of the cytoplasm on the nucleus was performed in a way that a young nucleus was isolated and implanted into the cytoplasm of an old cell. Such a young nucleus after implantation into an old cytoplasm will change its ultrastructure and within 3 to 5 days will adapt to the characteristics of an old nucleus. In a similar way, after implantation of an old nucleus into a young cytoplasm, the old nucleus will within a couple of days adapt to the morphology of a young nucleus.

These experiments, together with corresponding controls, underline the extraordinary role of the cytoplasm in determining the ultrastructure of the cell nucleus. Since, however, the structure reflects the functional state of the nucleus it means that the cytoplasm exerts a pronounced regulatory influence on the nucleus.

A second example which is to be discussed in this connection again demonstrates that the cytoplasm has a profound influence on distinct steps of gene expression in the nucleo-cytosol compartment. During recent years we have become interested in the regulation of those enzymes which supply the nucleus with substrates for replication, i.e., with deoxyribonucleotides. In order to describe the role of these enzymes it is essential to recall the life cycle of *Acetabularia.* During the vegetative phase of the life cycle an *Acetabularia* cell undergoes highly species-specific morphogenesis. During this phase the cell is uninucleate and the nucleus is located in the rhizoid. The vegetative phase ends with the formation of a maximum diameter cap. At that time the primary nucleus starts dividing and the secondary nuclei migrate

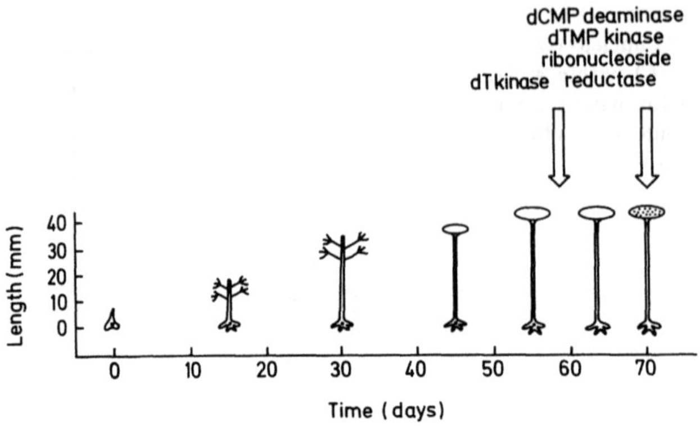

Fig. 2. Regulation of enzymes during the life cycle of *Acetabularia. Arrows* indicate the maximum of enzmye activities

through the stalk into the cap. Around each secondary nucleus a cyst is formed. In the cysts the secondary nuclei continue to divide, and eventually gametes are formed. The gametes are released from the cysts, they fuse and form zygotes.

During the generative phase the number of nuclei increases from one to 10^6 or even more. This increase requires a high rate of replication. The precursors which are used for replication originate in the metabolism of the cell. Therefore enzymes must be available which catalyze the formation of the precursors. One possibility of fulfilling this requirement would be the de novo synthesis of all the enzymes involved or at least some critical ones.

The experiments which were performed were aimed at the elucidation of the activities of the corresponding enzymes during the late vegetative but also during the generative phase. The following enzymes were studied: a thymidine kinase, a thymidylate kinase, a dCMP deaminase and a ribonucleoside reductase (Bannwarth and Schweiger 1975; Bannwarth, Ikehara and Schweiger 1977, 1982; de Groot and Schweiger 1981, 1983a,b and in preparation). It has been demonstrated that the activities of all four enzymes are subjected to pronounced regulation during the early generative phase (Fig. 2). Furthermore, it could be made probable that the increase in enzyme activity during this phase is due to de novo synthesis of the corresponding enzymes. These conclusions are based on estimation of the activity, on inhibitor studies and on mixing experiments.

The use of inhibitors which can distinguish between translation on 80S and on 70S ribosomes, and therefore indicate in which compartment translation occurs, have shown that the thymidine kinase, the dCMP deaminase and the ribonucleoside reductase are translated and therefore also transcribed in an organelle compartment, i.e., in the chloroplast. From these results one can conclude that the precursors which are essential for replication are produced by reactions which occur in chloroplasts. This means that we are dealing with a true cooperation between two compartments. It may be added that replication of organelle genomes can be excluded during the early generative phase since the number of chloroplasts and of mitochondria does not increase during this period, and it can also be excluded that the number of genomes per chloroplast is increased.

Table 1. Sensitivity of different enzymes to compartment specific translational inhibitors CX: cycloheximide; CAL: chloramphenicol

Enzyme	CX	CAL	Translation (ribosome type)
Thymidine kinase	–	+	70S
Thymidylate kinase	+	–	80S
dCMP deaminase	–	+	70S
Ribonucleoside reductase	–	+	70S

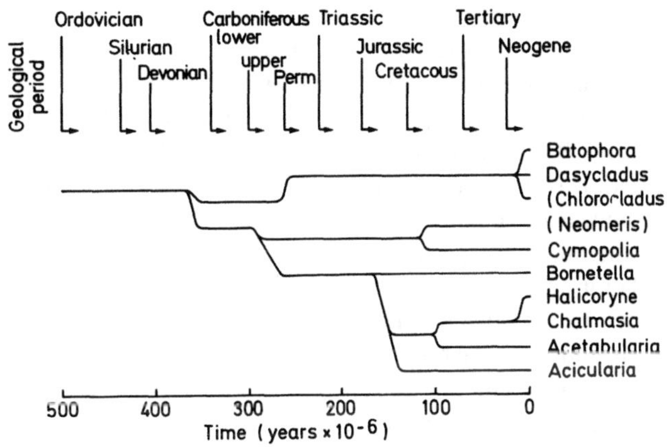

Fig. 3. Evolution of *Dasycladaceae*. (Based on Kamptner 1958)

A surprising result was that the situation with the thymidylate kinase is different. In this case the experimental results indicate that the enzyme is translated on 80S ribosomes of the nucleo-cytosol compartment of *Acetabularia* (Table 1). It is very hard to see why three of the enzymes which are involved in the deoxyribonucleotide metabolism are transcribed and translated in organelles, while a fourth closely related one is expressed in the nucleo-cytosol compartment. One possible explanation which has recently gained in weight has to do with the interesting fact that *Acetabularia* and the family of Dasycladaceae to which *Acetabularia* belongs can be traced back by around 500 million years (Fig. 3). The Dasycladaceae, therefore, represent an old family (Kamptner 1958). This situation is reflected also by the chloroplast genome which is clearly different from all other chloroplast genomes studied so far (Tymms, unpublished results). In particular, on the basis of spreading experiments, one may conclude that the size of the chloroplast genome in *Acetabularia* is significantly larger than other chloroplast genomes (Woodcock and Bogorad 1970; Büchsenstein, unpublished results). In fact it may be larger by a factor of 6 to 10.

It is an interesting speculation that the gene for the thymidylate kinase in the ancestors of *Acetabularia* was also located in the chloroplast genome and only during evolution has jumped out of the chloroplast genome into the nuclear genome. If this is correct then there should be a chance to find a relative of *Acetabularia* in which the gene for the thymidylate kinase is still in the chloroplast genome or in other words in which the gene is still expressed by translation on 70S ribosomes. Presently representatives of eight of the ten subfamilies of Dasycladaceae are being cultured. A systematic study of representatives of these different subfamilies is aimed at the question whether there is any species within the family of Dasycladaceae in which this enzyme is coded in the chloroplast genome. A positive result would support the hypothesis that genes can jump from the chloroplast genome into the nuclear genome and possibly also vice versa.

Experiments on enzymes which are involved in the deoxyribonucleotide metabolism in addition have indicated that the cell nucleus has a controlling effect on the regulation. The question has been studied of whether the enzymes which have to do with nuclear replication are also regulated when the cell nucleus is removed. Since this question could be positively answered one may conclude that the regulation of the synthesis of the enzymes which supply the nucleus with deoxyribonucleotides is not directly induced by a signal originating in the nucleus.

However, the regulation in anucleate cells cannot be directly compared. If the nucleus is removed a couple of weeks prior to the expected regulation, then the regulation occurs significantly earlier albeit not as pronouncedly as in nucleate cells (Bannwarth, Ikehara, and Schweiger 1982; de Groot and Schweiger 1983b). Removal of the nucleus is performed by amputation of the rhizoid. The advanced onset of regulation in the absence of the nucleus might therefore be an effect of cell surgery. In order to investigate this possibility control experiments were performed in which a rhizoid, immediately after surgery, was recombined with the stalk by grafting (Bannwarth and Schweiger 1983). Under these conditions the regulation occurs essentially as in the nucleate cell (Fig. 4). From this experiment it can be excluded that the suspension of the suppression of the regulation is due to the cell surgery. It can rather be concluded that the nucleus plays a regulatory role.

More recent experiments indicate that this suppressing function of the nucleus is species-specific. This conclusion is based on experiments in which the rhizoid of a cell of a comparable stage of development was grafted which did not come from the same but from a closely related species. In this case the suppressing effect could not be observed.

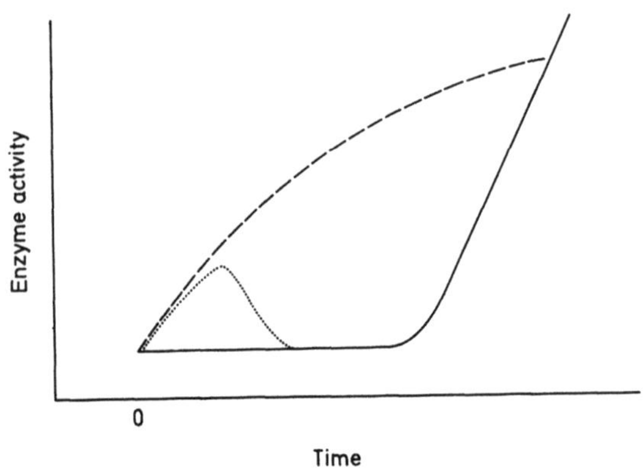

Fig. 4. Scheme of the time course of dCMP deaminase activity in nucleate (——), anucleate (- - -) and retransplanted cells (. . . .)

Another question which is closely related to the compartmentation of the cell is in how far the gene expression machinery of the nucleo-cytosol compartment is species-specific. One may ask whether the transcription which is localized in the nucleus as well as replication are specific for genes of *Acetabularia;* another question would be whether the expression in the cytosol, i.e., translation can be performed only with species-specific transcription products of the cell nucleus. In order to approach this problem a number of techniques have been adapted or developed. These included the isolation of a cell nucleus, the microinjection of such an isolated nucleus and the reimplantation of a microinjected nucleus into the cytoplasm. A similar technique included the fusion of a cytoplast with a loaded nucleus.

In order to investigate the specificity of the gene expression machinery of *Acetabularia,* foreign genomes and cloned genes were injected into an isolated nucleus and after implantation into a cytoplasm gene expression products were looked for by means of the indirect immunofluorescence method. Such experiments were performed with tobacco mosaic virus (TMV) RNA (Cairns et al. 1978c). In this case a comparison with control cells revealed that TMV RNA can be expressed in *Acetabularia.* Similar results were obtained when mengo virus RNA (Cairns et al. 1978b) was injected. Finally, it should be mentioned that experiments with a DNA, in this case adenovirus 2 DNA (Cairns et al. 1978a), were also successful. In all three cases the genomes were expressed and this clearly shows that the gene expression system of *Acetabularia* does not require highly species-specific genomes. This is true for the translation as well as for the transcription step.

These results were corroborated by more recent experiments in which cloned genes were used (Fig. 5). In a first experiment the gene for the small subunit of the ribulose 1,5-bisphosphate carboxylase from wheat integrated in a plasmid was injected (Broglie et al. 1983; Neuhaus et al. 1983). This experiment was supplemented

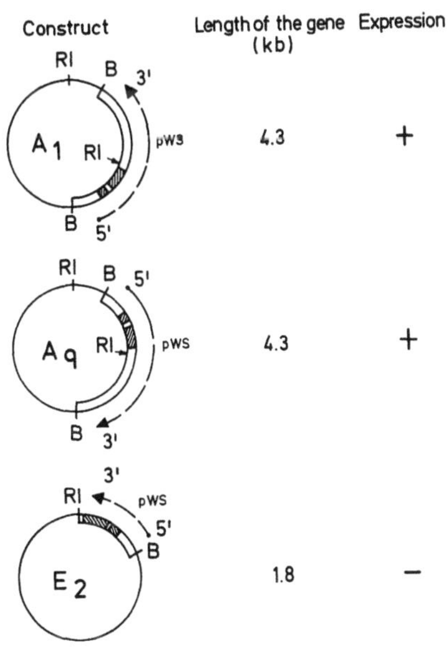

Fig. 5. Expression of cloned genes for the small subunit of the ribulose-1,5 bisphosphate carboxylase from wheat in *Acetabularia.* The genes were inserted in different orientations in pBR 322. These genomic clones were obtained from N.-H. Chua (Rockefeller University, New York). Abbreviations: **A1, A9, E2** code of the clones; **RI** restriction site of the restriction endonuclease EcoRI; **B** restriction site of the restriction endonuclease Bam HI; **pWS** name of the construct

by injecting constructs with different orientations of the inserted small subunit gene (Fig. 5). By means of these experiments it could be shown that the expression does not depend on the orientation of the insert but it does depend on the complete length of the gene. In another set of experiments a gene construct of the zein gene (Wienand et al. 1981) was injected (Langridge, Neuhaus-Url, Neuhaus, Feix and Schweiger in preparation). A comparison between a plasmid which carries a gene and an empty plasmid showed that the zein gene is expressed. Besides the fact of the expression, these experiments have revealed that the immunofluorescence in this specific case exhibits a characteristic structure which might be explained by the extreme hydrophobicity of the gene product.

Finally, a number of experiments should be mentioned in which the DNA of the simian virus 40 was injected (Neuhaus, Neuhaus-Url, Gruss and Schweiger in preparation). Two major observations are remarkable. The first is that the gene expression product has a pronounced affinity to the cell nucleus as is known from animal cells. This might indicate that the gene expression product not only contains the immunological determinants but also those sequences which are responsible for the affinity to the cell nucleus. A second point is the time the immunological reaction needs to become detectable. In the experiments discussed so far it was shown that RNA genomes like the TMV RNA and the mengo virus RNA are expressed significantly faster than DNA like the adenovirus 2 genome and the genes for the small subunit of the carboxylase from wheat and also the gene for zein. In contrast to these genes and genomes respectively the immunofluorescence appears in the cell nucleus as early as 2 h after injection. This indicates that the regulatory sequences contained in the SV40 genome are also effective in the *Acetabularia* system.

Summarizing, one may conclude that the components of the gene expression system in *Acetabularia* are rather unspecific. This is true for that part of the gene expression which occurs in the nuclear compartment, i.e., transcription as well as for that which occurs in the cytosol. Furthermore, one may conclude that gene expression depends on regulatory sequences and that the *Acetabularia* system recognizes sequences even from those genes which are normally expressed in higher animals. A question which is presently being approached is whether the expression is complete and in another set of experiments presently the possibility of the replication of foreign genes and genomes in the *Acetabularia* cell is being studied.

References

Bannwarth H, Ikehara N, Schweiger HG (1977) Proc R Soc Lond B 198:177–190
Bannwarth H, Ikehara N, Schweiger HG (1982) Eur J Cell Biol 27:200–205
Bannwarth H, Schweiger HG (1975) Proc R Soc Lond B 188:203–219
Bannwarth H, Schweiger HG (1983) Cell Biology Int Rep 7:859–867
Berger S, Schweiger HG (1975) J Cell Sci 17:517–529
Broglie R, Coruzzi G, Lamppa G, Keith B, Chua NH (1983) Bio-Technology 1:55–61
Cairns E, Doerfler W, Schweiger HG (1978a) FEBS Lett 96:295–297
Cairns E, Gschwender HH, Primke M, Yamakawa M, Traub P, Schweiger HG (1978b) PNAS 75: 5557–5559
Cairns E, Sarkar S, Schweiger HG (1978c) Cell Biol Int Reports 2, No 6

De Groot EJ, Schweiger HG (1981) Eur J Cell Biol 24:5
De Groot EJ, Schweiger HG (1983a) J Cell Sci 64:13–25
De Groot EJ, Schweiger HG (1983b) J Cell Sci 64:27–36
Hämmerling J (1932) Biol Zentralbl 52:42–61
Hämmerling J (1955) Biol Zentralbl 74:545–554
Hämmerling J (1958) Z Naturf 13b:440–448
Kamptner E (1958) Ann Naturhist Museum, Wien 62:95–122
Mazza A, Bonotto S, Felluga B (1977) In: Woodcock CLF (ed) Progress in acetabularia research. Academic Press, London New York, pp123–136
Neuhaus G, Neuhaus-Url G, Schweiger HG, Chua NH, Broglie R, Coruzzi G, Chu N (1983) Eur J Cell Biol, suppl 2, 30
Schweiger HG, Berger S (1979) Int Rev Cytol 9:11–44
Wienand U, Langridge P, Feix G (1981) Mol Gen Genet 182:440–444
Woodcok CLF, Bogorad L (1970) J Cell Biol 44:361–375

Control of Gene Expression During the Early Phase of Chloroplast Development

K. KLOPPSTECH[1], G. MEYER[1], K. BARTSCH[1], J. HUNDRIESER[1] and G. LINK[2]

Introduction

Light induces changes in the expression of nuclear and chloroplast genes in higher plants [3, 4, 10, 27] and algae 13]. These alterations belong to the most impressive examples in the regulation of genes both with respect to velocity and quantity. The expression of roughly 10 to 20 prevalent mRNA's and their correspondent proteins is regulated during this process. These effects fall into two groups. In one group the proteins and mRNA's disappear with illumination [3, 25], whereas in the other, larger group light exerts a positive control. To the latter belong chloroplast membrane proteins like the light-harvesting chlorophyll a/b protein (LHCP) [3, 9], the 32,000 MW photogene [18, 19], as well as chloroplast stromal proteins like ribulose-1,5-bisphosphate carboxylase (RuBPCase) [28], and the isoenzmyes of chloroplast glyceraldehyde phosphate dehydrogenase (NADP-GAP-DH) [8]. These light-inducible processes involve the regulation of transcription [14] but also rather probably posttranscriptional and posttranslational events such as mRNA and protein degradation [4, 6, 31]. Finally, it has been also found that the assembly of membranes is under the influence of light [4, 6]. These events are presented in Fig. 1 for the two most impressive proteins in barley. It should be stressed, however, that there exist dramatic differences with respect to particular proteins and different species in the degree of light-dependent regulation [6, 8]; this indicates that the expression of distinct genes can be coupled to a different extent to the photoreceptors which are blue light receptors in the case of algae [29] and phytochrome primarily in higher plants [1, 2]. Although it is established at which levels the expression of genes is controlled, the entire mechanism of phytochrome control – as is the case for the blue light receptor(s) – is undeciphered and mostly descriptive. Only a few known facts have implications as to the evolution of working hypotheses like the important finding that phytochrome controls both nuclear and chloroplast genes [18]. So far it is unknown whether these actions are exclusively under the control of phytochrome, e.g., by second messengers, or whether additional events at the levels of transcription and translation may be involved.

1 Institut für Botanik der Universität Hannover, Herrenhäuser Straße 2, 3000 Hannover 21, FRG

2 Biologisches Institut II der Universität Freiburg, Schänzlestraße 1, 7800 Freiburg/Breisgau, FRG

Compartments in Algal Cells and Their Interaction
ed. by W. Wiessner, D. Robinson, and R.C. Starr

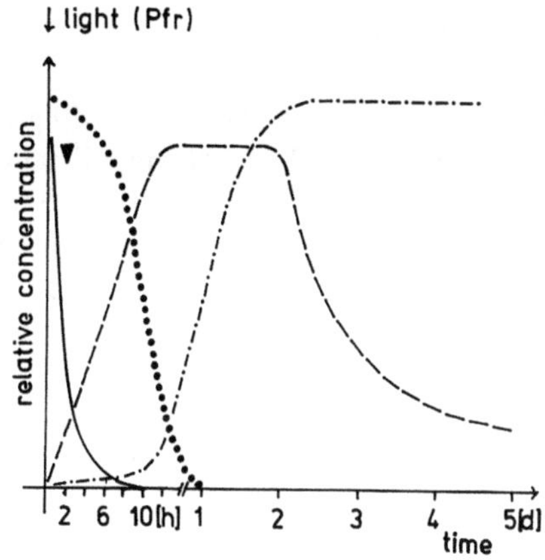

Fig. 1. Negative and positive control by light of two proteins and their correspondent mRNA's in barley. NADPH: Protochlorophyllide oxidoreductase mRNA (———) and the enzyme protein (. . . . . .) are turned off by light, whereas the positively regulated light-harvesting chlorophyll a/b protein (—·—·—) together with its mRNA (– – –) is induced. Semiquantitative representation of data [3, 25]. The *triangle* at 2 h indicates the maximum of the mRNA for the early light-induced protein

Here we provide evidence that in pea and barley additional gene products occur as well as disappear during the early phase of greening, which could indicate that newly synthesized proteins take part in the reorganization of the plastids during the process of greening, not only as structural but also as regulatory elements. These gene products are, however, again under the control of phytochrome.

Methods

Plant Growth

Pea seeds (*Pisum sativum* cv. Rosa Krone) were pre-soaked in tap water over night and grown for 4 d in moist vermiculite at 25°C in the dark and thereafter illuminated with 6000 lx for the indicated times. For the isolation of chloroplasts peas were grown for 7 d on soil under a 12 : 12 h light-dark cycle at 10,000 lx. Barley was cultured at 25°C for 5 d in the dark on vermiculite.

Extraction of RNA and in Vitro Translation

Six g plant material (the entire leaf in barley, the upper half of the stem in pea) was harvested at the indicated times, frozen in liquid nitrogen and the RNA extracted and purified as described [3].

Approximately 20 μg poly(A)RNA and 180 μCi [^{35}S] methionine were used for 240 μl translation assay, sufficient for one transport experiment. Following translation the ribosomes were removed by centrifugation (30 min at 100,000 *g*). Aliquots of 5 and 20 μl were taken for analyis of incorporation into hot trichloroacetic acid material and protein composition, respectively.

In Vitro Transport

Chloroplasts were isolated from 80 g pea plants according to Douce and Joyard [12] and Poincelot [23] with the modifications described [22]. The percoll-purified chloroplasts were washed twice with buffer C [22] and transferred into 1.5 ml transport buffer [15]. The transport protocol followed closely that developed by Grossman et al. [15], the assay consisted of 300 μl transport buffer containing 300–400 μg chlorophyll, 200 μl postribosomal translation supernatant (1.6 x 10^6 dis/min) and 100 μl of threefold concentrated transport buffer. The incubation lasted for 1 h at 25°C and 15,000 lx. Following transport the chloroplasts were pelleted at 2500 *g* for 5 min and the supernatant precipitated with 10% trichloroacetic acid. The precipitate was collected, washed once with acetone and dissolved in sample buffer [21]. The chloroplasts were treated and fractionated into stroma and membranes as described [15].

All fractions (1/5 volume of translation and transport supernatant in comparison to stroma and membrane fractions) were separated on 10 – 15% polyacrylamide gels [21] and treated for fluorography [7].

In Vivo Labeling

Four-d-old etiolated pea seedlings were transferred from the vermiculite to test tubes filled with water. Five plants for every estimation were labeled with 20 μCi[^{35}S] methionine in 0.05% Triton X 100 for the indicated times at 10,000 lx. Thereafter the plants were washed with 10 mM methionine and the upper 5 mm of the plants homogenized in 2 ml buffer (10 mM Hepes, 20 mM Tricine, pH 7.2 with KOH, 1 mM $MgCl_2$, 2% SDS) and proteins solubilized by heating at 65°C for 10 min. The clarified supernatant (10 min at 5000 *g*) was precipitated with 10% trichloroacetic acid, washed once with 2.5% trichloroacetic acid and dissolved in sample buffer. Polyacrylamide gel electrophoresis was performed as described [21].

Quantification of RNA by Northern Blotting

The RNA samples were denaturated for 1 h at 50°C in the presence of glyoxal [20] and separated in 1.1% agarose gels by horizontal electrophoresis [26]. The separated RNA was thereafter transferred with 1 x SSC buffer to gene screen membranes (New England Nuclear) by the capillary blotting method. After drying at 80°C for 2 h the filters were hybridized with nick translated plasmid DNA [24] at 42°C in the presence of 50% formamide (Gene screen manual. New England Nuclear [11, 32]) and washed. The quantification was performed by autoradiography as well as by counting the hybridization signals.

E. coli rRNA served as molecular weight marker.

Results

Quantification of mRNA by in Vitro Translation

The involvement of in the first instance one photoreceptor in higher plants during the light-induced changes in plastid differentiation raises the question as to whether all positively regulated proteins appear and turn over simultaneously or successively.

Examples for both types of regulation have been obtained as will be outlined below. Among the prevalent messengers – as detected by in vitro translation – there are, however, only a few examples of proteins that are repressed by light. To this category belong in pea a 26,000 MW protein (Fig. 2a) and a 21 kD protein of unknown functions as well as a protein precursor of 44,000 MW faintly reactive with the antibody against protochlorophyllide oxidoreductase (lower arrowhead on the right of Fig. 2b). More impressive are those effects on messengers coding for proteins

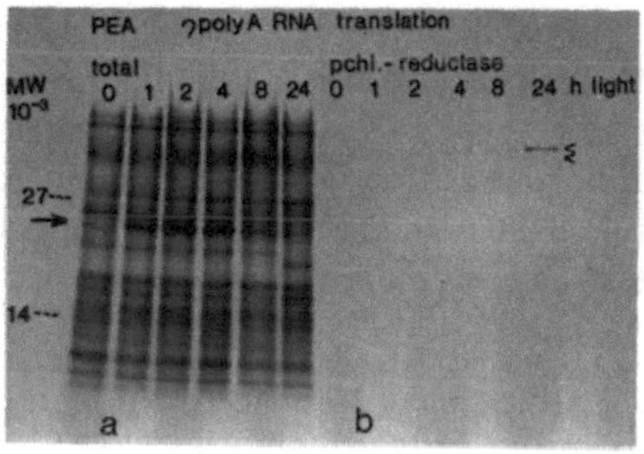

Fig. 2a,b

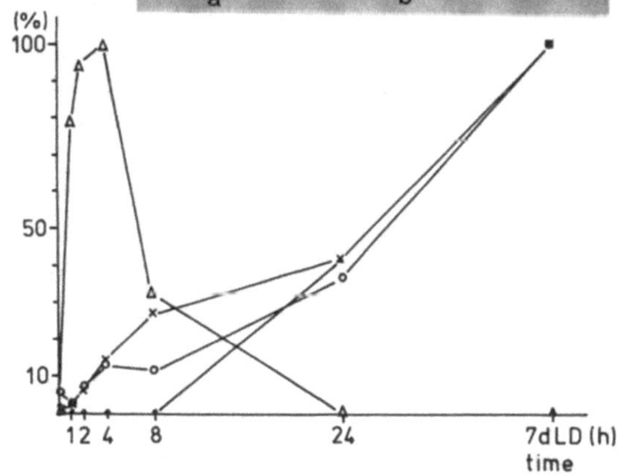

Fig. 3

Fig. 2a,b. Light-induced changes in the poly(A)RNA composition of pea. The total translation products obtained from poly(A)RNA of plants after different periods of illumination were either precipitated with trichloroacetic acid (**a**) or treated with the antibody against barley NADPH: protochlorophyllide oxidoreductase (**b**)

The two immunoreactive proteins are marked by the *arrow heads.* A product of 24,000 M_r culminates among the in vitro translation products at 2 – 4 h of illumination (*arrow* in a)

Fig. 3. Appearance of mRNA's for light-induced proteins as estimated from in vitro translation data. Poly(A)RNA's from plants after different periods of illumination were translated under limiting amounts of RNA and either separated directly or after immunoprecipitation. Autoradiograms were scanned and the linearity of density with radioactivity established.

Light-harvesting chlorophyll a/b protein (x——x), small subunit of RuBPCase (•——•), NADPH: protochlorophyllide oxidoreductase (o——o), early light-induced protein

that are induced by light. To these belong primarily the small subunit of RuBPCase, the LHCP, a protein reactive with the antibody against chloroplast GAP-DH and a protein precursor of 46,000 MW that can be precipitated with the antibody against barley protochlorophyllide oxidoreductase (Fig. 2b, upper arrowhead). The striking observation is, however, the appearance and disappearance of a protein of 24,000 MW within the first 24 h of greening. The translatable mRNA is not detectable in the dark, culminates between 2 and 4 h after illumination and disappears within less than one day. The data for these four proteins are summarized in Fig. 3 in which densitometer tracings of autoradiograms either directly or from immunoprecipitated proteins have been used for calculation (under criteria which ensure a linear relationship between radioactivity and density). This plot again demonstrates the unique kinetic behavior of the 24,000 MW protein. Furthermore the mRNA for two of these proteins, LHCP and the larger of the two oxidoreductase like proteins, are present in the dark and induced to about 20% of the light controls already after 8 h; at that time the mRNA for the small subunit of RuBPCase is present in traces, if at all.

Quantification of mRNA merely by in vitro translation might be a dangerous attempt, because it involves the formation of an initiation complex and competition between mRNA's with different binding constants. Attempts have been made to avoid this artifact by the translation of limiting amounts of RNA and we are convinced that by this method the values obtained at different times for but one protein are comparable (see also below).

Quantification of Messenger RNA by Hybridization

In order to exclude the above-mentioned possibility of competition during initiation of protein synthesis, we also used the method of hybridization of mRNA against cloned and radioactively labeled cDNA for different proteins. This method not only allows examination of the in vitro translation data presented above, but also gives access to the analysis of such proteins for which an antibody is not available like the 32,000 MW gene product of the chloroplast DNA. The data summarized in Fig. 4 confirm the data obtained by in vitro translation for the LHCP and the small subunit of RuBPCase with respect to the order of appearance and the extent of regulation.

In addition, we find that the light-dependent induction of the GAP-DH-like protein is even more slow than that of the other two proteins mentioned. The two chloroplast gene products are concerned, which reside in the poly(A) minus RNA fraction, they behave like their correspondent partners in the poyl(A)RNA fraction. The mRNA for the 32,000 MW photogene, which is in close contact to the LHCP within the thylakoid membranes, is already present in the dark, whereas the chloroplast coded large subunit of RuBPCase, the counterpart of the small subunit, is rather slowly induced and appears in parallel to the small subunit mRNA. These results indicate that chloroplast and nuclear genes which are functionally or structurally related are cooperatively expressed. It also shows that the chloroplast genes, which have been found to be under phytochrome control, are individually induced at least in pea.

Fig. 4. Quantification of light-induced mRNA's by Northern hybridization. Poly(A)RNA was denatured, separated on 1.1% agarose gels and transferred to nitrocellulose. The baked filters were hybridized to nick-translated, cloned cDNA as described in methods. The radioactive bands were localized by autoradiography, cut and counted. Data are expressed in percent of the maximal values. Light-harvesting chlorophyll a/b protein (x——x), small subunit of RuBPcase (•——•), GAP-DH like protein (o——o), 32,000 MW "photogene" (x- - - x), large subunit of RuBPCase (•- - - •)

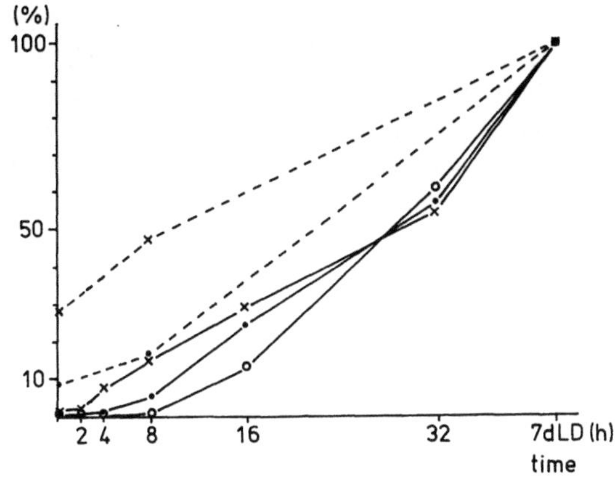

It should be stressed, however, that the chronological order in which these proteins occur in pea is by no means an absolute one in higher plants. The small and large subunit of RuBPCase, for instance, are expressed already in the dark to a considerable degree in the monocotyledonean species barley, oats and wheat ([8] and unpublished), whereas the LHCP and the 32,000 MW photogene are not. It seems that the differentitation of membranes is the essential step (see also below) to which the expression of RuBPCase is more or less tightly coupled.

Identification of the 24,000 Protein as a Nuclear Genome Coded Chloroplast Protein

A first step towards the establishment of the function of the 24,000 MW protein consists in the determination of its cellular location. Most of the light-mediated messengers code for chloroplast proteins. A useful tool to decide whether a protein belongs to this group is the in vitro transport system developed independently by Chua and Ellis and their coworkers [15, 17]. Figure 5 shows the results of a transport experiment. A number of membrane as well as soluble proteins appear with light. Among those identified are the GAP-DH and small subunit RuBPCase in the stroma

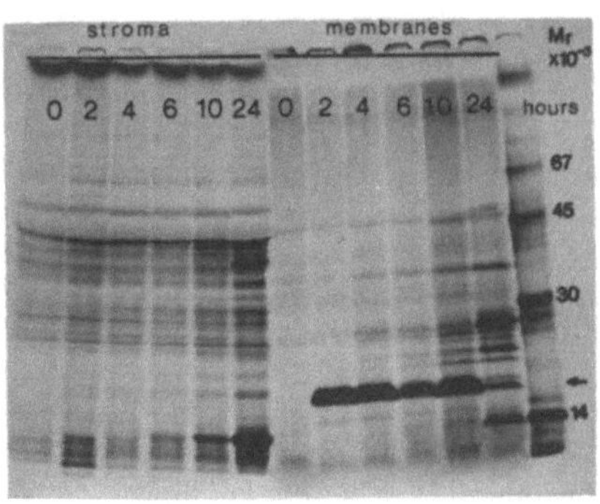

Fig. 5. Posttranslational transport of in vitro translation products obtained from pea mRNA. Translation products from mRNA of plants illuminated for the indicated periods were combined with intact chloroplasts as described in the methods. The washed and trypsinized chloroplasts were fractionated into stroma and membrane fraction and separated on a 10-15% polyacrylamide gel. *Arrow* indicates the 17,000 M_r protein

and the LHCP in the membrane fractions. A heavy band of 17,000 MW appears during the early phase of greening in the membrane fraction, this band is accompanied by a faint second band which is slightly larger. Finally a still fainter band resides in the 15,000 MW area. With translation products from RNA isolated after 24 h of illumination only a small amount of the 17,000 MW protein can be obtained in the chloroplast membranes. The massive 24,000 MW band before transport and the 17,000 MW band after transport show the same kinetics of appearance and decay. We therefore conclude that the 24,000 MW protein is transported into the chloroplasts and is thereby processed to the 17,000 MW protein which is then integrated into the membranes. There are a number of peculiarities in this connection; the first is that a protein whose mRNA has a maximum after about 3 h of illumination is integrated into the membrane fraction of chloroplasts taken from green, although actively growing plants. It seems to us that this binding is not an artificial one, since the binding is stable against the treatment with 1 mM sulfobetaine during 15 min at 0°C (data not shown).

Detection of the 17,000 Protein After Labeling in Vivo

In order to detect the 17,000 MW protein in vivo, plants were labeled for periods of 3 h throughout the first 24 h of illumination and a membrane fraction was isolated. The results are presented in Fig. 6. There is a faint band present when labeled from 4 to 5 h after illumination. This band migrates precisely with the 17,000 MW protein taken from membranes of an in vitro transport experiment. The data correlate with those described for the mRNA of the 24,000 protein, indicating that the mRNA for the 17,000 in vivo labeled protein is no longer present after about 24 h of illumination. From similar experiments a half live time of the 17,000 protein of approximately 5 h has been calculated.

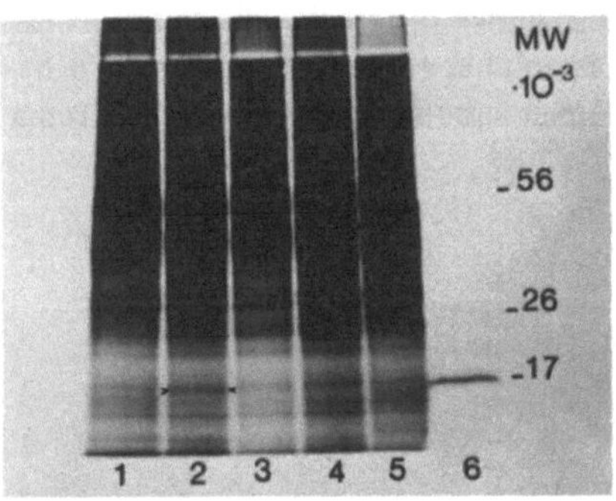

Fig. 6. In vivo labeling of the 17,000 M_r protein in pea. Etiolated pea plants were transferred to light and labeled as described in the methods section. The 17,000 M_r product of in vitro transport was used in a 10-15% polyacrylamide gel as a reference
The plants were homogenized after 3, 5, 6, 9, 24 h of illumination and applicated to slots 1-5, respectively. Incubation lasted for 3 h (slots 1, 3-5) and for 1 h (slot 2) prior to homogenization. The 17,000 M_r protein in slot 2 is marked by the *arrowheads*

Presence of the 17,000 Protein in Other Plant Species

The differences between different plant species in response to light are considerable. For this reason we tried to detect the 17,000 protein in another plant species. For obvious reasons barley was chosen; this species facilitates the analysis of phytochrome

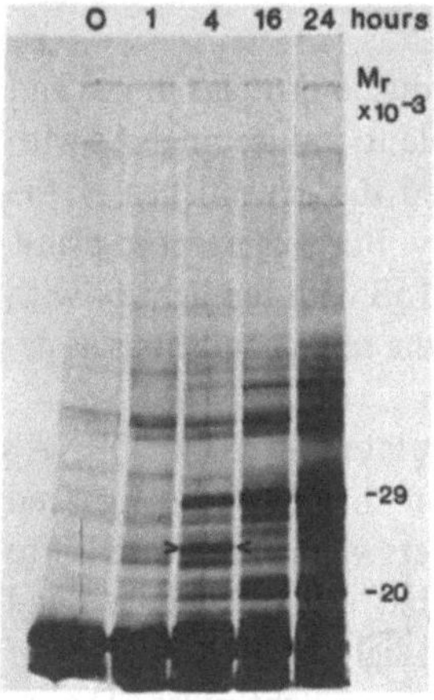

Fig. 7

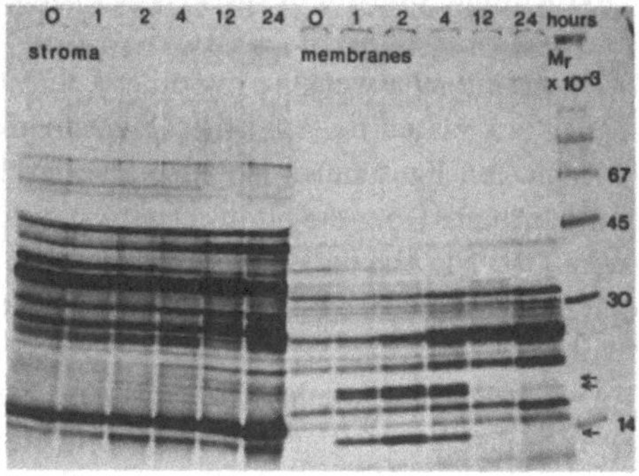

Fig. 8

Fig. 7. Light-induced changes in the mRNA composition of barley. In vitro translation products obtained from barley mRNA isolated after different times of illumination were precipitated with trichloroacetic acid and separated on a 10-15% polyacrylamide gel. *Arrowheads* indicate the position of the 24,000 MW translation product

Fig. 8. Heterologous transport into pea chloroplasts of translation products of mRNA isolated from barley plants after different times of illumination. After transport the chloroplasts were treated and separated as described in Fig. 5

action and offers the advantage of cells of different developmental stages within one leaf. Among the translation products from poly(A) RNA a protein of 24,000 MW shows up with a maximum of expression at 2 h of illumination (Fig. 7), indicating that the correspondent mRNA is occurring rapidly and disappearing from the poly(A) RNA fraction. Other peculiarities visible in this figure have been considered already [3]. Transport of these in vitro labeled products into pea chloroplasts yields similar results as obtained with the homologous pea system. In this case (Fig. 8) we obtain again a double band at about 17,000 MW but in this case of equal intensity as well as an additional band of approximately 14,000 again more prominent than that in the homologous system. For this reason it is hardly possible to establish a precursor product relationship, since we cannot decide whether the data represent incomplete or rate-limiting processing, or whether there are indeed two proteins of similar intensity; this question can only be decided by means of proteolytic fingerprinting. A similar argument is valid for the small band of 12,000-14,000 which appears and disappears in parallel to the double band; in this case we cannot exclude the possibility that this product stems from the processing step and is not digested in the heterologous system. In favor of the latter argument is the fact that no in vitro translated product of similar or higher molecular weight can be detected among the in vitro translation products. These studies have to be extended.

Entirely the same results as with the wild-type barley have been obtained for the mutant chlorina f-2, indicating that its genetic defect has nothing in common with these early proteins (results not shown).

Phytochrome Control of the 17,000 Proteins in Barley

We consider as a working hypothesis that the early light induced proteins in pea and barley are related to each other. Since in pea it is not possible to induce mRNAs with a single red light pulse to detectable levels, we investigated the dependence of the early proteins from the photoreceptor phytochrome in barley. For this purpose plants were grown in the dark and after 5 d either illuminated for 15S with red light or with red light followed by 5 min of far-red light entirely under the same conditions as described [1].

These plants and dark controls were harvested two hours later, the poly (A)RNA's were isolated and translated. The induction effect of phytochrome by a single red light pulse is rather weak, therefore the translation products were transported into chloroplasts in order to relatively enrich those proteins that are light-induced and membrane-bound. The results are presented in Fig. 9. Red light inducible bands of 32,000, 25,000 (LHCP), and 22,000 MW appear after the red light pulse. The influence of phytochrome is confirmed by the far-red light reversion which is not complete, as the dark controls show. Important is that also the bands at 17,000 and 12,000-14,000 MW appear induced by red light as is indicated by the arrows in Fig. 9. Again the induction can be partially reverted by far-red light treatment.

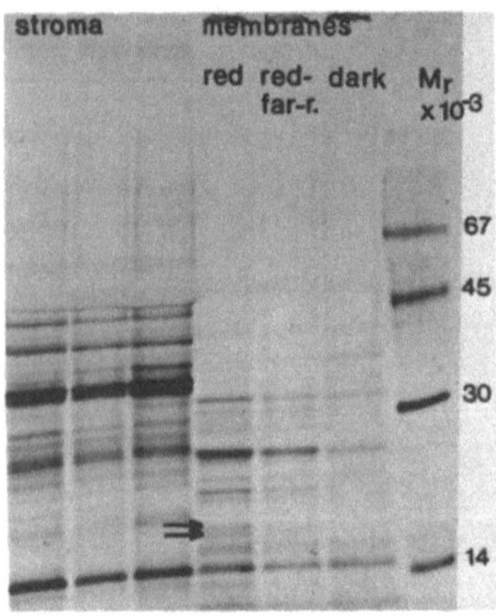

Fig. 9. Phytochrome effect on the appearance of the early proteins. Etiolated seedlings were either illuminated for 15 s with red light or with red light followed by 5 min far-red light and kept for another 2 hours in the dark, thereafter the poly(A)RNA was isolated and translated. Etiolated plants served as controls. The translation products were transported into isolated pea chloroplasts and these were fractionated into stroma and membranes as described in methods

Conclusions

One interesting aspect of the data reported and mentioned in this paper is the antagonistic regulation of particular light-induced proteins in mono- and dicotyledonous plants, like the LHCP, the 32,000 photogene and the two subunits of RuBPCase, although all of them seem to be under the control of phytochrome [1, 18, 28]. This could mean that despite the fact that the effects induced by light are rather impressive, their implication for plant development might be an overestimate; indeed, as the situation in gymnosperms and green algae shows, this development can occur in the absence of light and the differentiated structures can persist in darkness. Light control of plastid development seems a rather recent event, evolved to the highest in the monocotyledonean species.

Although there has been considerable progress in the knowledge of the light effects on gene expression during the last few years, the link between the photoreceptor phytochrome and the control of transcription at the molecular level is poorly understood. It should be pointed out, however, that the light-mediated processes resemble closely those animal systems which are under hormone control with respect to the rapidity of response, the number of genes involved and the extent to which these are regulated, and finally with respect to the primary mode of regulation at the level of transcription.

Of considerable importance is the finding that not only the nuclear genes but also the chloroplast genes are regulated via the phytochrome system. The protein described here could serve as a nuclear coded signal by which the expression of chloroplast genes might be regulated at the level of transcription.

This argument, however, seems unlikely since in both investigated species, pea as well as barley, one of the most prominent products of the chloroplast genome is expressed already in the dark, that is in the absence of both, light and the 17,000 MW early light-induced protein. The role in the regulation of transcription could, under these circumstances, be only a modifying one.

Another function during greening has been postulated to explain the specific disappearance of proteins [25] and other phenomena that occur when the normal process of greening is experimentally disturbed as is the case when plants are grown under intermittent light [9, 30], in far-red light [4] or redarkened [6]. In the first case specific proteins disappear rapidly, whereas in the other experimental attempts light-specific proteins do not accumulate although the latter are expressed, translated and transported into the organelles. Specific proteases would help to explain the observations.

There has been the description of a neutral protease in plants in connection with the light-induced development in oats [16]; this enzymatic activity appears and vanishes within a few hours; it has a maximum at about 4 h after light induction in oats. The protein has been postulated to function in the complete destruction of the prolamellar body providing amino acids for newly synthesized proteins. It is meanwhile known that the alterations in membrane proteins are limited to rather few proteins. If we assume that about 1 or 2 h are needed for the translation of sufficient amounts of enzyme protein, the data published [16] agree well with the information so far available for the early light-induced protein. We therefore propose as a working hypothesis that the 17,000 MW protein might function as a protease.

The rapidity in induction and disappearance of this protein could well help to understand the flexibility of light induction against alterations in the environment. Contrary to the original proposal of the authors [16] the function of the protease should be a rather specific one. This fact so far cannot be sustained by the published evidence because hemoglobin could be used as one of the substrates in vitro. One may expect that this finding probably will complicate the detection of the function under in vivo conditions.

Acknowledgements. We are grateful for the experienced experimental help of Mrs. Heidemarie Bliedung. Sacco de Vries, Groningen, kindly provided us with the clones for pea LHCP and small subunit RuBPCase. The study was supported by the Deutsche Forschungsgemeinschaft (Grant Kl 401/5-3).

References

1 Apel K (1979) Eur J Biochem 97:183–188
2 Apel K (1981) Eur J Biochem 120:89–93
3 Apel K, Kloppstech K (1978) Eur J. Biochem 85:581–588
4 Apel K, Kloppstech K (1980) Planta (Berl) 150:426–430
5 Bellemare G, Bartlett SG, Chua NH (1982) J Biol Chem. 257:7762–7767
6 Bennett J (1981) Eur J Biochem 118:61–70
7 Bonner WM, Laskey RA (1974) Eur J Biochem 46:83–88
8 Cerff R, Kloppstech K (1982) Proc Natl Acad Sci USA 79:7624–7628
9 Cuming AC, Bennett J (1981) Eur J Biochem 118:71–80
10 DeVries SC, Springer J, Wessels JGH (1982) Planta (Berl) 156:129–135
11 DeVries SC, Springer J, Wessels JGH (1983) Planta (Berl) 158:42–50
12 Douce R, Joyard J (1979) Adv Bot Res 7:1–116
13 Gilbert CW, Buetow DE (1982) Biochem Biophys Res Commun 107:649–655
14 Gollmer I, Apel K (1983) Eur J Biochem 133:309–313
15 Grossman AR, Bartlett SG, Schmidt GW, Mullet JE, Chua NH (1982) J Biol Chem 257: 1558–1563
16 Hampp R, De Filippis LF (1980) Plant Physiol (Bethesda) 65:663–668
17 Highfield PE, Ellis RJ (1978) Nature (Lond) 271:420–424
18 Link G (1982) Planta (Berl) 154:81–86
19 Mattoo AK, Pick U, Hoffman-Falk H, Edelman M (1981) Proc Natl Acad Sci USA 78: 1572–1576
20 McMaster GK, Carmichael GG (1977) Proc Natl Acad Sci USA 74:4835–4838
21 Neville DM (1971) J Biol Chem 246:6328–6334
22 Pfisterer J, Lachmann P, Kloppstech K (1982) Eur J Biochem 126:143–148
23 Poincelot RP (1980) Methods Enzymol 69:121–128
24 Rigby PWJ, Dieckman M, Rhodes C, Berg P (1977) J Mol Biol 113:237–251
25 Santel HJ, Apel K (1981) Eur J Biochem 120:95–103
26 Thomas PS (1981) Proc Natl Acad Sci USA 77:5201–5205
27 Tobin EM (1981) Plant Mol Biol 1:35–51
28 Tobin EM, Suttie JL (1980) Plant Physiol (Bethesda) 65:641–647
29 Vesk M, Jeffrey SW (1977) J Physiol 13:280–288
30 Viro M, Kloppstech K (1982) Planta (Berl) 154:18–23
31 Viro M, Kloppstech K (1983) Planta (Berl) 157:202–208
32 Wahl GM, Stern M, Stark GR (1979) Proc Natl Acad Sci USA 76:3683–3687

Aspects of Translational Coordination During Chloroplast Development

P. BRANDT[1]

The biogenesis of chloroplasts includes reproduction and differentiation of the organelles. Studies of this phenomenon in higher plants, as well as in unicellular algae, indicate that the biogenesis of the chloroplast is a complicated, highly organized, and well-regulated process depending on both chloroplast and cytoplasmic protein syntheses (Kirk and Tilney-Basset 1978; Herrmann, Börner and Hagemann 1980). Our present knowledge of the nature of the interaction between the nuclear and the chloroplast genetic systems was summarized recently in five principles by Ellis (1983). These principles include the site of encoding and of synthesis of chloroplast polypeptides, the translocation of the nuclear-coded polypeptides across the chloroplast envelope, the capacity of the chloroplast genetic system, the entire function of the chloroplast-encoded polypeptides within the chloroplast compartment, and the light stimulation of chloroplast and of cytoplasmic protein synthesis. These five principles reflect the required cooperation between the chloroplasts and the nucleocytoplasm in higher plants very well. There are, however, several lines of evidence which indicate that the proposed "cytoplasmic control principle" (Ellis 1977) in higher plants is contradicted by results from unicellular algae like *Euglena gracilis* (Pineau 1982), *Chlamydomonas reinhardii* (Iwanij, Chua and Siekevitz 1975) or *Chlorogonium elongatum* (Westhoff et al. 1981). In particular, in these organisms synthesis and assembly of the plastid-coded large subunit and of the nuclear-coded small subunit of the carboxydismutase are under strong control of the chloroplast genetic system. The degree of corregulation of the two genetic systems during the formation of the photosynthetic apparatus is much more complicated. The aim of the present investigations is to elucidate the mode of translational coordination in *Euglena gracilis*. Special reference will be made to the stage-specific formation of thylakoid proteins during the cell cycle of *Euglena gracilis*, which appears to be closely coupled, ensuring that the required constituents will be present at the right time and in the correct amounts to achieve an orderly assembly of the membranes and a suitable efficiency of the photosynthetic apparatus.

There are about 11 chloroplasts in autotrophically cultured *Euglena* cells which are located in the periphery of the cell. The chloroplast biogenesis in synchronized *Euglena gracilis* takes place during the light-time of the cell cycle, but neither the chloroplasts nor the *Euglena* cells divide during this period (Brandt and von Kessel 1983). It is important that a developmental period for the chloroplasts is provided

1 Abteilung für Experimentelle Phykologie, Pflanzenphysiologisches Institut, Universität Göttingen, FRG

Compartments in Algal Cells and Their Interaction
ed. by W. Wiessner, D. Robinson, and R.C. Starr

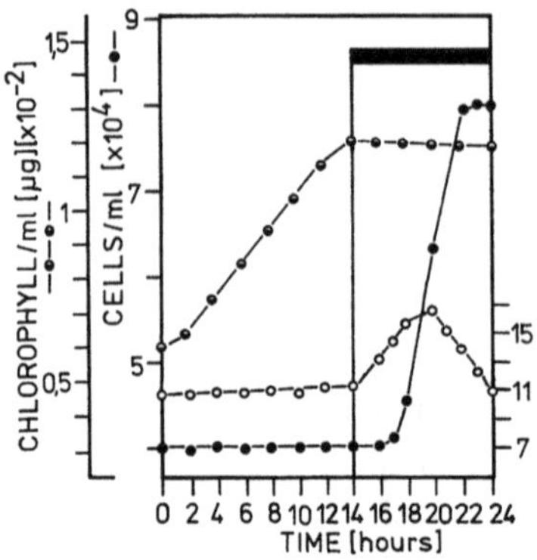

Fig. 1. Cells/ml (●), chlorophyll/ml (◒) and chloroplasts/cell (○) of a *Euglena gracilis* culture synchronized by a light-dark change of 14 h light and 10 h dark. (Brandt and von Kessel 1983)

by a light-dark change of 14 h light and 10 h dark in which division of the cells or of the chloroplasts does not occur (Fig. 1). Total protein and chlorophyll of *Euglena gracilis* double in a linear fashion during this light-time (Buetow et al. 1980). Synthesis of chlorophyll a, however, is especially favored between the 2nd and the 6th h of the light-time, whereas synthesis of chlorophyll b takes place with high intensity between the 6th and the 14th h of the light-time (Brandt 1980). In correlation with the consecutive syntheses of chlorophyll a and chlorophyll b the photosynthetic activity of whole *Euglena* cells (Brandt and von Kessel 1983), as well as of isolated thylakoid particles (Brandt 1981), changes remarkably (Figs. 2–4). In particular, the photosynthetic oxygen evolution does not reflect the chlorophyll accumulation and is low at the 6th and at about the 12th h of the light-time (Fig. 2). These stage-specific alterations of the photosynthetic oxygen evolution align with the cell cycle-dependent increase or decrease of the photosynthetic activity itself. To determine the changes of the PSI- and PSII activity of synchronized *Euglena gracilis* in more detail photosynthetically active thylakoid particles were isolated and separated into two fractions of PSII particles – named P2 and P3 – and one fraction of PSI particles. The photosynthetic activity of all of these particles shows maxima at about the 2nd and the 9th h and minima at the 6th and the 12th h of the light-time. It is of special interest that the mode of photosynthetic oxygen evolution coincides with the changes of PSI- and PSII activity (Figs. 2–4). Moreover, at the 6th h, where oxygen evolution, PSI- and PSII activity are low, the amount of the PSII particles P2 decreases and that of the PSII particles P3 increases (Fig. 3). This is the first hint of a drastic thylakoid reorganization at the 6th h, because the P3 particles are much smaller than the P2 particles.

For further elucidation of this phenomenon of thylakoid reorganization in the middle of the light-time, it was necessary to determine the exact time of synthesis of the photosystem II- and photosystem I-related proteins. These investigations were done with the chlorophyll-protein complexes CPI, CPa, and LHCP, which can be separated by gelelectrophoresis (Brandt 1980, Brandt and von Kessel 1983, Schuler et al. 1982). CPI is related to photosystem I and has an apparent molecular weight of 110,000. CPa is related to photosystem II and has an apparent molecular weight of 56,000. CPI and CPa contain chlorophyll a only. The light-harvesting-chlorophyll-

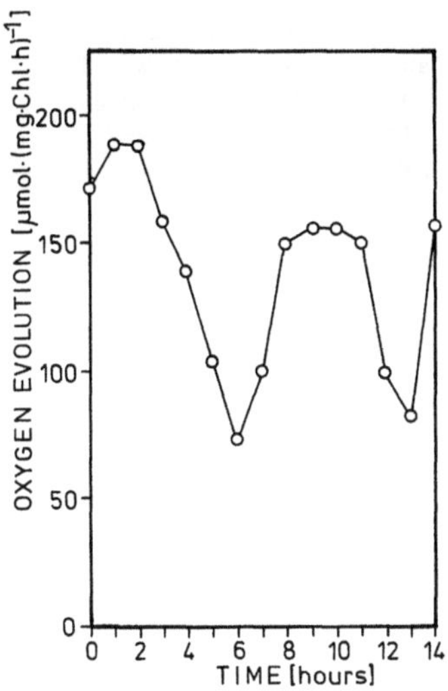

Fig. 2

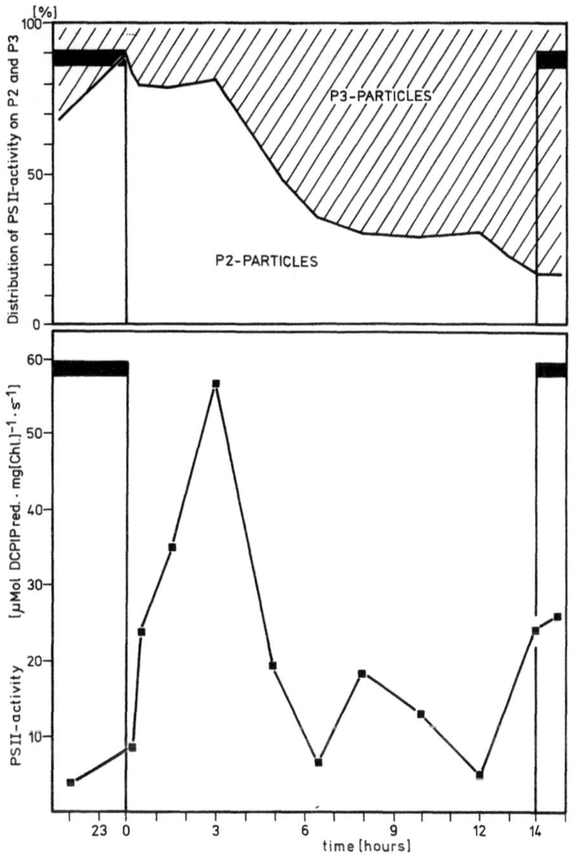

Fig. 3

Fig. 2. Photosynthetic oxygen evolution of synchronized *Euglena gracilis*, strain Z, during the light-time (14 h) of the cell cycle (24 h). (Brandt and von Kessel 1983)

Fig. 3. PSII-activity, measured as DCMU-sensitive DCPIP-reduction, of the P2- and P3 particles of *Euglena gracilis*, strain Z, during the light-time of the cell cycle, and percentual distribution among the two particle fractions. (Brandt 1981)

Fig. 4. PSI-activity, measured as methylviologen reduction of thylakoid particles of *Euglena gracilis*, strain Z, during the light-time of the cell cycle. (Brandt 1981)

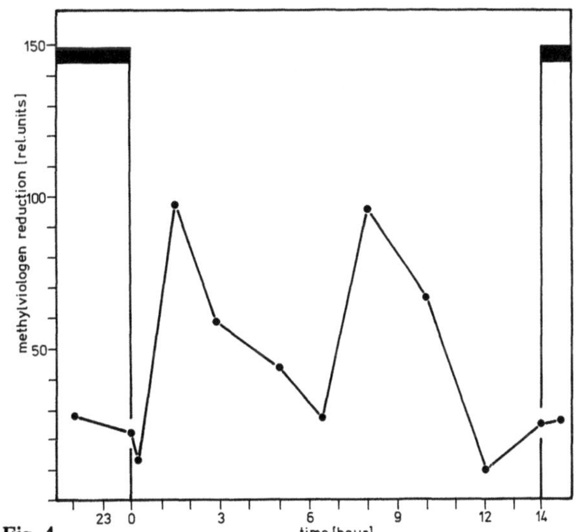

Fig. 4

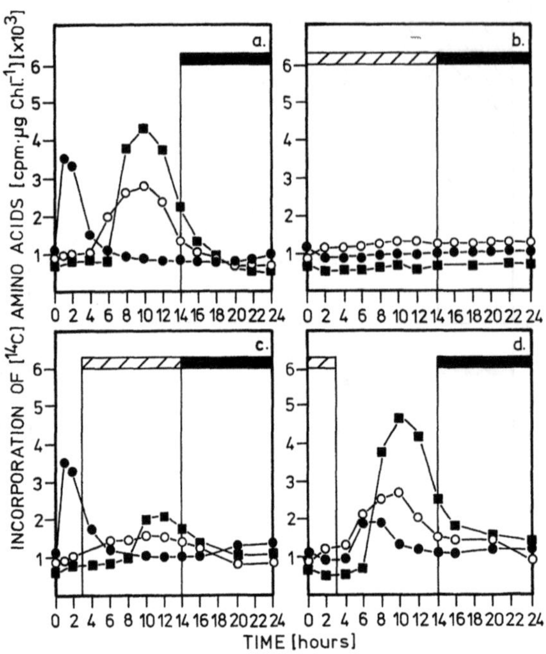

Fig. 5. Incorporation of [^{14}C] amino acids in the thylakoid chlorophyll-protein-complexes CPI (●), CPa (○), and LHCP (■) of synchronized *Euglena gracilis*, strain Z, during the cell cycle with 14 h light and 10 h dark (a), during the cell cycle with additional dark during the light-time (b), with additional dark at the 3rd to the 14th h of the light-time (c) or with additional dark at the beginning to the 3rd h (d). (Brandt and von Kessel 1983)

protein-complex LHCP, however, contains chlorophyll a and chlorophyll b and two apoproteins with the molecular weight of 24,000 and 26,000.

These special proteins are synthesized at distinct stages of the cell cycle only (Fig. 5; Brandt and von Kessel 1983).

a) Synthesis of CPI occurs at the 1st to the 2nd h of the light-time and is light-dependent. An additional dark period of 3 h at the beginning of the light-time causes a shift of the CPI synthesis to the time period between the 4th and the 7th h of the light-time and in addition a decrease in the rate of synthesis to about 50%.

b) Synthesis of CPa occurs at the 7th to the 12th h of the light-time and is also light-dependent. An additional dark period of 3 h at the beginning of the light-time causes no significant alterations; whereas, during an additional dark period of 11 h between the 3rd and the 14th h, the synthesis of the CPa-apoprotein decreases to about 10%.

c) The assembly of the two LHCP-apoproteins occurs at the 8th to the 12th h and is also light-dependent. In an additional dark period it behaves like the synthesis of CPa.

It is of special interest to consider that the syntheses of PSI- and PSII-related proteins are separated in time under synchronization conditions. These results align with reports that in greening barley seedlings PSI activity precedes PSII activity (Henningsen and Boardman 1973, Plesnicar and Bendall 1973). Moreover, these stage-specific syntheses of thylakoid proteins in *Euglena gracilis* are consistent with the stage-specific changes of the activity of PSI- or PSII particles (Figs. 3 and 4; Brandt 1981). The stage-specific decrease of the photosynthetic activities at the 6th h of the light-time can be explained by insertion of newly synthesized CPa and LHCP,

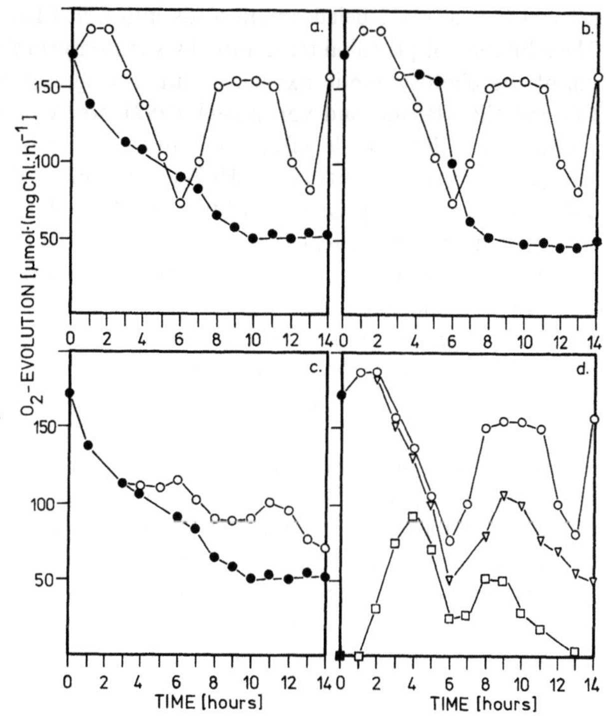

Fig. 6. Photosynthetic oxygen evolution (○) of synchronized *Euglena gracilis*, strain Z, during the light-time (14 h) of the cell cycle (24 h). Light intensity 32,026 Wm^{-2}. a (●), Additional dark period at the beginning to the 14th h of the light-time (LH). b (●), Additional dark period at the 3rd to the 14th h LH. c (●), Additional dark period at the beginning to the 3rd h LH. d oxygen evolution after treatment with cycloheximide (▽) or streptomycin (□) for 1 h. (Brandt and von Kessel 1983)

both related to PSII, and the occurrence of new PSII particles P3 necessarily accompanied by thylakoid reorganization.

The successive insertion of CPI, CPa, and LHCP is presumably necessary for the formation of an efficient photosynthetic apparatus, because simultaneous insertion is not efficient in photosynthetic oxygen evolution (Fig. 6c; Brandt and von Kessel 1983). The efficiency of oxygen evolution therefoe depends not only on the amount of the different components of the photosynthetic apparatus, but also on their optimal arrangement in the thylakoid system.

Before attempting any study of the photosynthetic efficiency of *Euglena* chloroplasts in vivo with respect to this succession of syntheses, it was important to investigate the behavior of the "water-splitting-enzyme" in *Euglena gracilis* chloroplasts. Especially in *Euglena*, the repair of water-splitting activity is accompanied by the synthesis of some polypeptides in the molecular weight range of 50,000 to 60,000 (Gurevitz et al. 1977). The thylakoid content of this group of polypeptides, as well as of two other polypeptides with molecular weights of about 23,000 and 16,000, presumably related to the "water-splitting-enzyme" (Akerlund et al. 1982) increases only very slightly during the light-time of the cell cycle, and therefore most probably is not responsible for alterations in photosynthetic oxygen evolution.

The direct connection between photosynthetic oxygen evolution and translation processes can be demonstrated by experiments with cycloheximide or streptomycin (Fig. 6d). In these experiments *Euglena gracilis* was treated with these translation-inhibitors for 1 h and then the photosynthetic oxygen evolution and the incorporation of ^{14}C-labeled amino acids into CPI, CPa, and LHCP was determined. The inhibition of plastidial translation by streptomycin strongly decreases both maxima of the photosynthetic oxygen evolution (Fig. 6d) and inhibits the synthesis of CPI and CPa (Brandt and von Kessel 1983), whereas the inhibitor of cytoplasmic translation, cycloheximide, effects only the second maximum of photosynthetic activity (Fig. 6d) and inhibits only LHCP-apoprotein synthesis (Brandt and von Kessel 1983). As reported by Ortiz and Stutz (1980), polypeptides related to CPI and to CPa are synthesized in *Euglena* chloroplasts and LHCP polypeptides are synthesized in the cytoplasm. Therefore, the first maximum of activity in oxygen evolution (Fig. 2) is directly related to plastidial synthesis of CPI proteins (Fig. 5) and the second maximum is related to plastidial synthesis of CPa proteins and to the assembly of LHCP proteins of cytoplasmic origin. The stage-specific synthesis of CPI in synchronized *Euglena gracilis* probably changes the efficiency of the photosynthetic apparatus resulting in the first maximum of oxygen evolution, whereas the stage-specific synthesis of CPa increases the amount of oxygen producing centers per cell itself and causes the second maximum of oxygen evolution. The decrease of oxygen evolution at the 4th to the 6th h of the light-time could be caused by a rearrangement of CPI and a beginning of insertion of CPa, because oxygen evolution activity is stable for a longer time when CPa protein synthesis is omitted (Fig. 6b).

However, what kind of regulation exists between the synthesis of LHCP precursor proteins in the cytoplasm and their simultaneous insertion into the thylakoids together with CPa proteins? One of the most important steps would be the passage of the LHCP precursor proteins through the *Euglena* chloroplast envelope. All hypotheses which explain the insertion of proteins into, and their eventual translocation through, biological membranes have in common a recognition step followed by internalization or externalization. Prior to translocation, modification is required of both the structure of the proteins to be translocated and of some areas of the membrane to recognize or to be recognized. Therefore the array of chloroplast envelope proteins would be a prerequisite for the regulation of the protein translocation.

The protein composition of the envelope of *Euglena* chloroplasts was established by chloroplast isolation and – after osmotic shock of the chloroplasts – by separation of an envelope fraction from the thylakoids. This method yields enough material to quantify 23 polypeptides of the inner envelope membrane after gelelectrophoresis and silver staining (Fig. 7; Brandt 1982). No data, however, are available about the polypeptide composition of the two outer envelope membranes. The amount of each polypeptide in the inner envelope membrane is not constant but changes markedly. These alterations are shown in more detail for five envelope polypeptides, which are also synthesized by isolated *Euglena* chloroplasts (Fig. 8). It should be recalled, however, that some of the other 18 polypeptides can also be plastid-coded and synthesized in vivo but not in organello, if their structural gene belongs to the chloroplast DNA while a regulatory gene of the nuclear DNA regulates its expression (Brandt 1976). Nevertheless, these five chloroplast-encoded polypeptides represent about 22% of the total envelope protein content during the whole chloroplast biogenesis. It is of special interest for the translocation of the LHCP precursor proteins, that at the 6th h of the light-time the amount of the 12,000 MW polypeptide is highest. But also the amounts of the four other polypeptide change remarkably at the 6th h of the light-time.

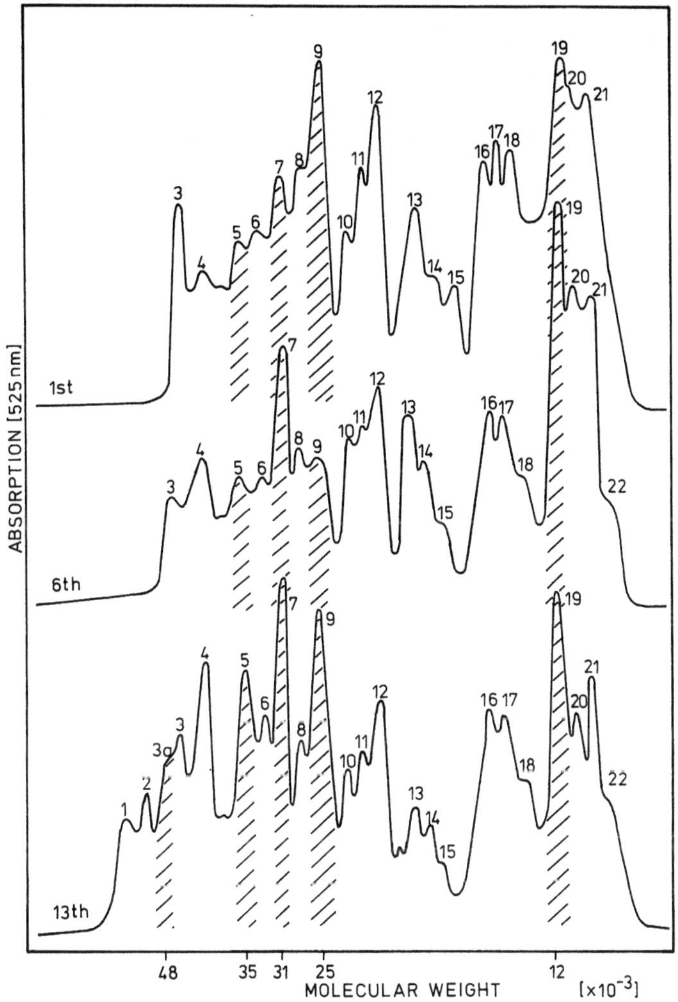

Fig. 7. SDS-PAGE of chloroplast envelope polypeptides of *Euglena gracilis*, strain Z, isolated at the 1st, 6th or 13th h of the cell cycle. (Brandt 1982)

In correlation with these alterations of the polypeptide composition of the chloroplast envelope the surface pK of isolated chloroplasts shifts slightly from 5.2 to 5.7 during the light-time (Brandt and Ernst 1982).

So far no experiments have been performed which indicate the direct dependency of LHCP precursor translocation on the stage-specific protein composition and arrangement of the chloroplast envelope by chloroplast-encoded polypeptides. Two experiments, however, hint at this presumed mode of regulation.

1. A few years ago it was demonstrated that the RNA polymerase of *Euglena* chloroplasts has its optimal activity at 28° and is inactive at 34°–35°, whereas the nuclear RNA polymerase are optimally active at 32°–33° and still active at 34°–35° (Brandt and Wiessner 1977). *Euglena* cultures are permanently bleached after about

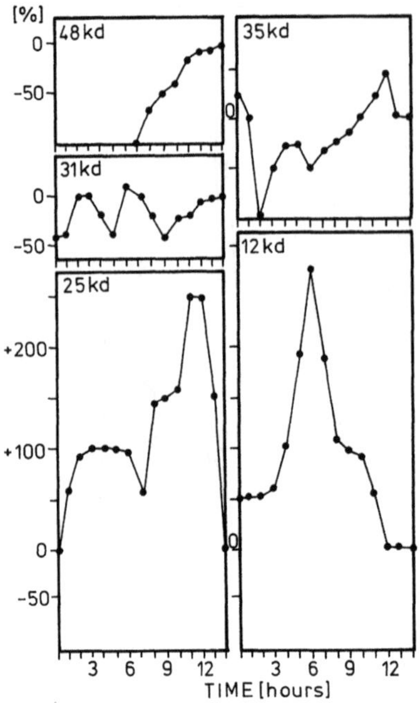

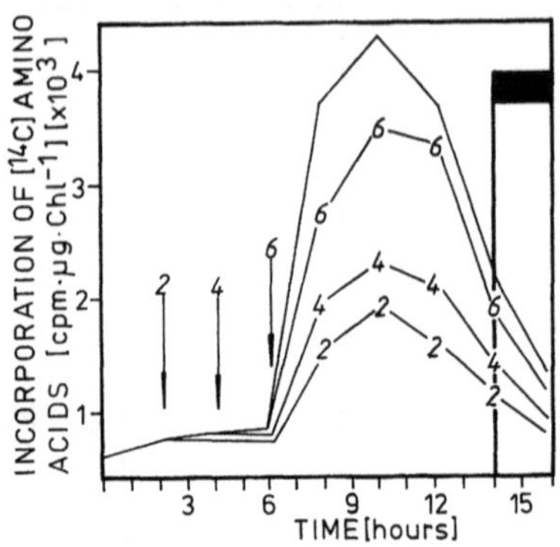

Fig. 9. Incorporation of (^{14}C) amino acids in the LHCP of synchronized *Euglena gracilis*, strain Z, during the light-time of the cell cycle at 27° (——) and after transfer to 35° at the 2nd h (—2—2—), at the 4th h (—4—4—) or at the 6th h (—6—6—) of the light-time

Fig. 8. Changes of the amounts of five polypeptides (MW 48,000, 35,000, 31,000, 25,000 and 12,000) in the chloroplast envelope during the light-time of the cell cycle of *Euglena gracilis*, strain Z. Their increase or decrease is given as a percentage of their amount in the envelope at the 14th of the light-time. (Brandt 1982)

five population doublings at 34°–35°. The inactivity of the chloroplast RNA polymerase at this temperature apparently leads to bleaching. This selective inhibition of the chloroplast transcription in *Euglena* by higher temperature can be used also for the study of the chloroplast-dependent LHCP-precursor translocation. The assembly of LHCP-apoproteins was determined during the chloroplast biogenesis after the transfer of *Euglena* cultures to 35° at the 2nd, 4th or 6th h of the light-time (Fig. 9). The time of incubation at 35° – that is the time of inhibition of the chloroplast transcription – correlates directly with the rate of the LHCP assembly. In particular, the transfer to 35° at the 2nd h yields the lowest rate of LHCP assembly.

2. It is well documented (Schiff 1973, 1980), that at least two photoreceptors control chloroplast development in *Euglena*: a blue-light receptor outside the chloroplast and a red-light receptor inside the chloroplast. The latter controls the formation of chlorophyll and the RNA synthesis in the chloroplasts. Moreover, the red-light photoreceptor in the *Euglena* chloroplasts can regulate the cytoplasmic synthetic reactions related to the chloroplast development. Accordingly red light alone can bring about normal chloroplast development in *Euglena*. Therefore experiments were performed in which normally synchronized *Euglena* cells were transferred to red light at the beginning of the light-time (Fig. 10). This does not modify the assembly of LHCP apoproteins. Treatment with chloramphenicol, however, at the 2nd, 4th or 6th h of the light-time decreases the LHCP assembly. The time of inhibition of chloroplast

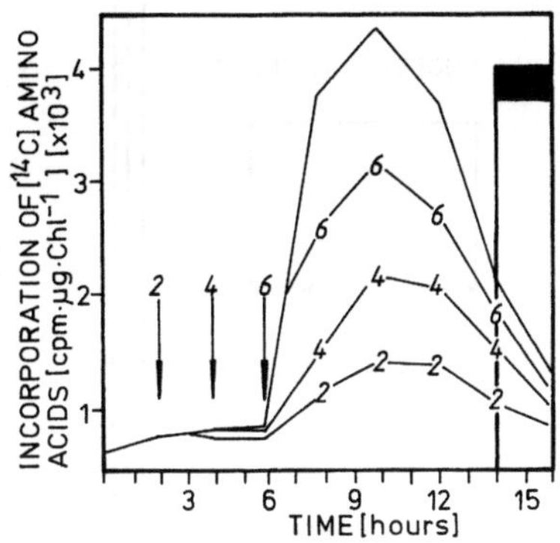

Fig.10. Incorporation of (^{14}C) amino acids in the LHCP of synchronized *Euglena gracilis,* strain Z, during the light-time of the cell cycle under red light conditions (——) and treatment with chloramphenicol at the 2nd h (–2–2–), at the 4th h (–4–4–) or at the 6th h (–6–6–) of the light-time

translation correlates directly with the rate of LHCP assembly. In particular, inhibition of the chloroplast translation after the 2nd h of the light-time yields the lowest rate of the LHCP assembly.

These two lines of experiments indicate that the transcription and the translation in the *Euglena* chloroplasts are essential for the protein translocation of LHCP-precursor proteins that are nuclear-coded and synthesized in the cytoplasm. This means in general that the *Euglena* chloroplasts can regulate the transfer of a product synthesized outside of themselves. This ability of transfer regulation may still be present in unicellular algae, but apparently is lost during evolution.

The interrelations between the *Euglena* chloroplasts and the nucleocytoplasm are summarized in Fig. 11:

1. The nucleocytoplasm and the chloroplasts both possess their own genetic system.

2. At 35° chloroplast transcription is selectively blocked.

3. Chloramphenicol or cycloheximide can be used to inhibit selectively chloroplast translation or translation in the cytoplasm.

4. There is an import of cytoplasmic-synthesized proteins into the chloroplasts. Logic demands that there will be a protein traffic also in the opposite direction, but it cannot be demonstrated so far. Some of these imported or exported proteins may have regulatory functions.

5. Proteins of different origin assemble to protein-complexes.

6. CPI and CPa are encoded and synthesized in the chloroplasts. The LHCP apoproteins, however, are encoded and synthesized in the cytoplasm.

7. Red light can be used to induce directly the genetic system in the *Euglena* chloroplasts. It induces indirectly the genetic system of the nucelocytoplasm, probably by unidentified chloroplast proteins transferred into the cytoplasm.

8. A set of chloroplast-coded and chloroplast-synthesized proteins is essential for the translocation of the LHCP-precursor proteins at the right time and in the right amount.

Acknowledgments. This research was supported by the Deutsche Forschungsgemeinschaft (Br 652/5-4).

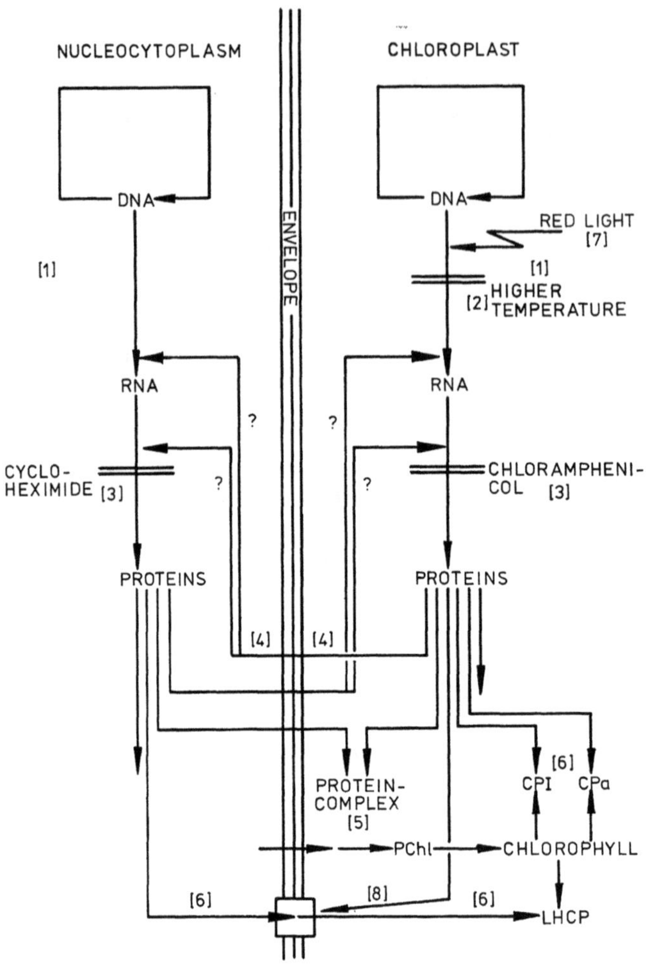

Fig. 11. Interrelations between the *Euglena* chloroplasts and the nucleocytoplasm. Detailed explanations are given in the text

References

Akerlund HE, Jansson C, Andersson B (1982) Biochim Biophys Acta 681:1-10

Brandt P (1976) Planta (Berl) 13:81–83

Brandt P (1980) Z Pflanzenphysiol 100:95–105

Brandt P (1981) Z Pflanzenphysiol 103:65–74

Brandt P (1982) Plant Sci Lett 28:237–244

Brandt P, Ernst D (1982) Electrophoresis 3:174–175

Brandt P, von Kessel B (1983) Plant Physiol (Bethesda) 72:616–619

Brandt P, Wiessner W (1977) Z Pflanzenphysiol 85:53–60

Buetow DE, Wurtz EA, Gallagher T (1980) In: Whitson GL (ed) Nuclear-cytoplasmic interactions in the cell cycle. Academic Press, New York, pp 9-55

Ellis RJ (1977) Biochim Biophys Acta 463:185–215
Ellis RJ (1983) In: Roodyn DB (ed) Subcellular Biochemistry, Vol 9. Plenum Publishing Corporation, pp 237–261
Gurevitz M, Kratz H, Ohad I (1977) Biochim Biophys Acta 461:475–488
Henningsen KW, Boardman NK (1973) Plant Physiol (Bethesda) 51:1117–1126
Herrmann FH, Börner T, Hagemann R (1980) In: Reinert J (ed) Chloroplasts. Springer, Berlin Heidelberg New York, pp 147–177
Iwanij V, Chua NH, Siekevitz P (1975) J Cell Biol 64:572–585
Kirk JTO, Tilney-Basset RAE (1978) The plastids. Their chemistry, structure, growth and inheritance. Elsevier/North-Holland Biomedical Press, Amsterdam
Ortiz W, Stutz E (1980) FEBS Lett 116:298–302
Pineau B (1982) Planta (Berl) 156:117–128
Plesnicar M, Bendall DS (1973) Biochem J 136:803–812
Schiff JA (1973) In: Abercrombie M (ed) Advances in morphogenesis, Vol X. Academic Press, London New York, pp 265–312
Schiff JA (1980) in: Senger H (ed) The blue-light syndrome. Springer, Berlin Heidelberg New York, pp 495–511
Schuler F, Brandt P, Wiessner W (1982) Z Naturforsch 37c:256–259
Westhoff P, Zimmermann K, Boege F, Zetsche K (1981) Z Naturforsch 36c:942–950

Genome and Gene Organization of the Cyanelle DNA from *Cyanophora paradoxa* in Relation to the Common Organization in Chloroplasts

H. J. BOHNERT[1] and W. LOEFFELHARDT[2]

A comparative discussion of cyanellar and chloroplast genome structure and gene organization might lead to a better understanding of plastid ancestry and to new concepts about how to trace plastid evolution. This was our hypothesis and initial aim when we decided to work with the cyanelle DNA from *Cyanophora paradoxa.*

The rationale for selecting this organism was based on several characteristics which had been revealed by cytological, biochemical and physiological studies. These will be covered in detail in several other contributions of this symposium, and this aspect has been reviewed at length in two recent articles already (Trench 1982a,b). In summary, these studies described cyanobacteria-like structures which even had retained a rudimentary cell wall of their own within a eukaryotic cell, the so-called host.

Biochemical investigations on accessory pigments, biliproteins, primary products of photosynthesis and thylakoid membrane structure did not yield a clear designation of cyanelles as either endosymbiotic cyanobacteria or chloroplasts. With such comparisons one always encounters the problem that, e.g., spinach or *Euglena* chloroplast physiology may not be analogous with the biochemical performance of an as yet undescribed "organelle" from an algal family whose systematic position itself is a matter of controversy (Skuja 1954, Kies 1979, 1980, Seckback et al. 1981, and see Cavalier-Smith 1982).

Disregarding these discussions we wanted to provide facts concerning the DNA of one of these cyanelle-bearing organisms. Convincing evidence had already been provided that the cyanelle DNA from *C. paradoxa* was of a surprisingly low kinetic complexity (Herdman and Stanier 1977) as measured by quantitative renaturation kinetics. The complexity reported of 1.17×10^8 dalton is by at least one order of magnitude lower than the corresponding values for free-living cyanobacteria (Herdman et al. 1979). This low value was confirmed using restriction endonuclease digestion of DNA from isolated cyanelles (Mucke et al. 1980, Jaynes et al. 1981). We were at first concerned with the possibility that the DNA component of low complexity might be an amplified fraction of cyanelle DNA only. Quantitative extraction of DNA from cyanelles (Mucke et al. 1980) and later from increasingly purified cyanelles

1 MPI für Züchtungsforschung, Köln-Vogelsang and Department of Biochemistry, University of Arizona, Tucson, Arizona, USA

2 Institut für Allgemeine Biochemie and Ludwig-Boltzmann-Forschungsstelle für Biochemie, Universität Wien, Wien, Austria

Compartments in Algal Cells and Their Interaction
ed. by W. Wiessner, D. Robinson, and R.C. Starr

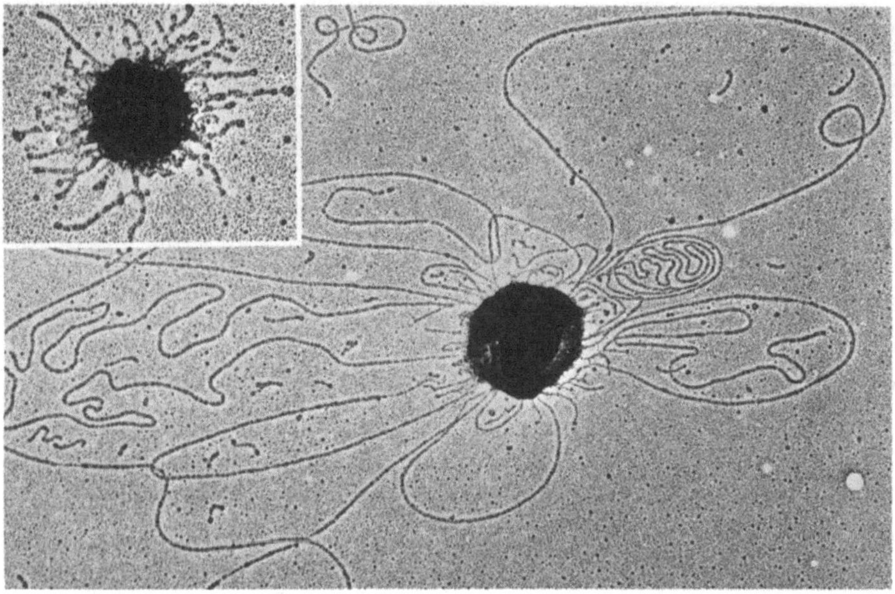

Fig.1. Electron micrograph of DNA/protein complex obtained after mild lysis of lysozyme-treated cyanelles and column chromatography of the lysate over Sepharose CL-23. The insert shows an example of the complex after shearing of DNA. Ca. x 24,000. (Bohnert et al. 1983a)

(Loeffelhardt et al. 1980, Bohnert et al. 1982) showed, however, that the DNA located in cyanelles was exclusively of this low kinetic complexity. Recently, in an attempt to obtain more information on the physical arrangement of cyanelle DNA, mild conditions of lysis with lysozyme, then SDS, followed by column chromatography (Koller and Delius 1982a) showed that the low complexity DNA is arranged surrounding the central body in the cyanelle (Bohnert et al. 1983) embedded in a matrix surrounding a crystal-like center (Fig. 1).

The "Real" Cyanelle DNA

Some contradictory reports concerning the real cyanelle DNA have to be considered. Discrepancies existed as to which buoyant density component out of several components found in distributions of total DNA from *C. paradoxa* had to be assigned to cyanelles. Our interpretation of many isolations which were checked in the analytical ultracentrifuge leaves no doubt that a buoyant density component of 1.692 g cm^{-3}, or 1.695 g cm^{-3} in another subculture (Table 2) is the cyanelle DNA. Several other subfractions of *Cyanophora* total DNA have meanwhile also been assigned to subcellular compartments (Table 1). As a suggestion, the different strains in culture, especially those strains for which a cyanelle DNA buoyant density of 1.716 g cm^{-3} was reported might be checked in a different laboratory once more.

Table 1. DNA components from *Cyanophora paradoxa* cells

Buoyant density (g cm^{-3})	Compartment	Complexity[a] (kbp)
1.688	cyanelle, not DNA, particulate	–
1.689–1.691	cytosol[b]	48
1.692	cyanelle	126.5
1.715–1.718	mitochondrial[c]	>200
1.710–1.712	nuclear (rDNA)[d]	15–17
1.720–1.730	nuclear, containing several subfractions	not resolvable

[a] Complexity (kbp) according to restriction endonuclease generated DNA fragment pattern
[b] Contaminates cyanelles not properly treated with DNAase; can also be removed by percoll-gradient centrifugation of not DNAase-treated cyanelles
[c] After removal of cyanelles from the supernatant, 10,000 *g* pellets were used as the source of DNA extraction
[d] *Cyanophora* cytosolic rRNA hybridizes to these DNA fragments (Bohnert, unpublished)

Different Strains of *Cyanophora paradoxa*

Another possibility to explain differing results from different groups might be that several strains of *Cyanophora* existed. Our attempts to look into this possibility are summarized in Table 2. (Loeffelhardt et al. 1983, Bohnert et al. 1983a).

Table 2. Characteristics of cyanelle DNA from different isolates or sources of *Cyanophora paradoxa*

Source	A R. Starr (LB555UTEX)	B L. Provasoli	C L. Kies (1555)	D L. Kies (1556)
Buoyant density (g cm^{-3})	1.692	1.695	1.692	N.D.
Molecular size (kbp)	126.5 ± 0.5	126.5 ± 0.5	138 ± 1	138 ± 1
Form	Circular	Circular	Circular	N.D.
Restriction endonuclease pattern -identical with	B	A	D	C
-different from	C,D	C,D	A,B	A,B
Size of repeat unit (kbp)	9.8–10.5	9.8–10.5	10–11	10–11
Suggested strain classification	Pringsheim	Pringsheim	Kies	Kies

It appears that at least two strains of *Cyanophora* may be distinguished on the basis of the different size and restriction fragment pattern of their cyanelle DNA's. We suggest that these two isolates should be referred to as the Pringsheim-B strain (Pringsheim 1958) and the Kies-strain (Kies 1979) although morphologically there appear to be no differences between them (Kies, personal communication).

The Physical Map of Cyanelle DNA

Restriction endonuclease digestion of cyanelle DNA from either strain results in a series of DNA fragments after separation according to size in agarose gels. By using several approaches a physical map was constructed. The techniques used were digestions with two restriction endonucleases, partial digestion with one enzmye, redigestion of isolated DNA fragments with a second restriction endonuclease and DNA/DNA hybridizations.

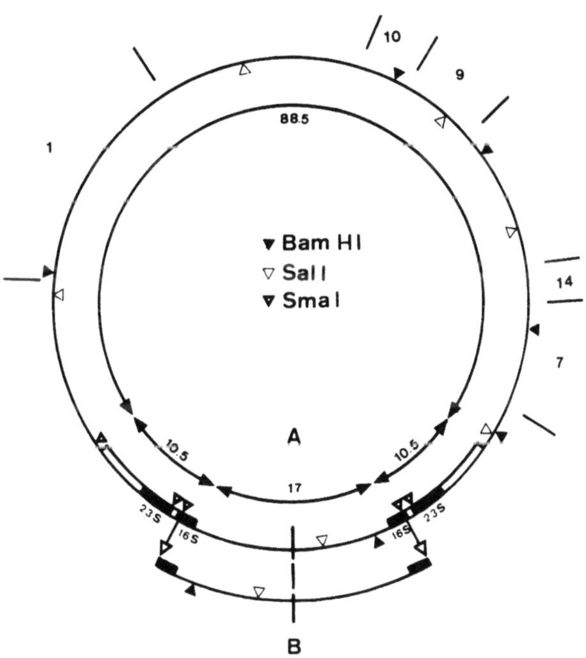

Fig. 2. Physical map of *Cyanophora paradoxa* (LB555UTEX) cyanelle DNA. The restriction endonuclease cleavage sites for Sal I, Bam HI and Sma I, and some sites for Bgl II (identified by fragment numbers) are shown. The box represents the two invertedly arranged repeat units, containing the rRNA genes. The two orientations of the single-copy DNA regions relative to each other as a result of intramolecular recombination are indicated (**A**,**B**)

The results showed, as expected from electron microscope studies of cyanelle DNA, that a circular map of restriction endonuclease cleavage sites could be constructed. The size of the molecule is 126.5 ± 0.5 kbp for the Pringsheim strain. For the circular alignment of DNA fragments eight restriction endonucleases have been used: Sal I, Bam HI, Sma I, Xho I, Bgl II, Hind III, Eco RI and Pst I. For the first five enzymes, which gave rise to between 5 and 19 DNA fragments a circular map is established (Bohnert et al. 1983b). For the restriction endonucleases EcoRI, HindIII and PstI each of which produces more than 30 DNA fragments, the map is not complete.

The outstanding characteristic of the circular map (LB555UTEX) is a twice-repeated DNA segment between 9.8 and 10.5 kbp in size (Figs. 2,3). The two repeat units are separated from each other by a 17 kbp small single-copy DNA region and by a 88.5 kbp large single-copy DNA region. The repeat unit is arranged in inverted orientation.

Comparison of Cyanelle and Chloroplast Genome Organization

This genome organization can be compared with the organization of the chloroplasts from higher plants (Bohnert et al. 1982b; as a review), from the liverwort *Marchantia polymorpha* (Ohyama et al. 1983), from the green alga *Chlamydomonas reinhardii* (Rochaix 1981) and from the euglenoid alga *Euglena gracilis* (Hallick et al. 1978). Besides *Euglena* whose chloroplast DNA includes either three or five tandem repeat units (Koller and Delius, 1982b) the other chloroplast DNA's are in principle of the same structure and of the same size as the cyanelle DNA (Fig. 3): two inverted repeats separate two single-copy DNA regions. Differences between the chloroplast DNA's and between chloroplast and cyanelle DNA concern mainly the size of the repeat unit which varies between approximately 10 kbp in cyanelle DNA and the chloroplast DNA from *Marchantia,* and ca. 28 kbp in some plants which have well-developed chromoplasts (see: Bohnert et al. 1982b for a review). Although it is noteworthy that the repeat unit size increases with organizational complexity of plastids, it is definitely too early to make this a strong point. Most angiosperm chloroplast DNA's are in the range of 21-23 kbp for the size of their inverted repeat unit.

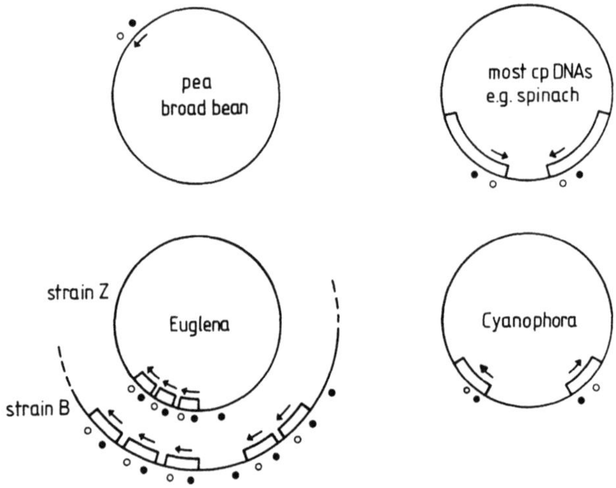

Fig. 3. Schematic drawings of the morphology of different types of plastid DNA's. The repeat units are depicted as boxes. *Solid circles* 16S rRNA genes; *open circles* 23S rRNA genes; *arrows* give directions of transcription of the rRNA operons

Table 3. Hybridization of chloroplast gene probes with cyanelle DNA

Probe	Protein coded	DNA fragments
16S rRNA		Bgl 2, Bgl 3
23S rRNA		Bgl 2, Bgl 3
spacer tRNA's		Bgl 2, Bgl 3
rbc L	RuBPcase LSU	Bgl 7, Bgl 14
atp B	CF_1, βSU	Bgl 7
atp E	CF_1, εSU	Bgl 7
atp A	CF_1, αSU	Bgl 9
psb A	herbicide binding	Bgl 14
atp H	CF_0, DCCD-binding	Bgl 1
pet B	cytochrome b6	Bgl 10
c4 pet D	cytochrome b6/f complex, SU IV	Bgl 10
rbc S	RuBPcase SSU	Bgl 7

In the chloroplast DNA's the rRNA operon contained in the repeat unit is always transcribed toward that single-copy DNA segment (Fig. 3, arrows) in which, up to now, no gene coding for a chloroplast protein was found. Contrary to this, in cyanelle DNA the direction of transcription of the rRNA operon is toward the single-copy DNA which contains the protein genes which are mapped up to now (Bohnert et al. 1983b and Table 3). This might mean that during evolution the ancestors of the extant (chlorophyll a- and chlorophyll b-containing) green algae and higher plants reversed the orientation of their inverted repeat units. Probably independently of this event, both in the green algae and in the lineage which gave rise to the angiosperms, the actual size of this inverted repeat unit increased.

Genes on Cyanelle DNA

We used the mapped, identified (Alt et al. 1983a,b) and, in part, sequenced genes from spinach chloroplast DNA as probes in the search for corresponding sequences on cyanelle DNA from *C. paradoxa*.

The probes used were cleaved from inside genes of spinach chloroplast DNA cloned in a variety of different vectors. The desired DNA fragment was isolated from agarose gels, nick-translated using α^{32} P-dCTP. The radioactive probe was first hybridized to filters containing immobilized vector DNA and subsequently to cyanelle DNA fragments.

In each hybridization, a single, or two at most, DNA fragments were labeled (Table 3). The positions of these sequences corresponding to the chloroplast genes and the nuclear rbcS gene could be included in a map (Bohnert et al. 1983b) of the cyanelle DNA. The rRNA genes including the ubiquitous trnI and trnA genes in the rDNA spacer (Kuntz et al. 1983) are present in one copy each in the approximately 10 kbp inverted repeat segment. The spacer rDNA is only between 300 and 400 kbp long and presumably no introns within the tRNA genes will be present. The cyanelle DNA shares this characteristic with eubacterial (*E. coli*), cyanobacterial (*Synechococ-*

cus 6301) and *Euglena* chloroplast DNA (Morgan et al. 1977, Tomioka et al. 1981, Graf et al. 1980, Williamson and Doolittle 1983). In contrast, the higher plant and, possibly, *Chlamydomonas* chloroplast DNA's have the spacer tRNA genes interrupted by introns (Koch et al. 1981, Rochaix 1981, Takaiwa and Sugiura 1982). The 16S rRNA gene in cyanelle DNA starts, at maximum, 600 bp inside the repeat unit (Fig. 2). This is similar to *Marchantia* chloroplast DNA (Ohyama et al. 1983) but unlike in the other chloroplast DNA's, where this gene is inside the middle part of the unit (see Bohnert et al. 1982b).

In addition to rRNA genes the cyanelle tRNA complement was also studied. For 16 out of 20 amino acids tRNA species were identified, some occurring in up to 4 isoaccepting molecules (Kuntz et al. 1983). Twenty-five tRNA genes were located on the physical map of cyanelle DNA by hybridizing the homologous tRNA to filter-immobilized cyanelle DNA fragments. Besides the genes tRNA and trnI in the spacer region, these tRNA genes are located in the single-copy DNA regions, as in chloroplast DNA's. Some tRNA genes appear to be clustered both in the cyanelle DNA and in the chloroplast DNA's in the same order. We are tempted to suggest that this positional conservation of genes between evolutionary long-separated genomes (trn M,V,A) represents an ancient feature, and that some constraint fixes this position. It occurs in a position of the spinach chloroplast DNA molecule which is near to a putative replication origin which was identified by its ability to serve as an origin in yeast (Bohnert HJ, Breunig K, unpublished). Such an origin of replication has, however, not yet been studied in cyanelle DNA.

Genes coding for proteins (see Table 3) are located in the large single-copy DNA region, as they are, e.g., in spinach chloroplast DNA. As in this molecule the atp B and atp E genes as well as the pet B and pet D genes are close together on cyanelle DNA. In addition, the order rbc L, atp B/E appears to be conserved in spinach chloroplast and cyanelle DNA, although these genes may be further apart in cyanelle DNA. One major difference between cyanelle and chloroplast DNA up to now may be that in the "higher plants" the gene rbc L is chloroplast-located while gene rbc S is in nuclear chromosomes. Both these genes appear to be close together on cyanelle DNA (Bohnert et al. 1983a), contained on a 2.2 kbp HindIII generated DNA fragment. Two points have to be made concerning this observation. Firstly, the same rbcS probe gives a faint signal also with spinach chloroplast DNA, with DNA fragment Bam H 1,6 which contains most of the 23S rRNA gene, the 4.5S and 5S rRNA genes and 1.5 kb downstream of these genes. Additionally, using RNA from purified cyanelles in Northern type hybridizations, we recently detected the signal for the rbcL transcript, though not for the rbcS mRNA. Although these experiments are in a very early stage and do not permit a rejection of the earlier observation, it should be emphasized that only the detection of a complete rbcS sequence and the demonstration that it is an active gene are the final proof. Still, it is possible that a pseudogene or scrambled sequences caused the observed signal. An example for this has recently been given since inactive, scrambled yeast mitochondrial DNA pieces have been found integrated in yeast nuclear chromosomes (Farrelly and Buetow 1983).

As a summary, chloroplast DNA's and the cyanelle DNA from *C. paradoxa* (Pringsheim) share organizational and gene sequence homologies which are, if we also take conserved positions of genes into account, too great to be dismissed as fortuitous.

Beyond that, when strengths of hybridization signals are considered too, we would like to suggest that cyanelle and, e.g., spinach plastid chromosomes are more closely related than each of these is related to *Euglena* chloroplast DNA. There appears to be no reason to regard the cyanelle as an endosymbiont: our interpretation speaks for the cyanelle being a type of plastid which has gone a different way during evolution, probably like (other) rhodophycean algae. Our results have up to now only shown that, and which, genes which are essential for plastids as photosynthetically active organelles are common. It will be interesting to find the genes that distinguish both types of plastids, like the genes responsible for cell wall synthesis or for the different type of protein for light-harvesting.

A Model for Plastid Origin and Evolution

It is tempting to try to include the increasing knowledge about movable DNA elements with the knowledge on chloroplast genome structure into a model. Such models, which will have to be revised as progress is made, might help in designing experiments in order to trace plastid DNA origin and evolution by comparing DNA sequences. DNA transposition and insertion is ubiquitous in bacteria and also eukaryotic chromosomes are subject to transpositional rearrangements (Bukhari et al. 1977, Shapiro 1983). There is no argument which would exclude the presence of such mobile genetic elements from plastid DNA. In fact, the known genome structures of plastids from those found in *Cyanophora, Chlamydomonas, Marchantia* (Ohyama et al. 1983), broad bean and pea and spinach or maize, as well as *Euglena,* suggest that these chromosomes might have been created by the fusion of (at least) two independent replicons as it is formally depicted in Fig. 4. In *Euglena* (Fig. 4, top) a replicon containing a rRNA-operon and a transposase might have transposed into another replicon leading to cointegration with a replicatively duplicated rRNA operon in tandem.

In the case of *Euglena* an alternative mode would have been unequal crossover of two replicons containing rRNA operons an event which also leads to tandem repeats. The fact that in *Euglena* chloroplast DNA transposon-like structures have been demonstrated (Koller and Delius 1982a) is in support of this model. In contrast, the presence of only one origin of replication in *Euglena* will have to be explained.

Following this particular model the existence of inverted repeats might be explained by an inversion of the once created tandem repeats which would involve prediction of small inverted repeats (as they are found in, e.g., transposon TN3; Shapiro 1983) in addition to the repeated rRNA operons. Such transposition events are drawn (Fig. 4, middle and bottom) in a way which would explain the two alternatives of how the inverted repeat units are arranged with respect to the single-copy DNA segments containing certain protein genes. The model, at present, is compatible with many observations, e.g., the more than one origin of replication in higher plant chloroplasts (Kolodner and Tewari 1975a,b) and the observation of head-to-head circular dimers as the result of recombination (Koldner and Tewari 1979), and the demonstration of intramolecular recombination (Palmer 1983, Bohnert and Loeffelhardt 1982). The advantage of a first model is that it can be refined, when it leads to

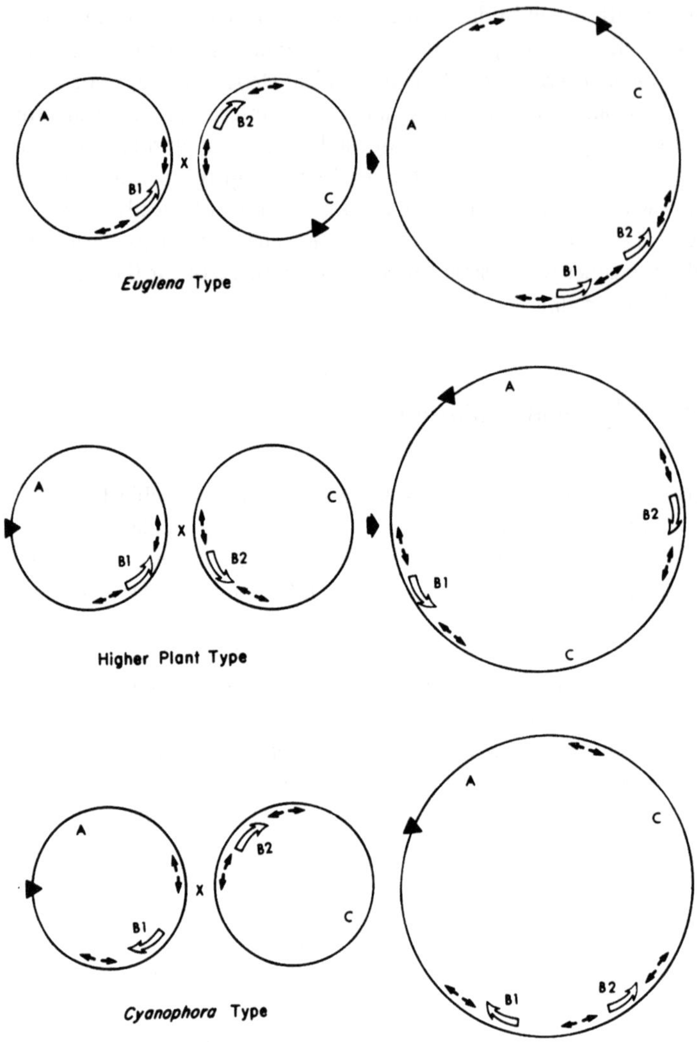

Fig. 4. Schematic explanation of the generation of three different types of plastid genome organization by recombination of two replicons with partial homology and/or containing a transposon-like segment

new experiments such as the demonstration of transposase and resolvase activities in chloroplasts, the structural analysis of replication origins or sites which recombine (e.g., even in *E. coli*).

When large parts of the plastid genome can be mobilized in such a way, this may have implications not only for our understanding of chloroplast evolution but also for (even experimentally induced) gene transfer from plastids into the nucleus or from plastids into mitochondria (Timmis and Scott 1983; Stern and Lonsdale 1982) or how gene transfer was accomplished in an established endosymbiosis.

Acknowledgments. Supported by grants from the Deutsche Forschungsgemeinschaft (HJB) and from the Fonds zur Förderung der Wissenschaftlichen Forschung (WL). We wish to thank our collaborators, identified in the various references, for their help. HJB was recipient of a Heisenberg Fellowship which is gratefully acknowledged.

References

Alt J, Winter P, Sebald W, Moser JG, Schedel R, Westhoff P, Herrman RG (1983a) Curr Genet 7: 129–138

Alt J, Westhoff P, Sears BB, Nelson N, Hurt E, Hauska G, Herrman RG (1983b) EMBO J 2: 979–986

Bohnert HJ, Loeffelhardt W (1982) FEBS Lett 150:403–406

Bohnert HJ, Crouse EJ, Pouyet J, Mucke H, Loeffelhardt W (1982a) Eur J Biochem 126:381–388

Bohnert HJ, Crouse EJ, Schmitt JM (1982b) In: Parthier B, Boulter D (eds) Encyclopedia Plant Physiol 14B. Springer, Berlin Heidelberg New York, pp 475–530

Bohnert HJ, Michalowski C, Koller B, Delius H, Mucke H, Loeffelhardt W (1983a) In: Schenk HEA, Schwemmler W (eds) Endocytobiology II. De Gruyter, Berlin New York, in press (1983)

Bohnert HJ, Michalowski C, Mucke H, Loeffelhardt W (1983b) Manuscript in preparation

Bukhair AI, Shapiro JA, Adhya SL (eds) (1977) DNA insertion elements, Plasmids, and Erisomes CSHL. Cold Spring Harbor

Cavalier-Smith T (1982) Biol J Linn Soc 17:289–306

Farrelly F, Buetow RA (1983) Nature (Lond) 301:296–301

Graf L, Koessel H, Stutz E (1980) Nature (Lond) 286:908–910

Hallick RB, Gray PW, Chelm BK, Rushlow KE, Orozco EM (1978) In: Akoyunoglou G, Argyroudi-Akoyunoglou JH (eds) Chloroplast Development. Elsevier, Amsterdam, pp 619–622

Herdman M, Stanier RY (1977) FEMS Lett 1:7–12

Herdman M, Janvier M, Rippka R, Stanier RY (1979) J Gen Microbiol 111:73–85

Jaynes JM, Vernon LP, Klein SM, Strobel GA (1981) Plant Sci Lett 21:345–356

Kies L (1979) Ber Dtsch Botan Ges 92:445–454

Kies L (1980) In: Schwemmler W, Schenk HEA (eds) Endocytobiology I. De Gruyter, Berlin New York, pp 7–19

Koch W, Edwards K, Koessel H (1981) Cell 25:203–213

Koller B, Delius H (1982a) FEBS Lett 139:86–92

Koller B, Delius H (1982b) Mol Gen Genet 188:305–308

Kolodner R, Tewari KK (1975a) J Biol Chem 250:8840–8847

Kolodner R, Tewari KK (1975b) Nature (Lond) 256:708–711

Kolodner R, Tewari KK (1979) Proc Natl Acad Sci USA 76:41–45

Kuntz M, Loeffelhardt W, Mubumbila M, Mucke H, Crouse EJ, Burkard G, Bohnert HJ, Weil JH (1983) Submitted

Loeffelhardt W, Mucke H, Bohnert HJ (1980) In: Schwemmler W, Schenk HEA (eds), Endocytobiology I. De Gruyter, Berlin New York, pp 523–530

Loeffelhardt W, Mucke H, Crouse EJ, Bohnert HJ (1983) Curr Genet 7:49–55

Morgan EA, Ikemura T, Nomura M (1977) Proc Natl Acad Sci USA 74:2710–2714

Mucke H, Loeffelhardt W, Bohnert HJ (1980) FEBS Lett 111:347–352

Ohyama K, Yamano Y, Fukuzawa H, Komano T, Yamagishi H, Fujimoto S, Sugiura M (1983) Mol Gen Genet 189:1–9

Palmer JD (1983) Nature (Lond) 301:92–93

Palmer JD, Thompson W (1982) Cell 29:537–550

Palmer JD, Thompson W (1981) Proc Natl Acad Sci USA 78:5533–5537

Pringsheim EG (1958) Stud Plant Physiol (Praha), p 165

Rochaix JD (1981) Experientia (Basel) 37:323–332

Seckbach J, Homersand IS, Hanania J (1981) Ann NY Acad Sci 361:409–425
Skuja H (1954) In: Melchior H, Wederman E (eds) Syllabus der Pflanzenfamilien, vol I. Borntraeger, Berlin, pp 56f
Shapiro JA (ed) (1983) Mobile Genetic Elements. Academic Press, New York London
Stern DB, Lonsdale DM (1982) Nature (Lond) 299:698–702
Takaiwa F, Sugiura M (1982) Nucleic Acids Res 10:2665–2675
Timmis JN, Scott NS (1983) Nature (Lond) 305:65–67
Tomioka N, Shinozaki K, Sugiura M (1981) Mol Gen Genet 184:359–363
Trench RK (1982a) In: Schiff JE (ed) On the origins of chloroplasts. Elsevier, Amsterdam, pp 55–76
Trench RK (1982b) In: Round FE, Chapman DJ (eds) Prog Phycol Res, vol 1. Elsevier, Amsterdam, pp 257–288
Williamson SE, Doolittle WF (1983) Nucleic Acids Res 11:225–235

The Organelle Versus Endosymbiont Problem of *Cyanophora paradoxa*

S. MARTEN and P. BRANDT[1]

Introduction

Cyanophora paradoxa is an apochlorotic unicellular flagellate containing one to eight blue-green compartments called cyanelles, a term introduced by Pascher (1929) with respect to intracellular symbiontic blue-green algae. Some criteria, such as the presence of a rudimentary cell wall (Trench 1978), which is sensitive to lysozyme (Schenk 1970), indicate the cyanophycean nature of the cyanelles. Because of their genome size, however, their guanine-cytosine content and the copy number of circular DNA molecules, the cyanelles can also be considered as chloroplasts (Herdman and Stanier 1977). According to Lee (1980) cyanelle-containing algae including *Cyanophora paradoxa* form the separate division, Glaucophyta. When we started work on *Cyanophora,* we focused our attention on three questions: first, how similar is the photosynthetic behavior of the cyanelles to the one of free-living blue-green algae? Second, if the cyanelles can be regarded as chloroplasts, does a similar relation between them and the host cell exist as between the chloróplast and the cytosol? Third, if the host cell is able to influence the metabolism of the cyanelles, is that a more or less general influence of a surrounding medium or does the relation between cyanelles and host cell depend on the developmental stage of the cyanelles? Consequently we will report (a) on the physiological and cytological development of the cyanelles during the life cycle of *Cyanophora* and (b) results which characterize the cyanelles as blue-green algae. Other data (c) hint at the chloroplastic nature of the cyanelles.

Materials and Methods

Cultivation of Algae. Cyanophora paradoxa (algal culture collection of the Institute of Plant Physiology, University of Göttingen, FRG) was cultivated autotrophically at 24°C (Marten et al. 1981) with a light-dark change of 14 h light and 10 h dark with dilution of the culture at the beginning of the light time.

1 Abteilung für Experimentelle Phykologie, Pflanzenphysiologisches Institut, Universität Göttingen, Untere Karspüle 2, 3400 Göttingen, FRG

Determination of Pigment Content. The content of chlorphyll was determined by the method of MacKinney (1941) and of phycobilines by the method of Yamanaka and Glazer (1981).

Isolation of Cyanelles and Separation of Cyanelle Proteins. Cyanelles were isolated according to Herdman and Stanier (1977). The cyanelle proteins were separated into one soluble and three membrane fractions and subjected to SDS-PAGE as described by Marten et al. (1983).

Measurement of Photosynthetic O_2-Evolution. Net oxygen evolution of whole algal cells was measured directly with a Clark type electrode at saturating light intensity (32 Wm^{-2}). Translation inhibitors were given in usual concentrations: Cycloheximide (3 μg/ml), chloramphenicol (300 μg/ml) and streptomycin (2.3 mg/ml).

Results and Discussion

Cell division of synchronized *Cyanophora paradoxa* takes place at the end of the dark period (Fig. 1). The syntheses of chlorophyll and phycobilins are strictly light-dependent (Fig. 2). Their rates increase after a lag phase of about 3 h in a linear fashion and stop with the beginning of the dark period. An extended dark period of 8 h at the beginning of the light period prevents the synthesis of both pigments. These characteristics of cell development in *Cyanophora paradoxa* are similar to those of other synchronized green algae like *Euglena, Chlamydomonas* and *Chlorella* (Buetow 1980). However, under the same culture conditions the blue-green alga *Synechococcus leopoliensis* behaves differently. In this alga the highest rates of chlorophyll and phycobilin synthesis take place in the dark period (Marten et al. 1983).

The physiological changes of the developing cyanelles are best illustrated by the change in the amounts of the chlorophyll-protein complexes and of the light-harvesting biliproteins as well as by their binding to the thylakoid membranes. For this investigation the thylakoidal proteins of the cyanelles were separated by deoxycholate treatment into one soluble fraction and three membrane fractions (MF) subsequently subjected to SDS-PAGE (Brandt 1980). The presence of the proteins in one or the other of the three membrane fractions is related to their binding in or on the thylakoids (Fig. 3).

The chlorophyll-protein complexes of photosystem I (CP I) and of photosystem II (CP a) (Ortiz and Stutz 1980) are present in the second and third MF. The phycobilin subunits can be detected in all fractions. The changes in thylakoidal amounts of these proteins have been determined by taking up the proteinspectra of the cyanelles for each hour of the cell cycle. The amounts are calculated as percent of the total protein of each fraction.

The most remarkable effect is seen in the distribution of CP I between the 2nd and 3rd MF (Fig. 2). During the first 2 h of the light period the CP I content of the 2nd MF strongly decreases, paralleled by a simultaneous increase in the content of this protein in the 3rd MF. This might signify an alteration of the organization of CP I

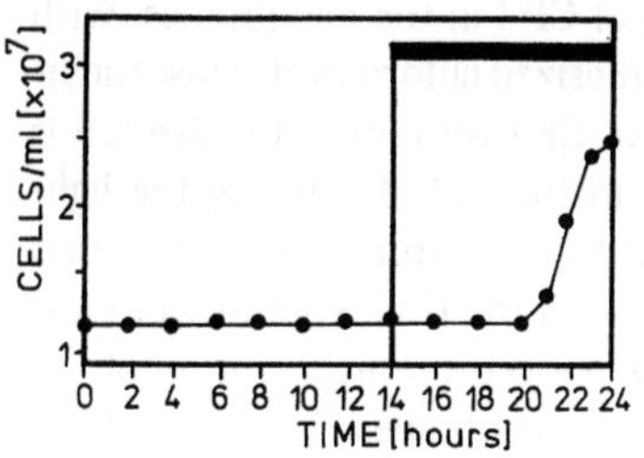

Fig. 1. Autotrophic growth of *Cyanophora paradoxa.* Cell division synchronized by a light-dark change of 14 h light and 10 h dark

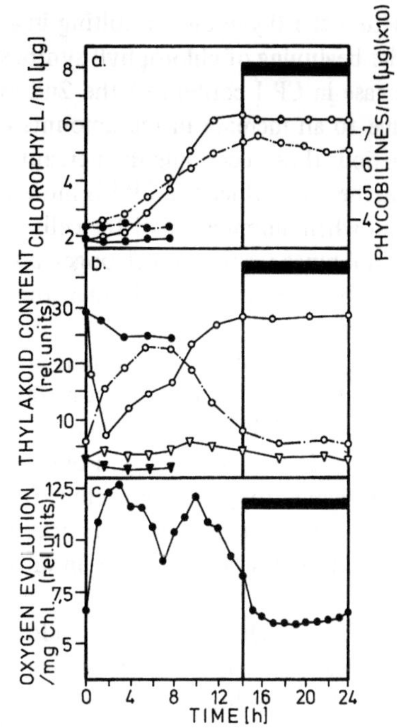

Fig. 2a-c. *a* Chlorophyll content (o– . –) and phycobiline content (o——) during the cell cycle of synchronized *Cyanophora paradoxa* cultures. Chlorophyll content (● – . –) and phycobiline content (●——) after an additional dark period. *b* Thylakoid content of CP I in the second (o——) and in the third (o– . –) membrane fraction and CP a in the second membrane fraction (– ∇–) isolated from endocyanelle thylakoids during the cell cycle of synchronized *Cyanophora paradoxa* and after additional dark (CP I – ● –, CP a – ▼–); *c* Photosynthetic oxygen evolution of synchronized *Cyanophora paradoxa*

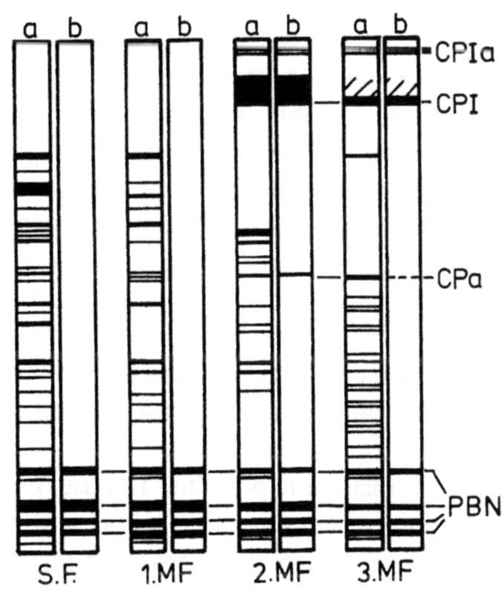

Fig. 3. Polypeptide pattern of the soluble protein fraction (*S.F.*) and of the first, second, and third membrane fraction (*1.MF, 2.MF, 3.MF*) of endocyanelle thylakoids separated by deoxycholate treatment and subsequent SDS–PAGE. Identification of the bands by pigments (*b*) or by staining with Coomassie blue (*a*)

within the thylakoids, resulting in a stronger binding of CP I in the membranes. With the beginning of chlorophyll synthesis the newly synthesized chlorophyll causes an increase in CP I content of the 2nd MF. The decrease in CP I content in the 3rd MF is due to an increase in the amounts of other proteins in the 3rd MF during the light period, thus decreasing the percent amount of CP I. It follows from these results that the newly synthesized CP I is more weakly associated with the thylakoid membranes. But when illumination of the cells starts after the dark period, a reorganization of the membranes results in a stronger binding of CP I to the membranes. This reorganization process is prevented when there is an additional dark period instead of the light period.

Cytologically the dark period is characterized by the presence of a stack of thylakoids called Thylakoid Organization Centers (TOC) in the endocyanelles which disappears rapidly within 10 min after the beginning of the light period (Marten et al. 1983). These "dark" thylakoids are of different structure than normal thylakoids with which they are connected. It is possible that the appearance of these TOC's results from a dark-dependent accumulation of thylakoidal polypeptids, the final insertion or position of which in or on the membranes occurs in a light-dependent process.

The development of the endocyanelles during the cell cycle of *Cyanophora paradoxa* is also characterized by changes in the photosynthetic activity. Oxygen evolution increases during the first 2 h of the light period (Fig. 2), corresponding to the reorganization of the thylakoid membranes. The further course of oxygen evolution shows a clear stage-specificity. After the first increase in oxygen evolution there is a strong decrease until the 7th hour of the light period. Then again oxygen evolution increases until the 10th hour and then decreases until the first hour of the dark period. In the dark period only minimal changes in oxygen evolution take place. Although these values for oxygen evolution are related to chlorophyll content of the cyanelles, there is no direct relationship between chlorophyll synthesis and oxygen evolution. The first increase in oxygen evolution is not due to chlorophyll synthesis which is still in a lag phase at that time. Further increases and decreases of oxygen evolution in the light period occur while the course of chlorophyll synthesis is rising in a linear way.

The question as to the physiological characteristics of cyanelles and free-living algae is inseparable from the question as to similarities in their life cycle dependent developments. Figure 4 shows striking similarities in the course of oxygen evolution of *Cyanophora paradoxa,* the eukaryotic *Euglena gracilis* and the blue-green free-living alga *Synechococcus leopoliensis.* In all three cases the maxima and minima of photosynthetic activity appear at the same time in the cell cycle. Obviously the development of photosynthetic activity of the cyanelles of *Cyanophora,* as well as of the chloroplasts of *Euglena* and of free-living blue-green algae, is similar. In this respect the prokaryotic algal character of the cyanelles is preserved.

The independence of the cyanelles from the host cells and their integration into the latter can be checked by experiments on the translation of the prokaryotic and cytoplasmatic proteins. The use of specific antibiotics enables us to determine the influence of cytoplasmic protein synthesis on different processes of development of the cyanelles. The main points of these investigations should be firstly, the stage-spe-

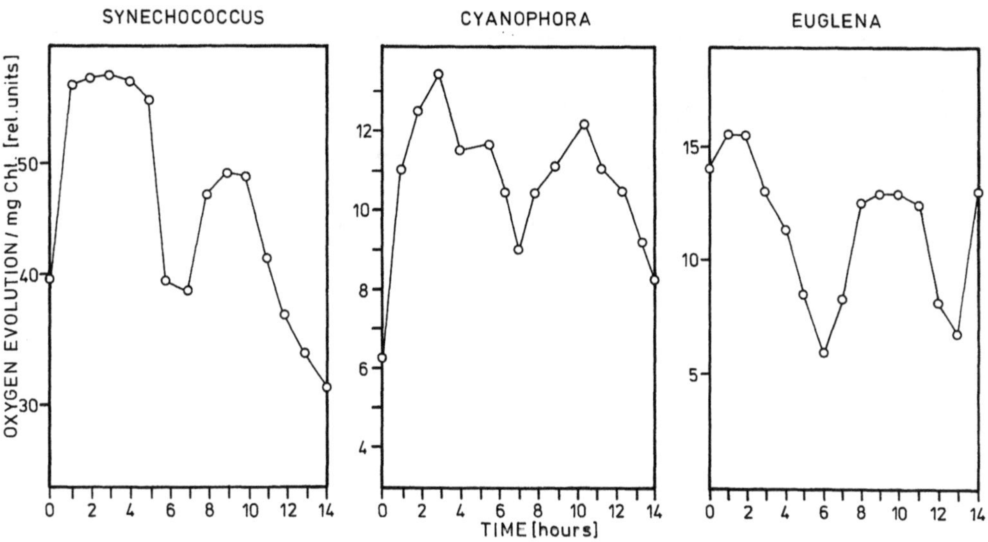

Fig. 4. Comparison of the photosynthetic oxygen evolution of *Synechococcus leopoliensis, Cyanophora paradoxa* and *Euglena gracilis* during the light time of cell cycles of 14 h light and 10 h dark

cificity of protein synthesis, secondly the development of the structure of the cyanelles and finally the development of photosynthetic activity. Here we stress the last point: investigation of photosynthetic activity is intended to offer a simple characterization of the influence of cytoplasmic protein synthesis. This provides further information on some aspects of the stage-specificity of this effect and on the degree of dependence of photosynthetic activity on the continuity of protein synthesis. The behavior of oxygen evolution is tested in presence of cycloheximide and chloramphenicol. A 1-h exposure to cycloheximide has no effect on oxygen evolution during the entire cell cycle, as there is no difference in the course of oxygen evolution in the cycloheximide-treated and untreated cells (Fig. 5). On the contrary, a 1-h treatment with chloramphenicol decreases oxygen evolution in a strongly stage-specific manner. Comparable experiments were done with *Euglena gracilis* and, additionally, the synthesis sequences of the chlorphyll-protein complexes were determined (Brandt and von Kessel 1983). The inhibition of prokaryotic protein synthesis shows in both algae the same stage-specificity. However, an inhibition of the eukaryotic protein synthesis of *Euglena gracilis* with cycloheximide results in a marked difference when compared with *Cyanophora paradoxa* (Fig. 5). In *Euglena* cycloheximide strongly decreases oxygen evolution in the second part of the light period, which is that period in the cell cycle in which the synthesis of LHCP in the cytoplasm and its assembly into the thylakoids takes place (Brandt and von Kessel 1983). Obviously when this synthesis is blocked by cycloheximide, the missing assembly of LHCP into the thylakoid membranes of *Euglena gracilis* causes a decrease in oxygen evolution. Thus we conclude that protein synthesis in the prokaryotic systems has an immediate effect on photosynthetic activity in *Cyanophora* as well as in *Euglena,* but that only in the

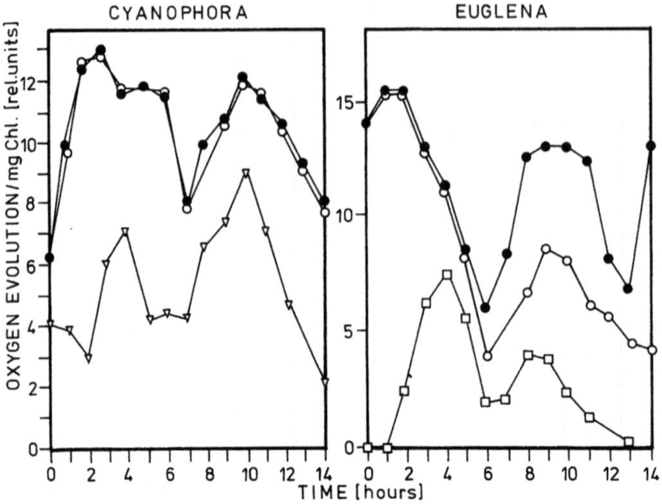

Fig. 5. Photosynthetic oxygen evolution of sychronized *Cyanophora paradoxa* and *Euglena gracilis* during the light period of the cell cycle (●) and after incubation with cycloheximide (○) or with chloramphenicol (▽) (*Cyanophora*) or streptomycin (□) (*Euglena*) for 1 h

case of LHCP synthesis in *Euglena* is a comparable direct effect on cytoplasmic protein synthesis recorded.

When *Cyanophora paradoxa* was treated with cycloheximide at the onset of light for a longer period of time, both chlorophyll synthesis and oxygen evolution are influenced. During the first 2 h of treatment oxygen evolution rises nearly as fast as in the unpoisoned cells (Fig. 6). Afterwards oxygen evolution strongly decreases with continuous incubation. The values for oxygen evolution are not related to chlorophyll, since chlorophyll synthesis is also affected very strongly by cycloheximide treatment. Chlorophyll synthesis of the cycloheximide-treated culture does not reach more than 25% of the values of the untreated culture till the 10th hour of the incubation. The effect of a long-term incubation with cycloheximide in the dark period in which cycloheximide is added at the beginning is different. Chlorophyll content is the same at each hour of the dark period in the cycloheximide and in the control cultures. This was to be expected because there is no chlorophyll synthesis in the dark. Oxygen evolution was also not very strongly affected (Fig. 6). There is only a slight decrease in the 10 h of incubation. But a remarkable effect is seen at the beginning of the next light period. The rise of oxygen evolution as a light-dependent process after the dark period which corresponds to the reorganization of the thylakoid membranes is prevented to a great extent by the incubation with cycloheximide in the dark. The same effect is seen when cycloheximide is given to the 4th, 6th or 8th hour of the dark period. An incubation time with cycloheximide of more than 24 h leads to the death of the algae. The long incubation experiments with cycloheximide have clearly shown that cytoplasmic translation has a strong effect on the development of the cyanelles. The elimination of cytoplasmic translation for longer times prevents the development of normal photosynthetic activity perhaps in connec-

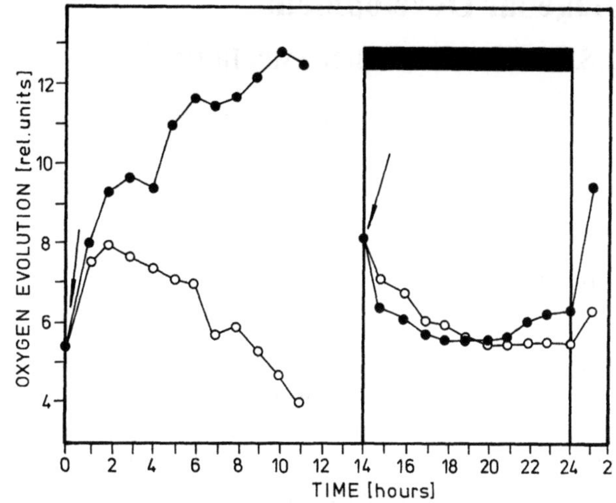

Fig. 6. Photosynthetic oxygen evolution of synchronized *Cyanophora paradoxa* in the light and dark period (●) and after incubation with cycloheximide (○). *Cycloheximide* was added at the beginning of the light or dark period

tion with thylakoid differentiation. In particular, chlorophyll synthesis is affected by cytoplasmic translation. The reasons for this are not clear from our experiments. There might be a breakdown of the regulative functions of the cytoplasm, or it might be due to the absence of proteins or other components from the thylakoids which are synthesized in the cytoplasm. It follows from our experiments that the cyanelles of *Cyanophora paradoxa* have preserved an algal cell-like behavior, as must be deduced from the parallel individual development of photosynthetic activity of free-living algae and endocyanelles. On the other hand, the cyanelles have lost their autonomy and – in a way which we do not yet know – the host cell directly influences the development and activity of the endocyanelles which makes them a cell-dependent cell organelle.

Acknowledgment. Support by the Deutsche Forschungsgemeinschaft is gratefully acknowledged (Br 652/6-1).

References

Brandt P (1980) Z Pflanzenphysiol 100:95–105
Brandt P, von Kessel B (1983) Plant Physiol (Bethesda) 72:616–619
Buetow DE, Wurtz EA, Gallagher T (1980) In: Whitson GL (ed) Nuclear-cytoplasmic interactions in the cell cycle. Academic Press, New York, pp 9–55
Herdman M, Stanier RY (1977) FEMS Lett 1:7–12
Lee RE (1980) Phycology. Cambridge University Press, Cambridge
MacKinney G (1941) J Biol Chem 140:315–322
Marten S, Brandt P, Wiessner W (1981) Ber Dtsch Bot Ges 94:727–732
Marten S, Brandt P, Robinson DG (1983) Eur J Cell Biol 30:18–24
Marten S, Stamm R, Brandt P (1983) Physiol Plant (Bethesda) 59:228–232
Ortiz W, Stutz E (1980) FEBS Lett 116:298–302
Pascher A (1929) Jahrb Wiss Bot 71:396–462
Schenk HAE (1970) Z Naturforsch 256:656
Trench RK, Pool RR, Logan M, Engelland A (1978) Proc R Soc Lond Biol Sci B202:423–443
Yamanaka G, Glazer AN (1981) Arch Microbiol 130:23–30

CER, Cell Surface-Flagellum Relationship During Flagellar Development

L. S. KAWANO and G. BENJAMIN BOUCK[1]

Introduction

Flagella of algae differ from those of most other eukaryotic organisms both in their point of attachment (anterior or bilateral) and in their extraordinary surface appendages which may include scales, spines and/or mastigonemes. Scales and mastigonemes have proven to be especially useful in clarifying the origin of surface materials in general since their synthesis and assembly can frequently be traced to intracellular compartments which include the Golgi apparatus (Manton 1966), chloroplast E.R. (CER, Mignot and Brugerolle 1982) and possibly even the mitochondrion (Moestrup 1982). However, while the structural pathway for some scales and mastigonemes may be reasonably well resolved, relatively little information is available on the biochemical composition and synthesis of individual components of these complexes.

One of the most clearly delineated intracellular pathways is that followed by the tubular mastigonemes of the Chrysophyte, *Ochromonas.* At least five morphologically distinct parts of the mature mastigoneme have been identified (the tubular shaft, the basal region, short lateral filaments, long lateral filaments and terminal filaments). The shaft and base are assembled in the CER and lateral filaments are later added in the Golgi complex (Bouck 1971). Thus at least two separate assembly compartments are utilized before mastigonemes are extruded to the surface. We have sought in this report to answer the question of whether each of these mastigoneme parts consists of a distinct glycoprotein or whether a single protein becomes variously modified in intracellular compartments to generate structurally different regions. Through the use of immunolabeling we have also attempted to relate mastigoneme parts with glycoproteins separated on acrylamide gels.

Materials and Methods

Mastigoneme Isolation

Ochromonas danica was cultured in defined media as previously described (Bouck 1971). Three days after innoculation cells were collected by centrifugation, mechani-

1 Department of Biological Sciences, University of Illinois, Chicago, Illinois, USA

Compartments in Algal Cells and Their Interaction
ed. by W. Wiessner, D. Robinson, and R.C. Starr

cally deflagellated and the flagella collected by differential centrifugation. Flagella were digested with the anionic detergent Sarkosyl NL 97 (sodium lauryl sarcosinate, ICN Pharmaceuticals, Inc., Plainview New York) and the detergent resistent mastigonemes were then purified by isopycnic centrifugation in cesium chloride.

Protease Digestion and Peptide Maps

Isolated mastigonemes were solubilized in 1% sodium dodecyl sulfate (SDS) and 0.018M β-mercaptoethanol in a boiling water bath and then applied to a 5% to 20% linear gradient acrylamide slab gel with a 3.5% stacking gel. Buffers and electrophoresis conditions were similar to those of Laemmli (1970). After electrophoresis the gels were stained with Coomassie blue to detect polypeptides or with the periodic acid Schiff (PAS) procedure to identify glycopeptides. Four prominent bands (designated as B, C, D and E) were cut from the gradient (preparative) gel. Each band was either eluted and used for generating antibodies (see below), or was applied to a second (analytical) gel for proteolytic digestion. Digestion followed the procedures of Cleveland et al. (1977) as modified by Tijssen and Kurstak (1983) using a long stacking gel and inserting the protein bands perpendicular to the direction of electrophoresis. Various concentrations of trypsin, chymotrypsin and Staph. A V-8 protease (Miles Laboratories, Elkhart, Indiana) were applied over the stacking gel and electrophoresis was conducted until the bromophenol blue tracking dye reached a few millimeters above the separating gel. Digestion for 30 minutes was followed by continuation of electrophoresis to separate the peptides. Gels were fixed, stained in Coomassie blue and scanned with a Zeineh soft laser scanning densitometer (Biomed Instruments Inc. Chicago, Ill.) for quantitation of bands.

Immunological Labeling

Proteins eluted from individual bands cut from gels of solubilized mastigonemes were injected subcutaneously into rabbits at weekly intervals. After the fifth injection rabbits were bled and the sera processed by standard methods. An IgG-enriched fraction was obtained after ammonium sulfate precipitation and DE-52 (Whatman) column chromatography. Each of the antisera was adsorbed against the heterologous antigens by passage of the sera through columns of agarose (Sepharose, Pharmacia) immobilized antigens.

For immunolabeling flagella were isolated after mechanical deflagellation and the flagella pellet fixed for 10 min in 4% glutaraldehyde buffered with 0.05M phosphate at pH 7.4. After several rinses with buffer and then buffer containing glycine (3 mg/ml) to quench free aldehydes, the flagella were incubated at 4°C overnight with the appropriate primary rabbit (anti-mastigoneme) antibody. Excess primary antibody was then rinsed out with several changes of buffer. Secondary antibody consisting of goat anti-rabbit IgG (GAR) conjugated to ferritin was applied for one hour at room temperature. GAR was coupled to ferritin by the method of Kishida et al. (1975) or purchased already coupled. Following secondary antibody application the flagella were rinsed several times in buffer, refixed in 4% glutaraldehyde and mounted on carbon stabilized, Formvar coated copper grids.

Results

Digestion

SDS gradient gel electrophoresis of isolated mastigonemes produced six well defined bands, all of which stained both with Coomassie blue (Fig. 1) and with the periodic acid Schiff (PAS) procedure (Fig. 3B). The largest of these glycopeptides (glycoprotein A) barely entered the gel and had an estimated molecular weight of 255,000. No further characterization of glycoprotein A was attempted. Glycoprotein B represented about 7% of the total Coomassie blue staining material and had an estimated molecular weight of 176,000. Glycoprotein C contained greater than 50% of the Coomassie blue staining proteins, and consisted of two closely migrating species designated C_1 and C_2 (Fig. 1). The two C bands were too closely associated to analyze separately, and were eluted together as one fraction. Band D and E had molecular weights respectively of 64,000 and 51,000. Each of these bands accounted for about 13% of the Coomassie blue staining material and each contained nearly identical amounts of glycoprotein. The fastest migrating band which was not further analyzed accounted for about 11% of the total stainable protein.

Whole intact mastigonemes were heated for 2 1/2 min in a boiling water bath and then directly digested with trypsin, papain or chymotrypsin (all at 25 μg/μl). Gels of the resulting digests (Fig. 2) revealed complete proteolysis of three fractions and nearly complete digestion of band C, indicating the sensitivity of all the fractions to protease. Limited proteolysis achieved with the modified (Tijssen and Kurstak 1983) Cleveland et al. (1977) procedure yielded additional information on the peptide fragments of each band. For these experiments four fractions (bands B, C, D and E) were individually removed from the separating gel and then applied to separate lanes of the analytical gel. As seen in Fig. 5 when chymotrypsin was added, all fractions were partially cleaved except fraction B. The SDS gel of the cleavage products revealed that peptide fragments from each of the four glycoproteins exhibited relative mobilities which were different from one another.

Similar but more extensive results were obtained with Staph. A V.8 protease. At 100 ng/channel three of the glycoproteins were digested to some extent (Fig. 3A and B) although substantial undigested protein remained in B, C and E. The peptide fragments obtained for each fraction showed no similarities when compared in gradient gels or in 8% gels (Fig. 4). Additional experiments were conducted to digest the glycoproteins as completely as possible in order to test the possibility that new peptides might appear at higher concentrations of enzyme. Glycoprotein B revealed no peptides at the highest concentration of protease tested (2 μg/channel), but after PAS staining slower migrating material was detected (Fig. 3b). Band C was ultimately completely digested but no new peptide fragments appeared that were not present at the lowest enzyme concentration (25 ng/channel). Band D is completely digested with 25 ng/channel of protease (Fig. 8), and produces a major cleavage fragment at low enzyme concentrations. This major fragment is itself digested at concentrations of enzyme above 100 ng/channel. Band D is clearly the most sensitive of all the glycoprotein bands, and is also more labile to heat denaturation (see below). Glycopro-

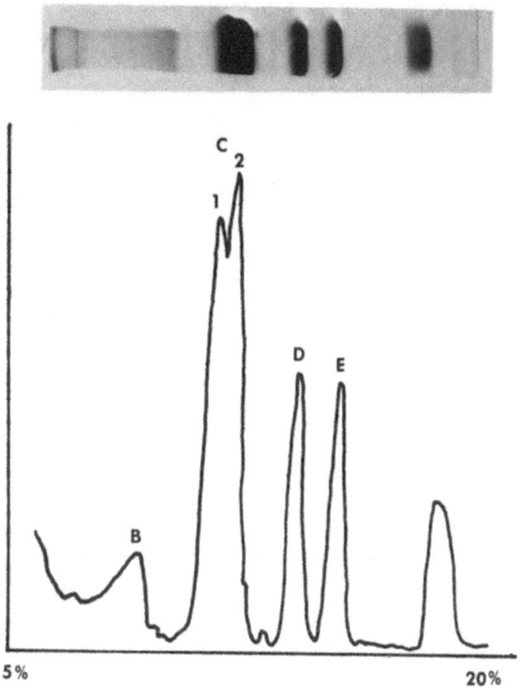

Fig. 1. Coomassie blue stained gel of isolated and solubilized mastigonemes of *Ochromonas*. The four bands labeled *B, C, D* and *E* were further analyzed by peptide digestion and immunolabeling. Laser scan of gel records quantitative differences in various bands

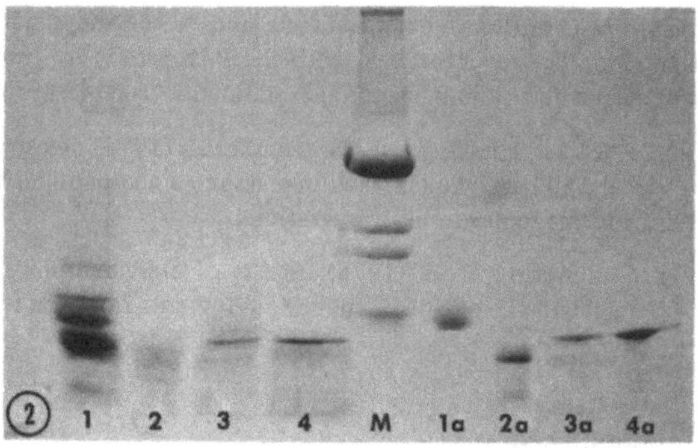

Fig. 2. Proteolytic digests of isolated mastigonemes using different proteases (25 μg protease/ 50 μg mastigoneme protein) electrophoresed through a 5-20% polyacrylamide gradient gel. *Lane* 1 Staph. a V.8 plus mastigonemes; *lane* 2 chymotrypsin plus mastigonemes; *lane* 3 trypsin plus mastigonemes; *lane* 4 papain plus mastigonemes; M, undigested mastigonemes. Lanes *1a-4a* are Staph. a V.8, chymotrypsin, trypsin and papain respectively all at 25 μg without mastigonemes

tein E seems to produce a new small peptide at very high concentrations (Fig. 9) of protease, but this band had no counterpart in digests from the other glycoproteins. Thus both partial and complete digestion of three (C, D, and E) of the four major mastigoneme glycoproteins yields fragments with no apparent homologies. The fourth glycoprotein (B) appears to be resistant to V8 protease when treated in gels.

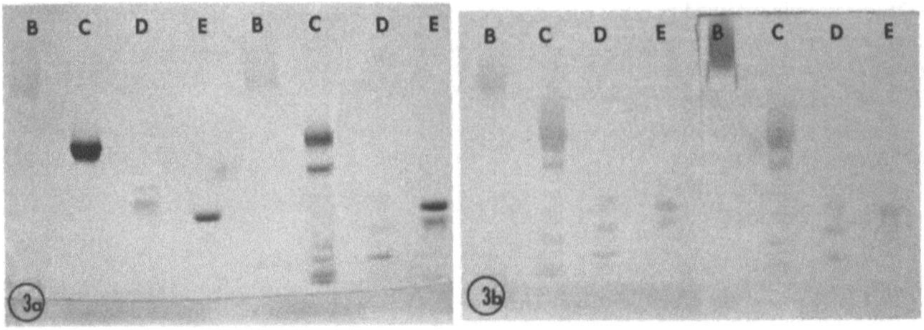

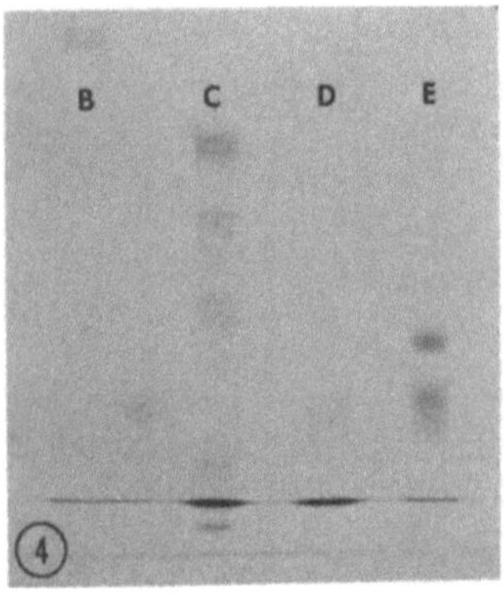

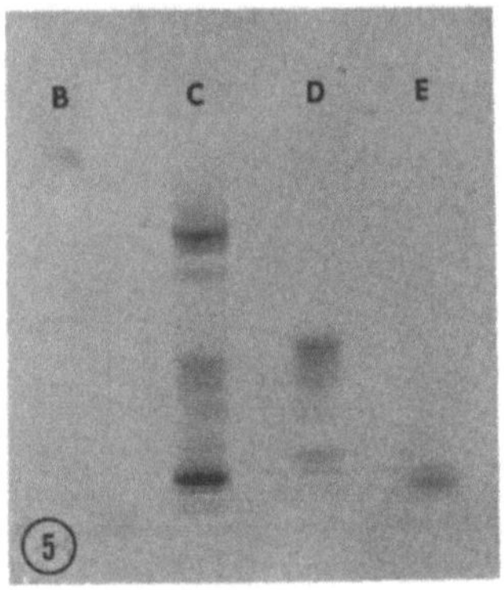

Figs. 3a,b. Staph a V.8 (100 μg/channel) digestion of individual bands (*B*, *C*, *D* and *E*) separated on a 5-20% gradient gel. In (**a**) left four lanes are undigested proteins, right four lanes are same proteins after Cleveland et al. (1975) digestion. Stained with Coomassie blue. In (**b**) a similar digestion was carried out but left four lanes were stained with Coomassie blue after digestion while right four lanes were stained with the PAS procedure for glycopeptides. Note that most of the Coomassie blue staining bands also stain with the PAS procedure

Fig. 4. 8% acrylamide gel of Cleveland et al. (1975) digested individual bands. Protease was Staph a V.8 at 150 ng/channel. Note that none of the peptides from the digests of the four glycoproteins appears to have similar migrations

Fig. 5. Cleveland et al. (1975) digests of individual mastigoneme bands using chymotrypsin at 500 ng/channel. None of the peptide fragments from the four glycoproteins appears to be similar

Immunolabeling

Attempts were made to relate each of the four glycoproteins (B, C, D, and E) with a specific region of the mastigoneme. Therefore after preparative gel electrophoresis individual bands were cut from the gels, eluted into buffer and injected into rabbits. The four antisera obtained from these injections were tested for specificity by Ouchterlony immunodiffusion against the heterologous antigens (i.e., antiserum against glycoprotein C was assayed against glycoprotein B, D, and E; other sera were similarly tested). Despite the apparent purity of the original SDS bands, each antiserum

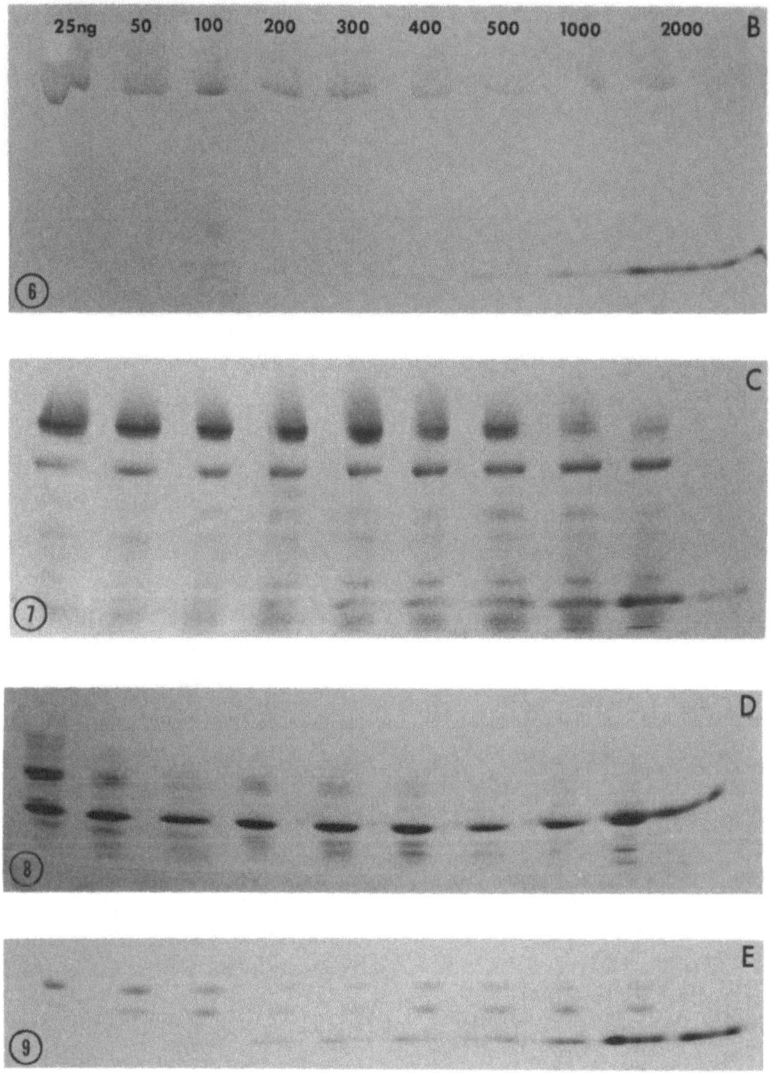

Figs. 6-9. Individual glycoproteins (*B, C, D,* and *E*) treated with increasingly high concentrations of Staph a V.8 protease. Note that with increasing concentrations of protease the number or position of peptide bands does not change except in *E* where a very low molecular weight band appears at the highest concentrations, and a band of higher molecular weight (below E) appears at a concentration of 25 μg protease but is eventually completely digested. Protease alone is electrophoresed in the last (right) channel. Numbers refer to amount in nanograms of enzyme added in each gel channel. All the gels are 5-20% gradient gels

cross-reacted with one or more of the heterologous antigens. Consequently it was necessary to adsorb each antisera with immobilized heterologous antigen. These adsorbed antisera were used for the experiments reported here. It was also found to be faster and technically easier to use isolated flagella with attached mastigonemes rather

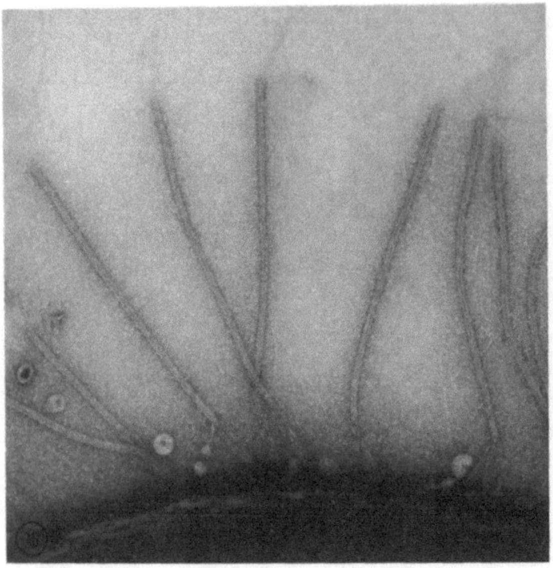

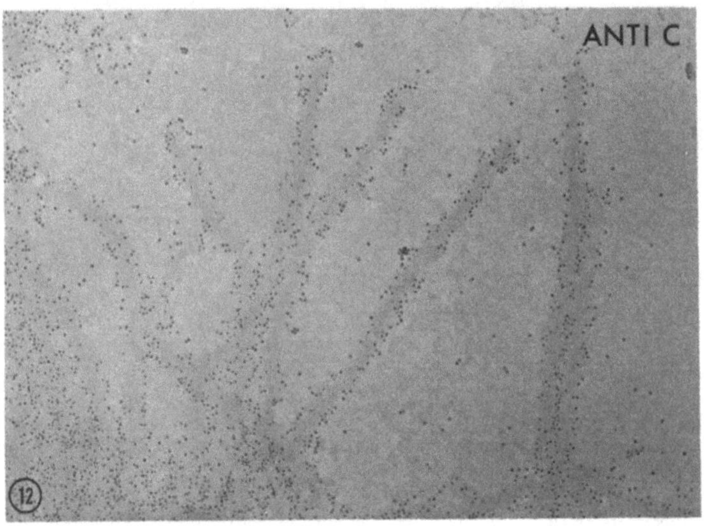

Fig. 10. Whole mount of flagella showing a portion of the laterally arranged mastigonemes with their various attached filaments. Negatively stained with uranyl acetate. x 50,000

Fig. 11. Single isolated mastigoneme illustrating position of different parts. *BA* basal region; *TS* tubular shaft; *LF* long lateral filaments; *SF* short lateral filaments. x 91,000

Fig. 12. Mastigonemes labeled with rabbit anti-C antibody. Secondary label is ferritin conjugated to anti-rabbit. Labeling is found surrounding the entire mastigoneme (*M*), a position suggesting that binding of anti-C may be to the short lateral filaments. x 73,000

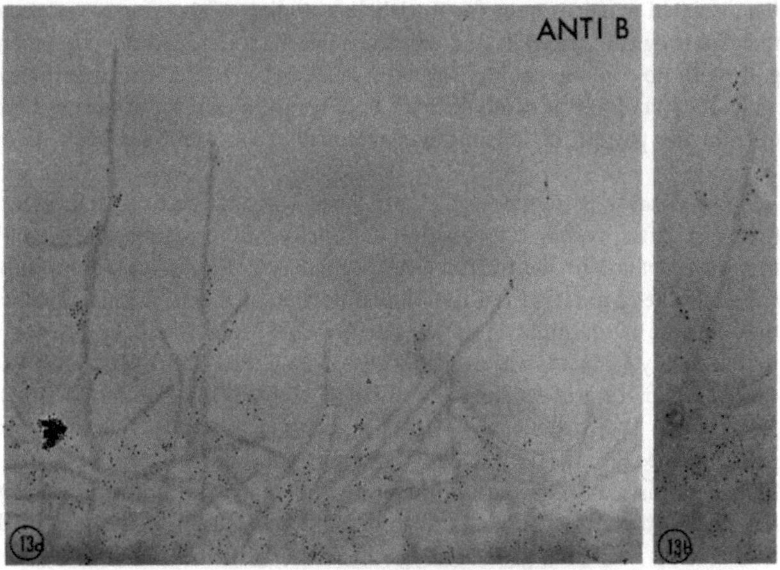

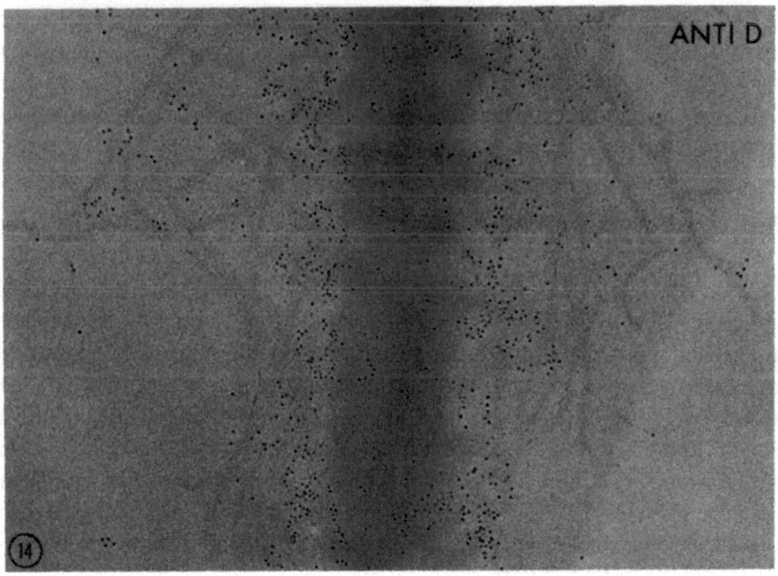

Fig. 13a,b. Mastigonemes labeled with adsorbed anti-B and ferritin GAR. Labeling seems to be found in patches thoughout the mastigoneme. Occasional labeling as in **b** suggests binding to the long lateral filaments. **a** x 55,000; **b** x 90,000

Fig. 14. Mastigonemes labeled with anti-D and then with ferritin-GAR. The pattern here suggests anti-D binds primarily to the basal attachment region of the mastigoneme. Scattered ferritin particles on other regions of the mastigoneme may be due to incomplete adsorbtion of the antisera. x 73,000

than isolated mastigonemes for the labeling studies. After primary and secondary immunolabeling (cf. Materials and Methods) flagella were dried directly on coated E.M. grids with no counterstaining. The latter tended to obscure ferritin particles, although without staining it was often difficult to judge the exact site of ferritin binding. Nonetheless the pattern of labeling was reasonably consistent for each of the antisera tested.

The clearest pattern was that produced with antibody against glycoprotein C (Fig. 12). After overnight incubation in primary antibody (unadsorbed rabbit anti-C) in most preparations the ferritin-GAR secondary antibody was found to be distributed along the entire length of the tubular portion of the mastigoneme. Since this portion consists of multiple short lateral filaments-Figs. 10 and 11), it was concluded that antibody against C was probably recognizing these short filaments. Another relatively clear pattern of labeling was obtained with adsorbed antibody directed against glycoprotein D. In this case GAR-ferritin bound to sites at the basal region of the mastigoneme (Fig. 14), i.e., at the attachment point of mastigoneme to the flagellum. Thus it seems likely that antiserum against D recognizes some portion of the carrot-like basal attachement region. Adsorbed antibody against glycoprotein B differed considerably from that generated against C and D. Anti-B binding was not localized at the base but was found throughout the mastigoneme, generally in small clusters (Fig. 13a). The spacing of these clusters as well as the occasional appearance of short linear arrays of ferritin (Fig. 13b) suggested that anti-B was recognizing the relatively sparse long lateral filaments found distributed along the mastigoneme shaft (see Fig. 11). Antibodies to glycoprotein E showed no consistent pattern of labeling and therefore its identity could not be determined.

Heat Disassociation of Mastigonemes

Purified mastigonemes incubated at 90°C for 5-15 min disassociated into filamentous structures (Fig. 15). These vary in length and seem to consist of elements similar to the lateral appendages found along the mastigoneme shaft. The base of the mastigoneme seems to be completely solubilized (Fig. 15). The heat stability of the filaments thus provided an independent method for determining the identity of the labile basal region. Therefore heat treated mastigonemes were centrifuged at high speed to pellet the filaments and the pellet and supernatant were both analyzed on SDS acrylamide gradient slab gels. Specifically reduced or altered in the pelleted material was glycoprotein D. The reduction of glycoprotein D is consistent with the tentative correlation of glycoprotein D with the basal mastigoneme region obtained by immunolabeling.

Discussion

Previous experiments have demonstrated the general pathway of mastigoneme development whereas the present report provides details on the nature of the various moieties that are assembled intracellularly. *Ochromonas* mastigonemes appear to consist of at least four unrelated glycoproteins which may, however, share some antigenic

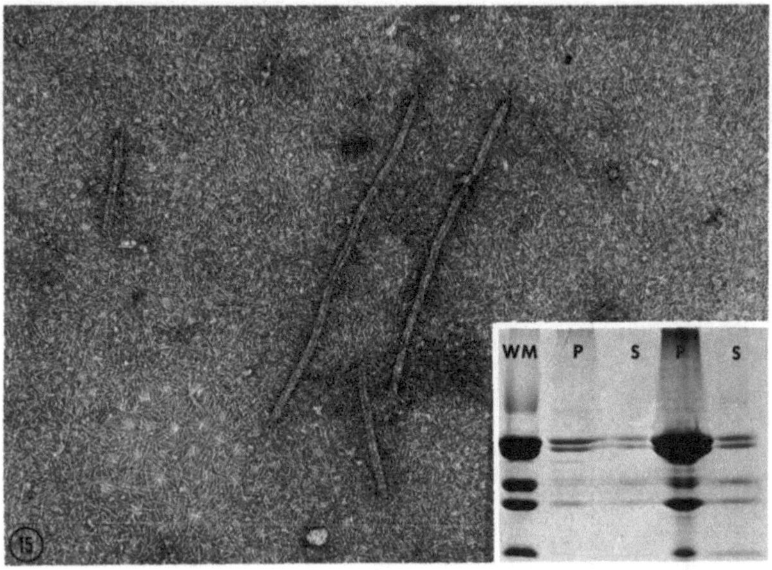

Fig. 15. Isolated mastigonemes heat treated at 90°C for 5 min. Most of the mastigonemes have disassociated into short filaments. The few intact mastigonemes have lost their basal regions. This preparation was made from the pellet obtained after heat treatment. *Inset* SDS gradient gel of mastigonemes after heat treatment similar to that of micrograph. The most obvious change in the heat resistent pellet (*P*) was a disproportionate loss of Coomassie blue staining material from band *D*. Micrograph x 66,000

determinants (see Table 1). Two additional glycoproteins have not been examined in detail, and one of the glycoproteins (glycoprotein C) which was examined actually consists of two closely migrating bands. The number of bands and their molecular weights differ significantly from earlier reports (Chen and Haines 1975) of mastigoneme composition and the molecular weight estimates of the constituent glycopeptides of the mastigoneme. The reasons for these differences are not clear. One explanation for the antigenic similarities of glycoproteins in the present report may be that these glycoproteins share common carbohydrates – a view supported in part by the positive PAS staining seen in all of the bands. Suitably absorbed, antisera will recognize only the original antigen. These results, together with proteolysis experiments, indicate that at least four of the mastigoneme glycoproteins are different biochemically from one another. Thus the assembly of mastigonemes in the CER and Golgi apparatus requires the sequestering of different peptides and glycopeptides at well defined intervals in mastigoneme ontogeny. The alternative explanation that one polypeptide is variously modified in different cellular compartments seems unlikely.

Mastigoneme development is apparently triggered by deflagellation. Newly formed "presumptive" mastigonemes appear first in the CER (Bouck 1971), the unique modification of the nuclear envelope (cf. Gibbs 1981) which seems to be involved in the assembly of several varieties of surface structures in different algae. In some cases such as in the production of mineralized surface scales, subsequent trans-

Table 1. Pathway of mastigoneme development

Glycoprotein	Estimated molecular weight	Percent of total mastig. protein	PAS[a]	Estimated sensitivity to digestion by *Staph. aureus* V8 protease	Sensitivity to heat treatment[c]	Cross-reactivities of unadsorbed antisera: Anti B	Anti C	Anti D	Anti E	Cell compartment where glycoprotein first appears	Suggested location on mastig.[b]
B	176,000	7%	++	No digestion at highest concentration of protease tested (2.0 μg)	Possibly solubilized	X	X			Golgi	Long lateral filaments
C (doublet)	C_1=115,000 C_2=100,000	+50%	+	Partial digest. at highest conc. of protease tested (2.0 μg/75 μg C)	Small amount solubilized	X	X X	X	X	Golgi CER?	Short lateral filaments. Tubular shaft?
D	64,000	13%	+	Complete digest. at 0.025 μg protease	Substantial amount solubilized		X	X	X	CER	Basal region
E	51,000	13%	+	Partial digest. at highest conc. of protease tested (2.0 μg/20 μg E)	Small amount solubilized		X	X	X	?	?

[a] Periodic acid-Schiff stain.

[b] Based on immunocytochemistry.

[c] Heat treatment =90°C, 5-15 min.

port to the Golgi complex does not appear to be necessary for extrusion to the cell surface (Mignot and Brugerolle 1982). In other cases formed structures cannot be recognized except in the Golgi complex (e.g., Brown et al. 1970). Mastigonemes of *Ochromonas,* however, are partially assembled in the CER, for they consist only of the basal region and tubular shaft in that compartment, and assembly is completed in the Golgi complex. From the present studies it tentatively can be concluded that the 64,000 MW glycoprotein D which forms all or part of the base must be accumulated or synthesized within the CER. The relative amount of glycoprotein D (13%) also compares favorably with the size of the basal region relative to other portions of the mastigoneme. In addition glycoprotein D is solubilized and/or altered in response to heat denaturation coincident with the loss of the basal region.

The tubular shaft of the mastigoneme is also found assembled in the CER but its biochemical identity is less certain. Glycoprotein C has been correlated with the short lateral filaments surrounding the shaft, but these are not present on presumptive mastigonemes in the CER. Because of its relative abundance (> 50% of the protein mass) and because of the fact that it consists of two closely migrating glycoproteins (C_1 and C_2), it seems likely that some component of glycoprotein C contributes to the tubular shaft.

As the mastigoneme progresses through the Golgi complex it acquires lateral appendages of which the shorter members have been correlated with glycoprotein C, whereas the longer and less numerous filaments apparently consist of the 176,000 MW glycoprotein B. Since glycoprotein B is strongly resistent to proteolysis and is PAS positive, it is probably heavily glycosylated, and hence is likely to be associated with the Golgi system. How these filaments are attached to the shaft is not yet clear.

The synthesis of *Ochromonas* flagellar mastigonemes seems to be a particularly well defined example of intracellular assembly in which various stages in development can be followed morphologically. This permits identification of the specific cell compartment where separate assembly steps take place. With some understanding of the glycoproteins associated with individual parts it would now seem appropriate to seek out precursors to these parts and attempt to resolve the signals which direct some proteins to the CER and others to the Golgi complex. Such studies might follow the successful and revealing studies on the synthesis and transport of lysozymal enzymes (Sly 1982). Supported by NSF Grant PCM 8203442 (GBB).

References

Bouck GB (1971) J Cell Biol 50:362–384
Brown RM, Franke WW, Kleinig H, Falk H, Sitte P (1970) J Cell Biol 45:246–271
Chen LL, Haines TH (1976) J Biol Chem 251:1828–1834
Cleveland D, Fischer SG, Kirschner MW, Laemmli UK (1977) J Biol Chem 252:1102–1106
Gibbs SP (1981) Int Rev Cytol 72:49–99
Kishida Y, Olsen B, Berg R, Prockop D (1975) J Cell Biol 64:331–339
Laemmli UK (1970) Nature 227:680–685
Manton I (1966) J Cell Sci 1:375
Mignot JP, Brugerolle G (1982) J Ultrastruct Res 81:13–26
Moestrup O (1982) Phycologia 21:427–528
Sly WS (1982) In: Horowitz MI (ed) The glycoconjugates. IV. Academic Press, New York
Tijssen P, Kurstak E (1983) Anal Biochem 128:26–35

Regulation of Tubulin Gene Expression in *Chlamydomonas reinhardii* Following Flagellar Excision

K. BRUNKE, J. ANTHONY, E. STERNBERG and D. WEEKS[1]

Introduction

Flagella play an important role in providing motility to many algal cells during the vegetative and/or gametic stages of their life cycle. In photosynthetic algae, flagella allow the organism to swim toward the environmental location where it can optimize photosynthetic energy production (e.g., [1]). In the gametic state, the flagella of certain species carry specific agglutination factors which facilitate initial contact between gametes of opposite mating types (e.g., [2]). Signals generated through the flagella by this initial contact cause rapid changes in the flagella and also in the specialized mating structures of the algal cell which allow the fusion of the mating pair [3]. Perhaps because of the essential nature of both of these functions (motility and sexual propagation) several types of alga are able to regenerate flagella that have been lost or shed.

One alga, *Chlamydomonas reinhardii*, has been the object of much research because of its ability to rapidly regenerate flagella that have been artifically removed by shearing [4] or by pH shock [5]. The pioneering work of Sir John Randall and his collaborators [4] demonstrated that *Chlamydomonas* cells could generate a new set of flagella in less than 2 to 3 h in a process that was highly temperature-dependent. From this followed the work of Rosenbaum et al. [6], who showed with the use of the protein synthesis inhibitor, cycloheximide, that *Chlamydomonas* contains a reserve of flagellar proteins that is adequate to allow regeneration of short flagella without the need for de novo protein synthesis. Subsequent work [7] with radioactively labeled arginine provided the first evidence that newly synthesized proteins are incorporated into the growing flagellum. This work and the earlier work of Randall's group [4] prompted us to ask whether or not flagellar excision triggered an increased synthesis of flagellar proteins. Because of their high abundance in the flagellum, we focused our attention on the synthesis of the α– and β–tubulin subunits which are the major building blocks of the flagellar axonemal microtubules.

As will be described below, we were able to demonstrate that the excision of flagella from *Chlamydomonas* results in a rapid induction of tubulin synthesis [8]. Subsequent work by our laboratory [9-15] and others [16] has shown the coordinate expression of the *Chlamydomonas* tubulin genes during flagellar regeneration. These

1 The Institute for Cancer Research, Philadelphia, PA 19111, USA
Zoecon Corporation, Palo Alto, CA 94304, USA

Compartments in Algal Cells and Their Interaction
ed. by W. Wiessner, D. Robinson, and R.C. Starr

investigations have established the tubulin induction system in *Chlamydomonas* as a model algal system in which the general problem of eukaryotic gene regulation can be fruitfully studied.

In the following sections a summary is presented of our research demonstrating the induction of tubulin synthesis in deflagellated cells, the effects of certain inhibitory reagents on this process, the post-translational modification of proteins during flagellar regeneration, and the possible involvement of a short nucleotide (consensus) sequence in the coordinate regulation of the four tubulin genes of *Chlamydomonas reinhardii.*

Flagellar Regeneration and the Induction of Tubulin Synthesis

The early experiments of Randall et al. [4] and Rosenbaum et al. [6] demonstrated that vegetative cells of *Chlamydomonas reinhardii* are capable of rapid flagellar regeneration. Within 60 to 90 min after deflagellation, cells are able to regrow flagella of nearly full (12-15 μm) length. In our studies with gametic cells of *Chlamydomonas* we have found similar kinetics of flagellar regeneration but only after a distinct 5-8 min lag period (Fig. 1).

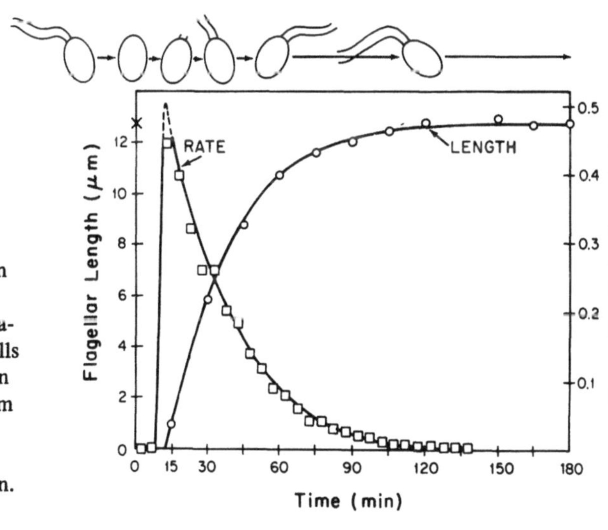

Fig. 1. Flagellar regeneration in gametes of *C. reinhardii.* Gametic cells were deflagellated by vigorously stirring cells at pH 4.7 for 1 min and then raising the pH of the medium to 7.0. X Initial flagellar length; O flagellar length; □ rate of flagellar elongation. (Weeks et al. 1977)

To determine if an induction of tubulin synthesis is required to support the rapid growth of new flagella, we undertook a set of in vitro translation experiments aimed at measuring the amount of tubulin mRNA in nondeflagellated and deflagellated cells. The results of translating both polyribosomes as well as mRNA in vitro demonstrated a large-scale induction of tubulin synthesis in deflagellated cells. This has also been shown in experiments where cells were labeled in vitro with $H_2{}^{35}SO_4$

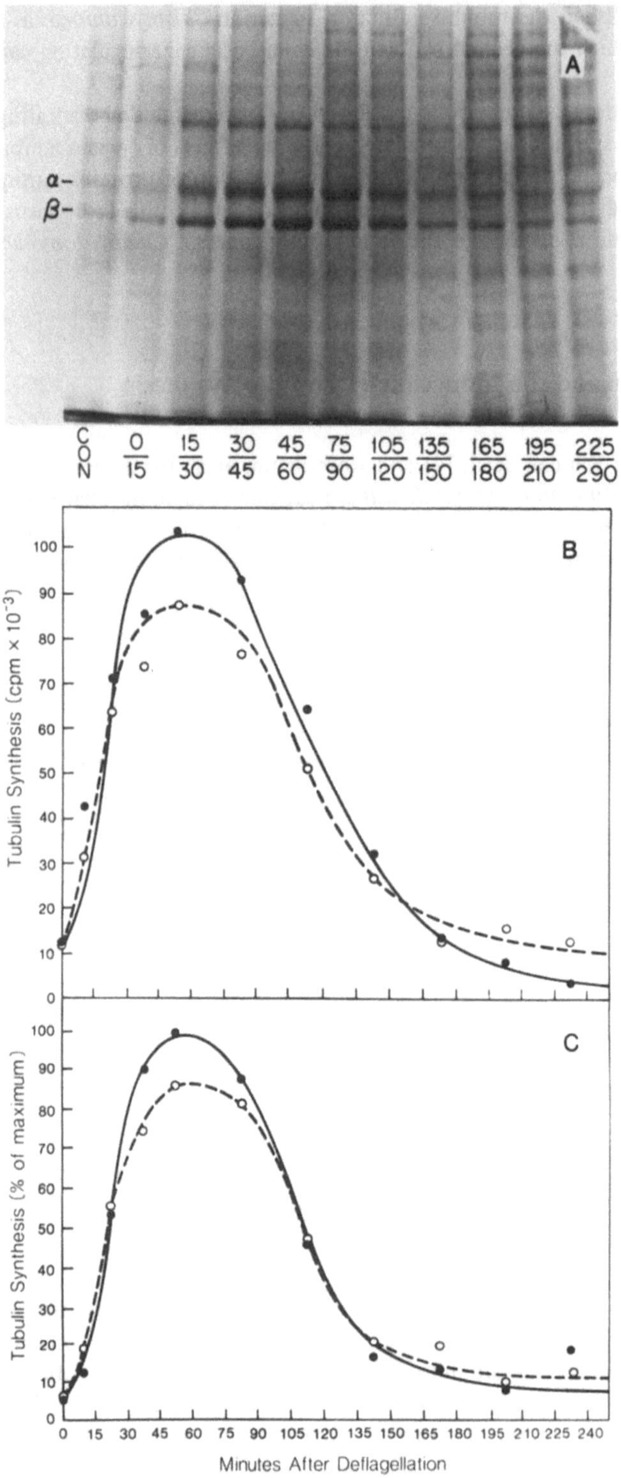

Fig. 2A-C. Time course of tubulin synthesis following flagellar excision. Cells were pulse-labeled with [^{35}S]-H_2SO_4 for 15 min during the interval after flagellar excision indicated below each lane. Cellular proteins were electrophoresed on 8% polyacrylamide gels and the radioactive proteins were detected by autoradiographic exposure of the dried gel (**A**). The relative amount of tubulin synthesis during each labeling interval was determined both by direct counting (**B**) of the portions of the gel containing α-(O--O) and β-tubulin (-- ●-- ●), and by densitometric analysis of the autoradiogram (**C**). Minami et al. 1981

at various intervals during flagellar regeneration. The kinetics of tubulin induction derived from these labeling experiments are presented in Fig. 2. The data shows a rapid and nearly equal increase in the amount of both α– and β–tubulin subunits. After a 45- to 60-min period of maximal synthesis, tubulin production declines rapidly to pre-deflagellation levels. The induction of tubulin synthesis is not only rapid, it is also extensive. During the peak period of synthesis, between 10 and 15% of total protein synthesis appears to be directed toward tubulin production.

Effects of Chemical Reagents on Flagellar Outgrowth and Tubulin Induction

One of the major unknowns that still exists in our studies of the tubulin induction system concerns the nature of the signal that is produced as a result of deflagellation and how this signal operates to cause the initiation of flagellar outgrowth and tubulin induction. However, the use of various chemical reagents has allowed us to learn more about the timing and the nature of the inductive signal. We know, for example, that the signal which prompts tubulin synthesis is not a feed-back signal generated from the utilization of tubulin pools for flagellar outgrowth. First, increases in tubulin synthesis in deflagellated gametic cells can be detected before flagellar regrowth begins (i.e., in less than 5 min). Secondly, if we use reagents, such as colchicine, which block flagellar outgrowth by blocking microtubule assembly, we [9] and, more recently, others [17] have shown that there is no inhibition of the rapid initial induction of tubulin synthesis.

Another reagent, amiprophosmethyl (APM), has also given us useful insights into the regulation of tubulin synthesis and flagellar regeneration. This compound which was first observed by Quader et al. [18] to prevent flagellar regeneration in *Chlamydomonas,* has a mode of action quite different than colchicine. For while, like colchicine, it blocks flagellar regeneration, APM has the ability to completely block tubulin induction when given to cells immediately prior to deflagellation [12]. The effects on tubulin synthesis are quite selective since the syntheses of no other major cellular proteins are affected. Even when given to fully induced cells midway through flagellar regeneration, it has the ability to stop flagellar outgrowth and halt tubulin synthesis. Both in vitro translation studies [12] and hybridization with cDNA probes specific to tubulin mRNA [13] have shown that tubulin mRNA disappears from fully induced cells within 10 to 15 min after administration of APM. Interestingly, the effects of the inhibitor are easily and rapidly reversed by transferring cells to fresh medium without APM.

Taken together, the results with the inhibitors cycloheximide, colchicine, and APM give us an insight into what appears to be a multi-step signaling process that leads to flagellar outgrowth and tubulin induction. The possible nature of the multistep process is summarized in the following schematic diagram.

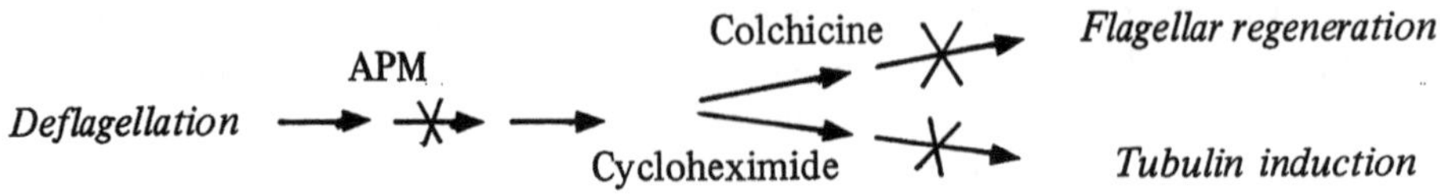

The diagram denotes the fact that, while flagellar regeneration may be blocked by reagents such as colchicine, this has no effect on the ability of the cells to support a burst of tubulin synthesis. Vice versa, the inhibition of tubulin production by cycloheximide has no effect on the ability to begin regeneration of new flagella. On the other hand, it would appear that APM blocks an earlier step in the signaling pathway that prevents both flagellar outgrowth and tubulin induction. The fact that cells can be treated for several hours with APM and still respond rapidly by regrowing flagella and initiating tubulin synthesis after the APM has been removed suggests that the inductive signal persists (even for long periods) until flagellar regeneration and tubulin synthesis is complete.

Analysis of the Coordinate Expression of the Tubulin Gene in *Chlamydomonas*

The simultaneous increase and decrease in the synthesis of α– and β–tubulin subunits during the induction and deinduction phases of tubulin synthesis (Fig. 2) suggested that there might be a concomitant increase and decrease in tubulin mRNA synthesis. By constructing and using cDNA clones to α and β tubulin mRNA's [13] we were able to show that this, indeed, is the case. By analysis of the levels of tubulin mRNA at short intervals after flagellar excision, we showed that there is a large and rapid increase in tubulin mRNA synthesis. Indeed, the increase and decrease in tubulin mRNA content in the cell during flagellar regeneration closely matched the kinetics of tubulin synthesis (Fig. 2). This demonstrated that there is no large pool of tubulin mRNA in the cell prior to deflagellation and that there are no major translational-level controls of tubulin production. The data also shows a clear increase in tubulin mRNA synthesis within only 3 to 5 min after deflagellation. This indicated that the signal generated as a result of flagellar excision rapidly triggers a set of subcellular reactions that lead to the increased production of tubulin mRNA.

When the tubulin mRNA's produced during flagellar regeneration are separated according to size in agarose gels containing the denaturing agent, methylmercury hydroxide, four distinct sizes of mRNA can be detected [15]. Two of these hybridize to α–tubulin cDNA and two hybridize to β–tubulin cDNA. Importantly, when mRNA's are isolated from cells at various points during the induction-deinduction cycle all four mRNA's are shown to increase and decrease together. These results suggest that there is a highly coordinated regulation of tubulin mRNA synthesis during tubulin induction.

Using the same tubulin cDNA's that were employed in the studies described above, we carried out Southern analysis of genomic DNA to determine how many tubulin genes are contained in the *Chlamydomonas* genome. By utilizing a variety of restriction enzyme cuts and portions of cDNA clones that selectively hybridized to only one of the four tubulin mRNA's, we gathered preliminary data that strongly indicates there are only four tubulin genes in *Chlamydomonas* – two encoding α-tubulin and two encoding β-tubulin [13]. Silflow et al. [19], using similar techniques, reached a similar preliminary conclusion. To confirm the gene number and to characterize the tubulin genes further we inserted fragments of *Chlamydomonas* genomic DNA into the bacteriophage cloning vector, Charon 30. From several hundred clones which hybridized with radioactive cDNA probes, we were able to select only four different classes of genes. Restriction mapping revealed that each of these classes corresponded to one of the four potential tubulin genes revealed by Southern analysis. From this and other data we and others [20] have been able to show that each of these four tubulin genes gives rise to one of the four tubulin genes that are produced in response to deflagellation.

During the past year we have begun analysis of the four tubulin genes to determine if they contain similar sequences which might allow for their coordinate transcription during tubulin induction (and during the increase in tubulin synthesis that we had earlier observed during cell division in synchronized cultures of *Chlamydomonas* [10]). We have focused on the region of the gene just 5' (upstream) of the initiation site for tubulin mRNA synthesis in each gene, since this promoter region has been implicated in several recent studies as the region in eukaryotic genes which may carry sequences important to the coordinate regulation of gene expression [21]. Our DNA-sequencing data has revealed that there are, indeed, shared regions of homology in the four genes [25]. We find a 16 base pair consensus sequence in multiple copies in the area just upstream of the TATA box in the promoter region of each gene. Although such a finding suggests this sequence as a prime candidate for involvement in the coordinate regulation of the *Chlamydomonas* tubulin genes, ultimate proof can only come from experiments in which foreign genes with these or altered sequences in appropriate positions are introduced into *Chlamydomonas* cells by transformation. The ability of these sequences to direct the transcription of the foreign gene in response to deflagellation can then be ascertained. The recent development of a genetic transformation system for *Chlamydomonas* [22] makes these experiments reasonable to consider in the near future.

Post-Translational Modification of α-Tubulin During Flagellar Outgrowth

Cells of higher eukaryotes contain multiple tubulin genes along with multiple forms of tubulin subunits. This suggests that microtubules with specialized function in different cellular compartments or organelles might gain that differential specialization because they are transcribed from different tubulin genes. In *Chlamydomonas* we have shown that there are only four different tubulin genes. Thus the idea seems less tenable in this organism that there are separate sets of α– and β–tubulin genes coding for tubulin in the flagellar apparatus, for the mitotic spindle, the cytoskeleton, etc. An alternative possibility is that tubulin subunits are modified for specific functions after translation and before (or during) their assembly into the microtubules of a spe-

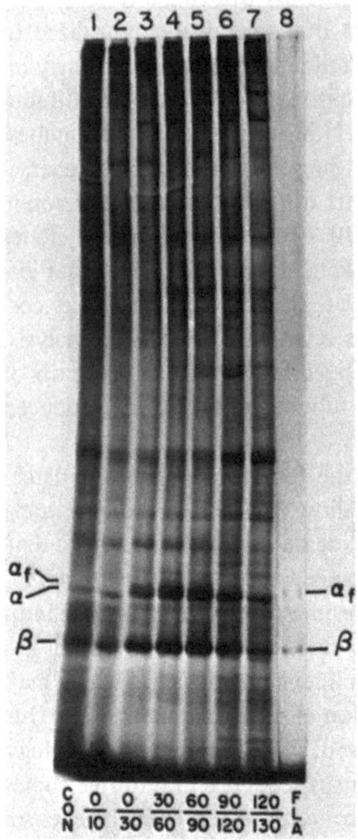

Fig. 3. Tubulin induction following flagellar excision in *C. reinhardi.* Autoradiograph of a SDS-polyacrylamide gel containing $H_2{}^{35}SO_4$-labeled proteins produced at various times before and after flagellar excision. Gametic cells were pulsed with $H_2{}^{35}SO_4$ before deflagellation (CON) and for the 10- or 30-min periods after flagellar removal shown below each track (for example, *30/60* denotes labeling from 30 to 60 min after deflagellation). α α–tubulin subunit which migrates as a protein with an apparent molecular weight of 56,000; α_f the predominant α–tubulin subunit of the flagellum; β the β–tubulin subunit of molecular weight 53,000. Track *8* contains radiactively-labeled flagellar markers. (Brunke et al. 1982)

cific cellular compartment. In the course of examining tubulin synthesis during flagellar regeneration we have found evidence that such modifications do take place in *Chlamydomonas.* We observed (Fig. 3) that a new species of α–tubulin, α_f, is synthesized during the period of active flagellar outgrowth [14]. In the period 0 to 10 min after deflagellation, and before flagellar outgrowth began, only a single species of α–tubulin can be detected on SDS-polyacrylamide gels. Likewise, the synthesis of α_f can not be detected during the latter period of tubulin induction (i.e., after 90 min) when flagellar outgrowth has been completed but tubulin synthesis still remains high (Fig. 3). Lefebvre et al. [23] had previously suggested a difference between the α–tubulin in isolated flagella and that produced during in vitro translation of tubulin mRNA. We have confirmed that observation [14] and have further shown by pulse-chase and inhibitor studies that a species of α tubulin that is produced and stored in the cell body is modified post-transcriptionally only during flagellar outgrowth into the α_f form of α–tubulin. These observations have recently received further confirmation by L'Hernault et al. [24]. The data thus strongly suggests that specialized microtubule function may arise at least in part from the post-translational modification of tubulin subunits at or near the site of organelle assembly and during the biogenesis of the organelle itself.

References

1. Foster K (1982) Biophys J 37:221A
2. Adair W, Hwang C, Goodenough U (1983) Cell 33:183–189
3. Goodenough U, Adair S, Caligor C, Forest C, Hoffmann J, Mesland D, Spath (1980) In: Gilular NB (ed) Membrane-Membrane Interactions (New York, Raven Press) pp 131–152
4. Randall J, Cavalier-Smith T, McVitte A, Warr J. Hopkins JM (1976) Dev Biol Suppl 1:43–83
5. Witman GB, Carlson K, Berliner J, Rosenbaum J (1972) J Cell Biol 54:507–539
6. Rosenbaum JL, Moulder JE, Ringo D (1969) J Cell Biol 41:600–619
7. Gorovsky MA, Carlson K, Rosenbaum JL (1970) Anal Biochem 35:359–370
8. Weeks DP, Collis PS (1976) Cell 9:15–27
9. Weeks DP, Collis PS, Gealt MA (1977) Nature (Lond) 268:667–668
10. Weeks DP, Collis PS (1979) Dev Biol 69:400–407
11. Gealt MA, Weeks DP (1980) Exp Cell Res 127:329–339
12. Collis PS, Weeks DP (1978) Science (Wash DC) 202:440–442
13. Minami SA, Collis PS, Young EE, Weeks DP (1981) Cell 24:89–95
14. Brunke KJ, Collis PS, Weeks DP (1982) Nature (Lond) 297:516–519
15. Brunke KJ, Young EE, Buchbinder BU, Weeks DP (1982) Nuc Acid Res 10:1295–1310
16. Siflow CD, Lefebvre PA, McKeithan TM, Schloss JA, Keller LR, Rosenbaum JL (1982) Cold Spring Harbor Symp Quant Biol 46:157–169
17. Lefebvre PA, Nordstrom SA, Moulder JE, Rosenbaum JL (1978) J Cell Biol 78:8–27
18. Quader H, Cherniak J, Filner P (1977) Plant Physiol (Bethesda) 59:19a
19. Silflow CD, Rosenbaum JL (1981) Cell 24:81–88
20. Silflow CD, Schloss JA, Rosenbaum JL (1982) J Cell Biol 95:477a
21. Davidson EH, Jacobs HI, Britten RJ (1983) Nature (Lond) 301:468–470
22. Rochaix JD, van Dillewijn J (1982) Nature (Lond) 296:70–72
23. Lefebvre PA, Silflow CD, Weiben ED, Rosenbaum J (1980) Cell 20:469–477
24. L'Hernault SW, Rosenbaum JL (1983) J Cell Biol 97:258–263
25. Brunke KJ, Anthony JG, Sternberg EJ, Weeks DP (1984) Molecular and Cellular Biology (in press)

Flagellar Root-Mediated Interactions Between the Flagellar Apparatus and Cell Organelles in Green Algae

M. MELKONIAN[1]

The Flagellar Apparatus – A Separate Compartment ?

The flagellar apparatus is a structurally and biochemically complex cell organelle, but it is not a separate compartment in the sense of Schnepf (this vol.). It is a local specialization of the cytoplasmic compartment involving a microtubular organelle (the axoneme), its associated basal apparatus and the ensheathing plasma membrane plus cell surface components. There is exchange of metabolites, macromolecules and ATP into and out of the flagellar apparatus without requiring passage through a membrane, therefore also in functional terms the flagellar apparatus is not a separate compartment. It nevertheless maintains structural and functional integrity and therefore qualifies as an important eukaryotic cell organelle. There is good evidence that passage of ribosomes into the free flagellum does not occur, although in some green algal taxa ribosomes may be found consistently inside the basal bodies (e.g., Melkonian and Berns 1983). The flagellar membrane is also specific and does not mix with other areas of the plasma membrane (review by Bloodgood 1982). The same holds true for cell surface components of the flagellar membrane (sexual agglutinin of green algal gametes, flagellar hairs and scales; Adair et al. 1983, Melkonian 1984). How lateral diffusion of integral membrane proteins from the flagellar membrane into other areas of the plasma membrane is prevented, is unknown, but the flagellar membrane covering the transition zone may be such a lateral diffusion barrier, since it is not susceptible to low concentrations of nonionic detergents unlike the rest of the flagellar membrane (Goodenough 1983). It should be further mentioned that the diffusible low molecular weight protein calmodulin occurs in much higher concentrations inside the flagellar apparatus in comparison to other regions of the cytoplasm (Gitelman and Witman 1980, Surek and Melkonian 1983). The amount of calmodulin inside the flagellum is probably governed by a source-sink relationship due to the presence of specific calmodulin-binding proteins in the flagellar apparatus. In summary it may be stated that although the flagellar apparatus is not a compartment sensu stricto it exerts some influence on the amount and quality of components entering or leaving it (ions, regulatory proteins, membrane proteins, etc.) reflecting its unique biochemical composition.

This short review concentrates on interactions of the mature flagellar apparatus of green algae with various cell organelles mediated by flagellar roots. The interactions

1 Botanisches Institut, Universität Münster, Schloßgarten 3, 4400 Münster, FRG

Compartments in Algal Cells and Their Interaction
ed. by W. Wiessner, D. Robinson, and R.C. Starr

between the flagellar apparatus and cell organelles (mainly the nucleus) during development of flagella are dealt with in other articles in this book (Bouck, Weeks, this vol.).

Nucleus-Flagellar Root Interactions

The flagellar apparatus of green algal motile cells has been the subject of several reviews over the last few years (Moestrup 1978, Melkonian 1980a, Moestrup 1982, Melkonian 1984). From these it can be seen that structural interactions between the flagellar bases and the nucleus do occur more frequently than previously anticipated. Improved methods of fine structural preservation and the investigation of more green algal taxa leads to the conclusion that principally two types of interactions exist: (a) the basal bodies are directly connected to the nuclear envelope or to an elongated branch of endoplasmic reticulum as an extension of the nuclear envelope (Figs. 1,8), (b) the basal bodies are indirectly linked to the nuclear envelope through a fibrillar flagellar root traditionally known as a rhizoplast (system II fiber according to Melkonian 1980a, see also Figs. 3,4). In some green algal motile cells both types of connections occur simultaneously (Watson and Arnott 1973, Melkonian and Berns 1983). In the latter case the system II fiber runs parallel and in close contact with the branch of endoplasmic reticulum. In a few rare cases microtubular flagellar roots are connected to the nuclear envelope (Fig. 2: in *Nephroselmis* and also in *Pedinomonas* species, Ettl and Manton 1964). It should be recalled that such connections between microtubular flagellar roots and the nucleus are a characteristic feature of sperm cells of higher plants (e.g., Myles and Hepler 1982). In the peculiar backward-swimming green flagellate *Scourfieldia* a special type of flagellar root, here provisionally called the lamellate root, connects the basal body with the nuclear membrane and runs inside a nuclear groove to the posterior end of the nucleus (Fig. 5).

If system II fibers (rhizoplasts) provide the linkage between basal bodies and the nucleus generally each basal body of the flagellar apparatus has its own fiber. In the octoflagellate species of *Pyramimonas* for example the number of system II fibers per flagellar apparatus is greatly increased. Exceptions to this rule are texa in which pairs of flagella beat coordinately (hydrodynamically coupled) as in the quadriflagellate *Tetraselmis* (Salisbury et al. 1981). Here the system II fiber branches proximally (near the basal bodies) and the branches are indirectly linked to at least two basal bodies.

The functional significance of the basal body-system II fiber-nucleus connection is largely unknown. In most scaly green flagellates the system II fibers pass the nucleus and do not terminate at the nuclear envelope (Robenek and Melkonian 1979). In *Tetraselmis* it has been shown that the system II fiber makes contact with the nuclear envelope preferentially in the region of the cross bars of the fiber (Robenek and Melkonian 1979). The nuclear envelope subtending the fiber is devoid of nuclear pores, lacks attached ribosomes and on its inner side is associated with condensed chromatin. A similar association of condensed chromatin with an overlying system II fiber is also apparent in other scaly green flagellates (e.g., *Pyramimonas amylifera,*

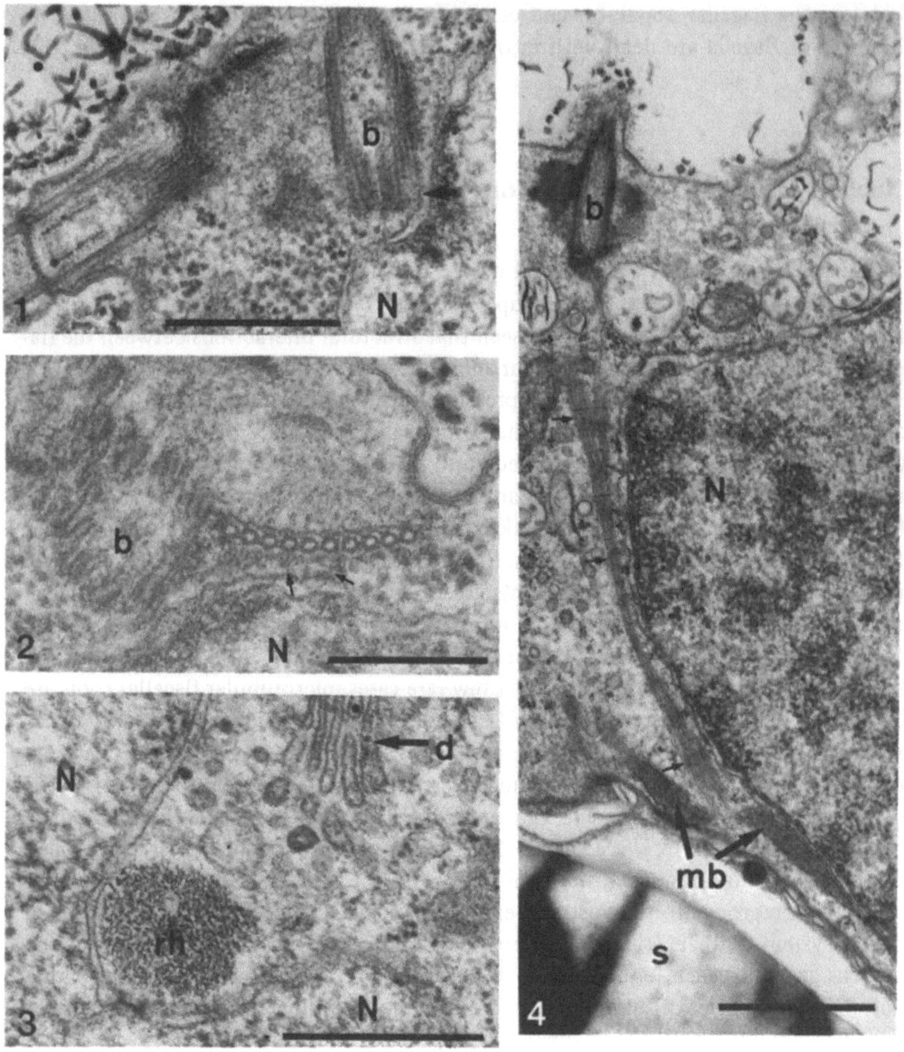

Fig. 1. Section through both basal bodies (*b*) of the scaly flagellate *Nephroselmis olivacea. Arrow* one of the basal bodies is attached to the nucleus (*N*). Bar = 0.5 μm

Fig. 2. Cross section through the broad microtubular flagellar root of *Nephroselmis olivacea.* The lower stratum of the root is attached (*small arrows)* to the nucleus (*N*). Bar = 0.25 μm

Fig. 3. Cross section through the system II fiber (rhizoplast, *rh*) of *Tetraselmis cordiformis* located in a groove of the nucleus (*N*). *d* dictyosome. Bar = 0.5 μm

Fig. 4. Longitudinal section through a system II fiber in *Pyramimonas amylifera* (*small arrows*). The fiber runs parallel to the nuclear envelope, which on its inner side is associated with condensed chromatin, *mb* microbody profiles; *N* nucleus; *b* basal body; *s* starch grain. Bar = 1 μm

Fig. 5. Longitudinal section through basal body (*b*) and lamellated roots (*long arrows*) of *Scourfieldia caeca. N* nucleus; *short arrow* condensed chromatin associated with one of the lamellated roots. Bar = 0.5 μm

Fig. 6. Longitudinal section through the scale ductus (*scd*) of *Pyramimonas amylifera.* A cross-striated flagellar root (*prominent arrow*) links the flagellar apparatus with th duct opening. *Small arrows* the scale ductus is ensheathed by filaments. Bar = 1 μm

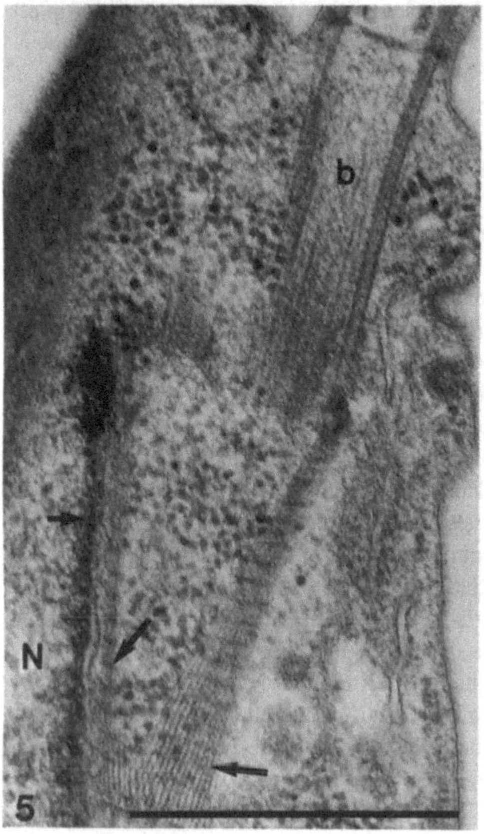

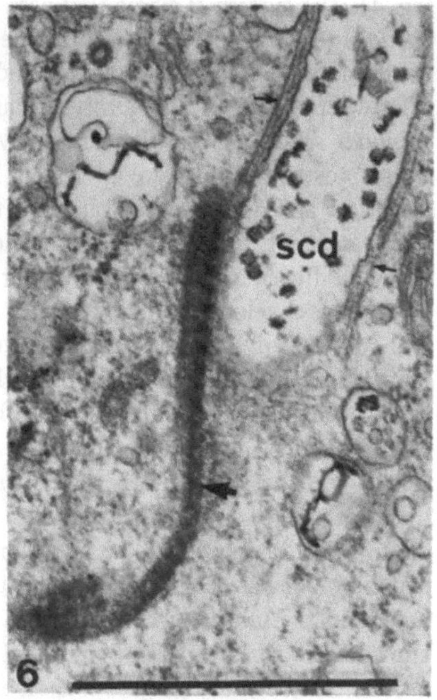

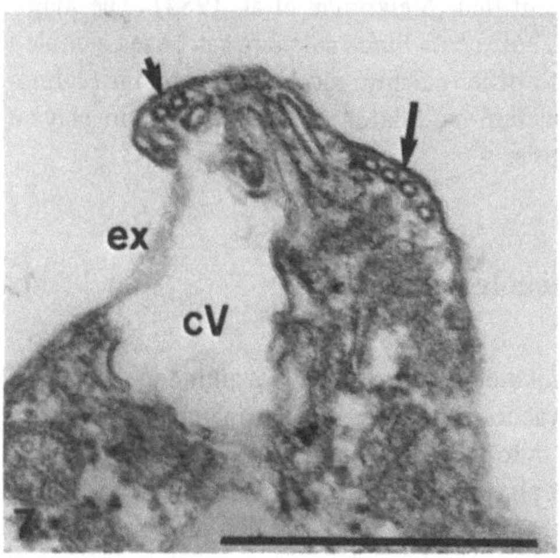

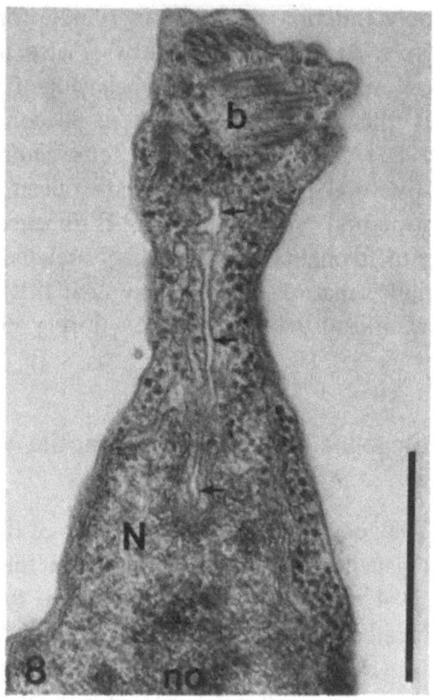

Fig. 7. Section through the single contractile vacuole (*cV*) of zoospores of *Friedmannia israelensis*. *ex* exocytosis site at the plasma membrane; *short arrow* cross section through two-stranded microtubular root; *long arrow* cross section through four-stranded microtubular root. Bar = 0.5 μm

Fig. 8. Longitudinal section through the anterior parts of the naked green flagellate *Spermatozopsis similis*. A branch of endoplasmic reticulum (*small arrows*) extends from near the basal body (*b*) to the nucleus (*N*). *no* nucleolus. Bar = 0.5 μm

Fig. 4). The system II fibers that pass the nucleus either terminate at the chloroplast or individual filaments of the fiber are attached to special sites at the plasma membrane (as in *Tetraselmis,* Robenek and Melkonian 1979). It is possilbe that even in cases where the fiber apparently terminates at the chloroplast (for example in *Pyramimonas* species) some filaments may reach the plasma membrane. In other green algae (mostly Chlorophyceae) system II fibers are less massive as in the scaly green flagellates and do not extend beyond the nuclear region. However, in no case has a termination of filaments at the nuclear envelope been conclusively shown. Most often the fiber terminates in a granular region of cytoplasm devoid of ribosomes near the nuclear envelope (as in *Carteria* for example; Lembi 1975).

The fate of the system II fiber during mitosis might offer some clues as to the function of this flagellar apparatus-nucleus association. In *Tetraselmis subcordiformis* it has been shown that the system II fiber dissolves during prophase to form a granular mass from which the newly organized spindle microtubules emanate (Stewart et al. 1974). It has therefore been concluded that the system II fibers are involved with the formation of spindle microtubules. During telophase system II fibers are reformed presumably originating at the basal bodies and they regain contact with the nuclear envelope (Stewart et al. 1974). The connection of the system II fiber to the basal body and the nuclear envelope might therefore be a reflection of both the organization of the fiber at basal bodies and its function in spindle formation during mitosis. In cases where the system II fibers terminate near the nucleus (see above) their only function might lie in spindle formation during mitosis, whereas in most scaly green flagellates they probably also perform additional functions in the interphase cell related to the motile behavior of the cell (see below). In *Tetraselmis* the system II fibers have been shown to be contractile (Salisbury and Floyd 1978, Salisbury 1983). System II fibers of *Tetraselmis* have been successfully isolated and the major structural protein (20,000 MW) been identified (Melkonian et al. 1984). The major structural protein of system II fibers of *Tetraselmis* binds calcium, can be reassembled into filaments and is probably related to other calcium-modulated proteins (calmodulin, spasmin). It is likely that this calcium-modulated contractile protein plays a significant yet unknown role during mitosis.

Flagellar Apparatus-Contractile Vacuole Interactions

The contractile vacuole complex of flagellated green algal cells is an independent cell organelle (recent survey of contractile vacuoles in algae by Hausmann and Patterson 1984, this vol.) and its position in the cytoplasm is usually well defined and fixed. Numbers of contractile vacuoles in interphase cells vary in the different taxa from 1 to 4. The few reports of a greater number of contractile vacuoles per cell scattered around the cell periphery presumably relate to preprophase stages of taxa where cell division is transverse not longitudinal (members of the Haematococcaceae including *Chlorogonium*). The position of the contractile vacuole complex is fixed with respect to the plane of beat of the flagella. In *Chlamydomonas*-type flagellated cells two contractile vacuoles are located close to the flagellar insertion perpendicular to the plane

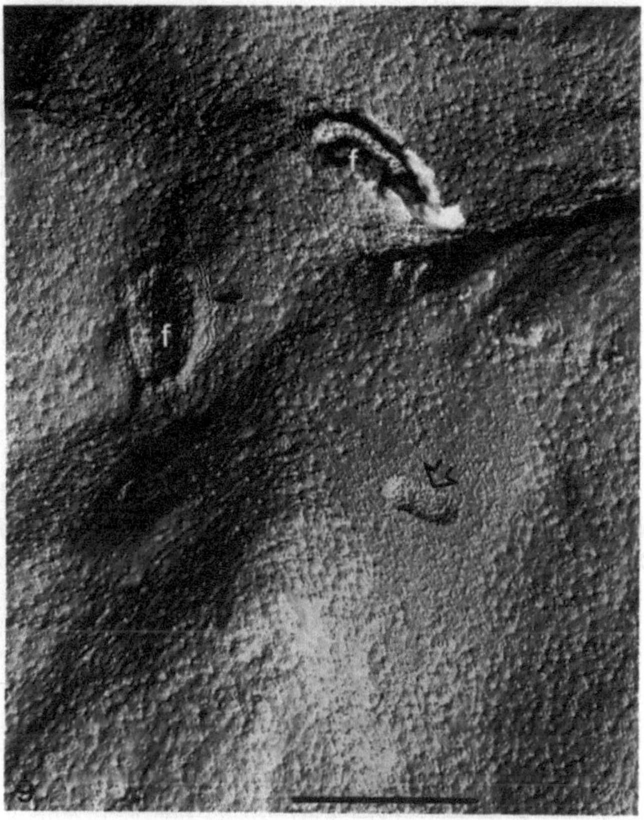

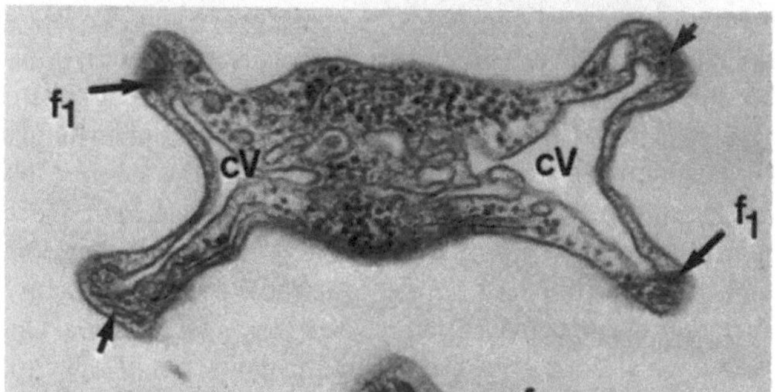

Fig. 9. Freeze-fracture image of the plasma membrane of *Chlamydomonas reinhardii* (CCAP 11/32 CW 15^+) after treatment with filipin according to methods described in Robenek and Melkonian (1981b). Filipin-sterol complexes cover most areas of the plasma membrane with exception of the contractile vacuole exocytosis site (*open arrow*) and the adjacent membrane area. *f* cross fractured flagellar bases; *arrow* flagellar bracelet. Bar = 0.5 μm

Fig. 10. Cross sections through the anterior part of the cell of *Spermatozopsis similis* revealing two contractile vacuole complexes (*cV*); *f* oblique section through one of the two flagella; *short arrow* five-stranded microtubular roots; f_1 system I fibers associated with two-stranded microtubular roots. Bar = 0.5 μm

of beat of the two flagella (Fig. 9). The exocytosis site of the contractile vacuole complex and the surrounding area of the plasma membrane is specialized (for example it does not bind the polyene antibiotic filipin which forms complexes with 3β-hydroxysterols; Fig. 9; see also review by Severs and Robenek 1983). How does the fixed position of the contractile vacuole complex in green algae arise and how is it maintained? Recent studies indicate that the membrane of the contractile vacuole is linked to two different microtubular flagellar roots (Figs. 7,10). In *Friedmannia*-type motile cells (with only one contractile vacuole per cell) the membrane of the contractile vacuole is connected to a two- and four-stranded microtubular root which originate from the same basal body (Melkonian and Berns 1983; Fig. 7), in *Chlamydomonas*-type organisms (with two contractile vacuoles per cell) the connection is provided again by linkage to a two- and four-stranded microtubular flagellar root, but this time both roots originate from different basal bodies (Melkonian and Preisig 1984a; Fig. 10). In Fig. 10 the left contractile vacuole is probably in the state of late systole and it can be seen that the membrane of the contractile vacuole is still connected to the two microtubular roots resulting in an irregular profile of the vacuole and indicating that the connection may also be a functional one. It may therefore be that the microtubular flagellar roots play a role in maintaining the position of the contractile vacuoles with respect to the flagellar apparatus in green algae. It should however be mentioned that in the freshwater prasinophyte *Tetraselmis cordiformis* the two contractile vacuoles are not associated with microtubular flagellar roots yet still occupy a definite position with respect to the plane of beat of the two pairs of flagella (unpubl. observations). It is therefore more likely that the microtubular roots are needed to establish or determine the correct positioning of the contractile vacuole during contractile vacuole morphogenesis rather than to maintain its position during interphase. It should be recalled that during cell division contractile vacuoles are distributed among daughter cells in the semi-conservative way, i.e., each daughter cell (if only two daughter cells are produced) receives one old contractile vacuole and forms the new one de novo, presumably in close association with the development of flagellar roots. In *Tetraselmis* it has been shown that the microtubular roots in recently divided daughter cells extend the entire length of the anterior lobes of the cells and only after theca formation retract to their interphase length and become associated with the rhizanchora (Domozych et al. 1981). They may therefore perform an organizing function for positioning of the contractile vacuole also in this genus, a direct evidence for this presumed function however is still lacking. Once the close contact between a small portion of the vacuole membrane and the exocytosis area of the plasma membrane is established (Weiss et al. 1977 and Fig. 10) there may not be a functional need for maintaining a close structural connection between microtubular flagellar roots and the contractile vacuole membrane (see example: *Tetraselmis cordiformis*). Why then should the exocytosis site of the contractile vacuole of green algae have a specific location in relation to the plane of beat of the flagella? It might be that the sudden extrusion of fluid from the exocytosis site would adversely affect the proper hydrodynamic function of the beat cycle of the flagella if both the exocytosis site and the plane of beat of the flagella were in the same plane. It is therefore interesting to note that even in *Friedmannia*-type green algae where there is only one contractile vacuole roughly located in the plane of beat of the two flagella the exocytosis site is still clearly lateral and therefore the fluid is expelled perpendicular to the plane of beat of the flagella (Melkonian and Berns 1983 and Fig. 7).

Flagellar Root-Plasma Membrane Interactions

Structural connections between the basal bodies and the plasma membrane in green algal motile cells always appear to be indirect, either mediated through the axoneme with its transition region (review by Melkonian 1984) or mediated through flagellar roots. One example, the system II fibers of the prasinophyte *Tetraselmis* have already been mentioned (see above), other system II fibers, especially in the Ulvophyceae sensu Stewart and Mattox (1978), run just below the cell periphery for several μm and probably always terminate at the plasma membrane. Their role has been evaluated during gametic fusion in *Ulva lactuca* gametes (Melkonian 1980b and Figs. 11, 12). The most common linkage of the flagellar apparatus with the plasma membrane is provided by microtubular flagellar roots (review: Moestrup 1978, 1982). Although one of the initially assumed functions of microtubular flagellar roots, namely that of a cellular cytoskeleton in green algal motile cells, does not seem to hold true for the majority of flagellated green algae (see Melkonian et al. 1980 and McFadden and Wetherbee, to be published), it is still likely that in motile cells lacking additional secondary cytoskeletal microtubules (for example the motile cells of *Friedmannia*-type algae, Melkonian and Berns 1983) the microtubular flagellar roots perform a cytoskeletal role. In green algal motile cells where abundant secondary cytoskeletal microtubules occur (in the *Chlamydomonas*-type flagellates and in motile cells of the Ulvophyceae sensu Stewart and Mattox) the microtubular flagellar roots function as microtubule-organizing centers (MTOC's) for these additional microtubules (Stearns and Brown 1981).

Mating structures are specialized regions of the plasma membrane of algal gametes where gametic fusion of opposite mating types is initiated (review: Melkonian 1980b, Fig. 12). Similar to the contractile vacuole complex mating structures always appear to have a constant positional relationship to the plane of beat of the flagella (Goodenough and Weiss 1978, Robenek and Melkonian 1981a,b amongst others). In several taxa it has been demonstrated that the specific position of the mating structure might be functionally important to guarantee the efficiency of gametic fusion. Until now all mating structures that have been investigated ultrastructurally are linked to one type of microtubular flagellar root, the two-stranded root. It might be one of the functions of this root type to determine the future site of mating structure differentiation during gametogenesis. It is interesting to note that only one mating structure occurs per gamete which appears to be in accordance with the view that only one basal body (the "younger" basal body of a pair) organizes the mating structure through its flagellar roots (similar to the proposal made for the development of the phototactic apparatus in green algal motile cells; Melkonian 1984). The molecular basis of this presumptive organizing function of flagellar roots is totally obscure at present. It is also interesting to mention that the two-stranded microtubular root in those taxa with mating structures appears to be always associated with a system I

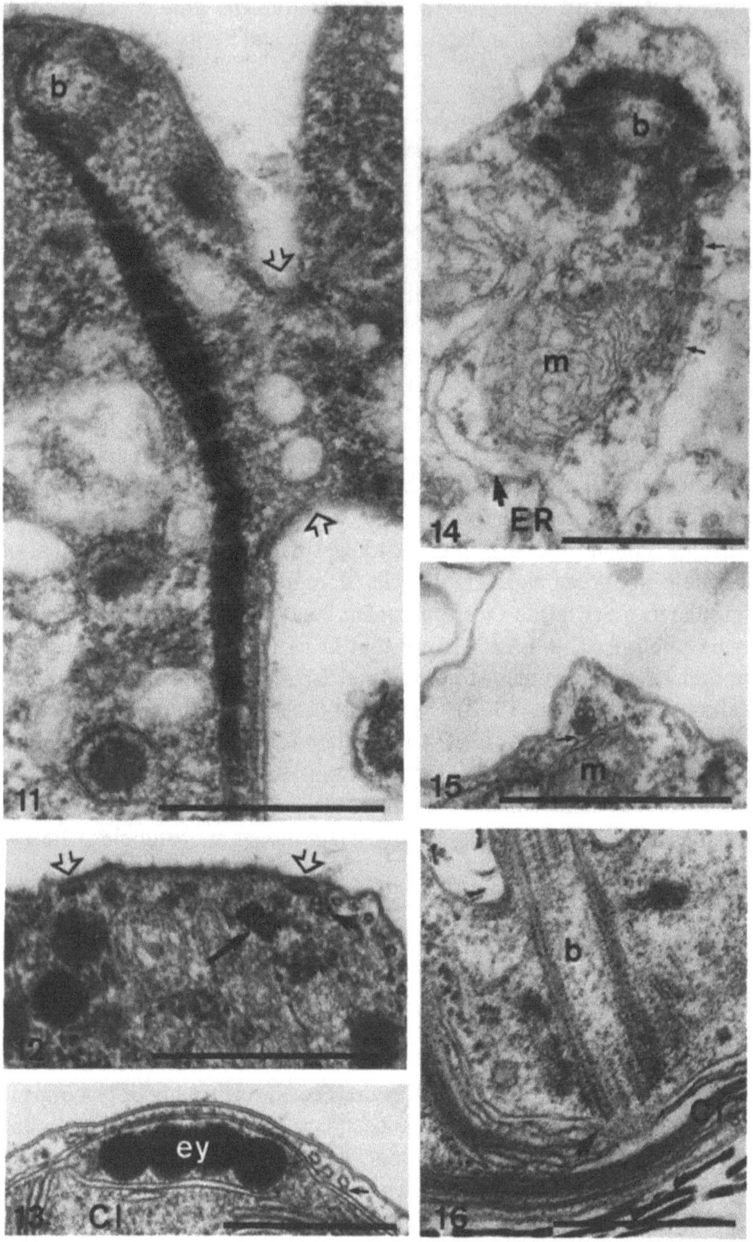

Fig. 11. Longitudinal section through two gametes of *Ulva lactuca* shortly after gametic fusion has taken place in the area of the mating structures (*open arrows*). From a basal body (*b*) a prominent system II fiber passes the fusion site. Bar = 0.5 μm

Fig. 12. Cross section through the mating structure (limits of the mating structure indicated by *open arrows*) of the female gamete of *Ulva lactuca. Long arrow* cross section through system II fiber; *short arrow* cross section through two-stranded microtubular root. Bar = 0.5 μm

Fig. 13. Cross section through the eyespot (*ey*) of zoospores of the green alga *Chlorosarcinopsis pseudominor*. *Small arrow* link between microtubular flagellar root and outer chloroplast envelope membrane; *Cl* chloroplast. Bar = 0.5 μm

Fig. 14. Longitudinal section through the anterior part of a cell of *Dunaliella salina*. *b* basal body; *m* mitochondrial profile; *ER* endoplasmic reticulum; *small arrows* a system II fiber connects the basal body with the mitochondrial profile. Bar = 0.5 μm

Fig. 15. Cross section through the two-stranded microtubular flagellar root of zoospores of *Chlorosarcinopsis dissociata*. *Small arrow* electron dense material underlying the root microtubules is attached to a mitochondrial profile (*m*). Bar = 0.5 μm

Fig. 16. Section through a basal body (*b*) of the scaly flagellate *Mesostigma viridie*. *Short arrow* the basal body is linked to the outer chloroplast envelope membrane. *Cl* chloroplast. Bar = 0.5 μm

fiber (for definition see Melkonian 1980a; possible exception *Bryopsis* ?). A function of the system I fiber as a signal transducer in mating of *Chlamydomonas* has been suggested (Goodenough and Weiss 1978).

Another interesting association between a flagellar root and a prospective plasma membrane is that of the scale ductus in some *Pyramimonas* species. The scale ductus which leads to the scale reservoir is connected to the basal bodies in some *Pyramimonas* species by a cross-striated fiber (Manton 1968 and Fig. 6). From this fiber filaments arise that ensheath the ductus (Fig. 6). The cross-striated fiber is associated with a microtubular flagellar root which in its absolute configuration corresponds to the two-stranded microtubular root of an X-2-X-2 flagellar root system (sensu Moestrup 1978). The scale reservoir is a singular structure and is distributed during cell division in the same way as the contractile vacuole complex (semi-conservative replication). The reason why the scale reservoir opening should have a fixed position with respect to the flagellar apparatus is at present unknown.

Flagellar Apparatus – Mitochondrion Interactions

In a greater number of flagellated green algal cells the flagellar apparatus is structurally connected to mitochondrial branches. This connection may be direct (i.e., mediated through the basal bodies; e.g., in zoospores of *Chlorokybus atmophyticus*, Rogers et al. 1980 and in the scaly flagellate *Pyraminonas obovata*, Melkonian 1981) or more often indirect, mediated by microtubular or fibrous flagellar roots. In the motile cells of all Charophyceae sensu Stewart and Mattox (Mattox and Stewart 1984) mitochondrial branches are linked to the flagellar apparatus by the broad microtubular root containing the multilayered structure (e.g., Moestrup 1974). The same apparently holds true for the sperm cells of the archegoniate land plants. In scaly green flagellates it may be a system II fiber that links the flagellar apparatus with mitochondrial branches (review on mitochondria of prasinophytes by Norris 1980). It appears most likely that in all the taxa where flagellar roots connect mitochondrial branches with the basal bodies there is only a single mitochondrial reticulum per cell, as has recently been demonstrated in *Pyramimonas gelidicola* (McFadden and Wetherbee 1982). In flagellates the association of flagellar roots with the mitochondrion may help to distribute daughter mitochondria regularly during cell division. In other green algal motile cells the relationship between mitochondria and the flagellar apparatus may not be so obvious, sometimes two-stranded microtubular flagellar roots may be close to mitochondrial profiles (Fig. 15), in other cases system II fibers pass along mitochondrial profiles on their way toward the nucleus as in *Dunaliella* species (Melkonian and Preisig 1984b and Fig. 14). An obvious functional role for the close association of mitochondrial branches with the flagellar apparatus is related to the energy producing function of the mitochondria and the energy requirement for flagella function.

Flagellar Apparatus – Microbody Interactions

In the Charophyceae sensu Stewart and Mattox and in the prasinophytes generally one large microbody occurs per motile cell. It is generally attached to the basal bodies either directly or via a broad microtubular root containing a multilayered structure (in zoospores of Charophycean algae) or a system II fiber (discussion of these relationships in Rogers et al. 1980). A functional relationship between the microbody and the flagellar apparatus is not apparent compared to the mitochondria-flagellar apparatus relationship and since the single microbody is divided during cell division into two daughter microbodies (Floyd et al. 1972) it is likely that its association with basal bodies is necessary for division and regular distribution of daughter microbodies. It is therefore interesting to note that in the zoospore producing charophyte *Klebsormidium* the centriole-microbody connection persists in the vegetative cells (Pickett-Heaps 1972). If the microbody is divided during cell division, (instead of one microbody retained and another one formed de novo) then the cleavage plane must pass through the microbody. This probably requires that the microbody be associated with a special area of a basal body (as in *Pyramimonas obovata,* Melkonian 1981) or with a particular microtubular flagellar root (as in *Mesostigma,* unpubl. observations). In most Chlorophyceae sensu Stewart and Mattox there are numerous small microbodies scattered in the cytoplasm and there is no apparent structural relationship with the flagellar appratus.

Flagellar Apparatus – Chloroplast Interactions

Finally the flagellar apparatus of motile cells of green algae may also be linked to the chloroplast. In a few rare cases basal bodies make direct contact with the chloroplast envelope (e.g., in *Mesostigma,* Fig. 16), but in general the chloroplast does not closely approach the basal bodies. A very common indirect linkage of the chloroplast envelope with the basal bodies is provided by microtubular flagellar roots which connect with a specialized area of the chloroplast housing the phototactic apparatus (eyespot apparatus, reviews Moestrup 1978 and Melkonian 1982). The evolution of the special positional relationship between a particular (the X-root) microtubular root and the green algal eyespot is probably related to the behavior of the eyespot during cell division in these green algae (semi-conservative replication, Melkonian 1984). In those prasinophytes in which the cleavage plane passes through the eyespot and divides the eyespot into two daughter eyespots an eyespot-microtubular flagellar root connection is not found (Melkonian and Robenek, in prep.).

The connection between system II fibers and the chloroplast in some scaly flagellates has been discussed above.

Conclusion

In flagellated cells of green algae the flagellar apparatus is structurally connected to all major cell organelles. Most often the connection is provided by flagellar roots. The function of these interactions has not been evaluated in most cases but present evidence suggests: (a) microtubular flagellar roots may determine the position of cell organelles with respect to the flagellar apparatus and more specifically the plane of beat of the flagella. The connection is presumably established during development of the flagellar apparatus (zoosporogenesis, gametogenesis, cell division). In some cases (eyespot, mating structure, contractile vacuole complex, mitochondrion) the positional relationship may be necessary for proper function, in others it may be necessary to ensure correct distribution of divided cell organelles (microbody, mitochondrion?) during cytokinesis. (b) In most green algal motile cells microtubular flagellar roots organize secondary cytoskeletal microtubules which interact with the plasma membrane to maintain the cellular cytoskeleton. (c) The basal bodies are organizing centers for calcium-modulated contractile filaments which play a role in intracellular movements, possibly including some events during mitosis.

Acknowledgments. I would like to thank B. Berns, H.R. Preisig, H. Robenek and B. Surek for help during various stages of the experimental work. This study was supported by the Deutsche Forschungsgemeinschaft.

References

Adair WS, Hwang C, Goodenough UW (1983) Cell 33:183–193
Bloodgood RA (1982) In: Amos WB, Duckett JG (eds) Prokaryotic and Eukaryotic Flagella. University Press, Cambridge, pp 353–380
Domozych DS, Stewart KD, Mattox KR (1981) J Cell Sci 52:351–371
Ettl H, Manton I (1964) Nova Hedwigia 8:421–451
Floyd GL, Stewart KD, Mattox KR (1972) J Phycol 8:176–184
Gitelman SE, Witman GB (1980) J Cell Biol 87:764–770
Goodenough UW, Weiss RL (1978) J Cell Biol 76:43–438
Goodenough UW (1983) J Cell Biol 96:1610–1621
Hausmann K, Patterson DJ (1984) In: Wiessner W, Robinson DG (eds) Compartments in algal cells and their interaction. Springer, Berlin Heidelberg New York, pp 000–000

Lembi CA (1975) J Phycol 11:219–221
Manton I (1968) Proc Linn Soc (Lond) 179:147–152
Mattox KR, Stewart KD (1984) In: Irvine DEG, John DM (eds) The systematics of the Gren Algae. Academic Press, London, pp 000–000
McFadden GI, Wetherbee R (1982) Protoplasma 111:79–82
Melkonian M (1980a) BioSystems 12:85–103
Melkonian M (1980b) J Cell Sci 46:149–169
Melkonian M (1981) Protoplasma 108:341–355
Melkonian M (1982) In: Amos WB, Duckett JG (eds) Prokaryotic and Eukaryotic Flagella. University Press, Cambridge, pp 589–606
Melkonian M (1984) In: Irvine DEG, John DM (eds) The systematics of the Green Algae. Academic Press, London, pp 000–000
Melkonian M, Berns B (1983) Protoplasma 114:67–84
Melkonian M, Preisig HR (1984a) Plant Syst Evol 000:000–000
Melkonian M, Preisig HR (1984b) Plant Syst Evol 000:000–000
Melkonian M, Kröger K-H, Marquardt K-G (1980) Protoplasma 104:283–293
Melkonian M, Surek B, Salisbury JL (1984) Protistologica (in press)
Moestrup Ø (1974) Biol J Linn Soc 6:111–125
Moestrup Ø (1978) BioSystems 10:117–144
Moestrup Ø (1982) Phycologia 21:427–528
Myles DG, Hepler PK (1982) Dev Biol 90:238–252
Norris RE (1980) In: Cox ER (ed) Phytoflagellates. Elsevier/North Holland, New York, pp 85–145
Pickett-Heaps JD (1972) Cytobios 6:167–183
Robenek H, Melkonian M (1979) Arch Protistenkd 122:340–351
Robenek H, Melkonian M (1981a) J Cell Sci 50:149–164
Robenek H, Melkonian M (1981b) Eur J Cell Biol 25:258–264
Rogers CE, Mattox KR, Stewart KD (1980) Am J Bot 67:774–783
Salisbury JL (1983) J submicrosc Cytol 15:105–110
Salisbury JL, Floyd GL (1978) Science (Wash DC) 202:975–977
Salisbury JL, Swanson JA, Floyd GL, Hall R, Maihle NJ (1981) Protoplasma 107:1–11
Severs NJ, Robenek H (1983) Biochim Biophys Acta 737:373–408
Stearns ME, Brown DL (1981) J Ultrastruct Res 77:366–378
Stewart KD, Mattox KR (1978) BioSystems 10:145–152
Stewart KD, Mattox KR, Chandler CD (1974) J Phycol 10:65–79
Surek B, Melkonian M (1983) Br Phycol J 18:210–211
Watson MW, Arnott HJ (1973) J Phycol 9:15–29
Weiss RL, Goodenough DA, Goodenough UW (1977) J Cell Biol 72:133–143

Flagella Development in *Chlamydomonas reinhardii*: Some Regulatory Aspects Concerning the Shortening Response of the Flagella

H. QUADER[1]

Introduction

Chlamydomonas reinhardii has been introduced as a test organism to study the development of flagella: the locomotive organelle of this alga (Ringo 1967, Rosenbaum et al. 1969, Witman et al. 1972, Cavalier-Smith 1974).

Vegetatively growing cells withdraw and regenerate their flagella before or after cell division, respectively (Bloodgood 1974). Both events, flagellar withdrawal and regeneration, are apparently controlled via calcium (Quader et al. 1978). Lowering the extracellular calcium concentration to <1 μM through the application of the highly selective calcium-chelator ethyleneglycol-bis(aminoethyl)-N,N'-tetraaceticacid (EGTA) or increasing the free calcium concentration of the cytosol through the herbicide amiprophosmethyl (APM) prevents the outgrowth or induces the cells to gradually shorten their flagella (Quader and Filner 1980, Hertel et al. 1980).

The axonemal microtubules are the main structural component of the flagella. Since tubulin assembly is affected by calcium in vitro (Olmsted and Borisy 1975) and microtubules are disintegrated by this cation (Schliwa 1976), it seems conceivable that the inhibition of flagellar regeneration and the induction of flagellar shortening by calcium antagonistic substances may simply be the result of the known effect of calcium on tubulin assembly and microtubule integrity.

To gain more insight into the role of calcium in flagellar regeneration and shortening some basic questions need to be resolved. Glycerol and deuterium oxide (D_2O) are known to promote and to stabilize microtubules in vitro and in vivo (Dustin 1978). Since the length of the flagella depends, at least to some extent, upon the tonicity of the medium (Solter and Gibor 1978) glycerol because of its osmotic potency appears less suitable for experiments to stabilize the axonemal microtubules in *Chlamydomonas*. Does, however, deuterium oxide reduce the effect of the calcium-antagonistic substances?

Sulfhydryl groups have been discussed as being involved in cellular responses which are the result of a change in calcium distribution (Rebhun 1976). Assuming that naturally occurring flagellar regeneration is preceded by a redistribution of cal-

1 Abt. Cytologie, Pflanzenphysiologisches Institut der Universität Göttingen, Untere Karspüle 2, 3400 Göttingen, FRG

Present address: Zellenlehre, Universität Heidelberg, Im Neuenheimer Feld 230, 6900-Heidelberg, FRG

Compartments in Algal Cells and Their Interaction
ed. by W. Wiessner, D. Robinson, and R.C. Starr

cium as in the case of the chemically induced flagellar shortening, sulfhydryl groups may also have a regulatory role in flagellar development.

Finally, is there any evidence for a compartmentalization of calcium during flagellar withdrawal and regeneration?

Materials and Methods

The methods used to measure flagellar length, to fix, to embed, and to section the material for electronmicroscopy have already been described (Quader et al. 1978a and b).

The isolation of cell wall-flagella-complexes was achieved using the procedure introduced by Hunt and Ellar (1974) to isolate the pellicle of trypanosomes.

Deuterium oxide was purchased from Janssen Pharmaceutica (Beerse, Belgium), APM was kindly provided by Bayer Company (Leverkusen, FRG), other chemicals were obtained from Sigma (Munich, FRG).

Results and Discussion

Effect of D_2O on Flagellar Regeneration and Withdrawal

Generally, it is believed that D_2O strengthens hydrophobic interactions (Siegel and Galun 1978), or restricts the transfer of protons (Wiberg 1955). Microtubule stabilization by D_2O could be the result of either one of the suggested mechanisms. In a number of studies the stabilizing effect of D_2O upon microtubules has been demonstrated (see Dustin 1978). In the green alga *Oocystis solitaria* D_2O affects the microtubule-controlled cellulose microfibril deposition (Quader 1981). Are the axonemal microtubules stabilized by D_2O in a way that the effects caused by EGTA or APM are, at least, reduced?

After detaching the flagella from the cell body by the pH-shock-method *Chlamydomonas reinhardii* regenerates its flagella in about 90-120 min. In the presence of 50% D_2O the cells are able to regenerate their flagella but at the different time intervals the flagella are shorter than the ones of the control cells or the ones treated with 25% D_2O (Table 1). This is probably due to a prolonged adaptation period to the higher D_2O concentration because the elongation rate is similar to that of untreated cells. Unicellular organisms seem to need some time to adapt to D_2O (Flaumenhaft et al. 1965), which is supported by the results obtained with 50% D_2O (Tables 1,2). The outgrowth is, however, completely inhibited by 75% D_2O. This effect cannot be explained by an extension of the period of adaption.

D_2O given to cells with fully developed flagella show similar concentration-dependent responses. In the presence of 25% and 50% D_2O the flagella shorten slightly within the first 30 min of treatment but then recover to almost the length shown by untreated cells (Fig. 1). In contrast, the induction of flagellar shortening by 75%

Table 1. Flagellar regeneration in the presence of different D_2O concentrations. Deflagellation was achieved by the pH-shock-method

Treatment	Flagellar length (µm) after	
	90 min	180 min
none	7.9 (1.2)[a]	9.9 (0.7)
D_2O (25%)	8.1 (1.1)	10.0 (0.8)
D_2O (50%)	1.9 (1.0)	5.2 (1.3)
D_2O (75%)	0.0	0.0

[a]Standard deviation

Table 2. Flagellar shortening induced by the herbicide amiprophosmethyl (APM) in the presence of 50% D_2O. Flagella were detached by the pH-shock-method and were allowed to regenerate in the presence of D_2O. After 60 min the D_2O-treated cell suspension was divided into two parts to one of which APM (10^{-5} M) was added

Treatment	Flagellar length (µm) at different time intervals (min) after deflagellation			
	30	60	120	180
None	5.6[a]	8.9	9.8	10.1
D_2O (50%)	3.1	6.5	7.9	8.5
D_2O + APM (50%) (10^{-5} M)	-	-	6.9	3.8

[a]The standard deviations were within the range of 0.8 - 1.1 µm

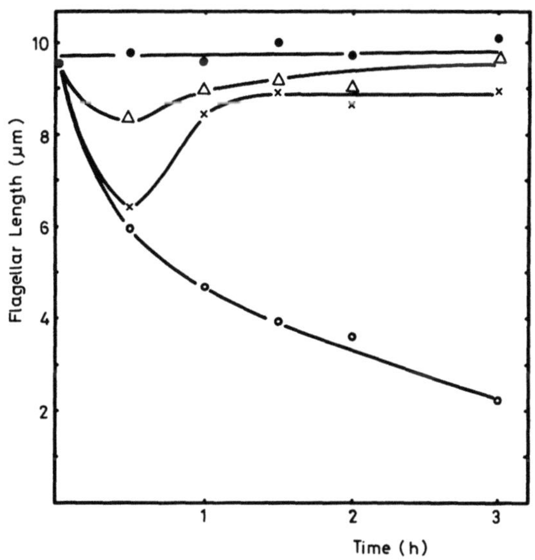

Fig. 1. Change in the flagellar length after treatment with different D_2O concentrations. Cells with fully developed flagella were treated with D_2O. ●——● no treatment; △——△ 25% D_2O; x——x 50% D_2O; ○——○ 75% D_2O

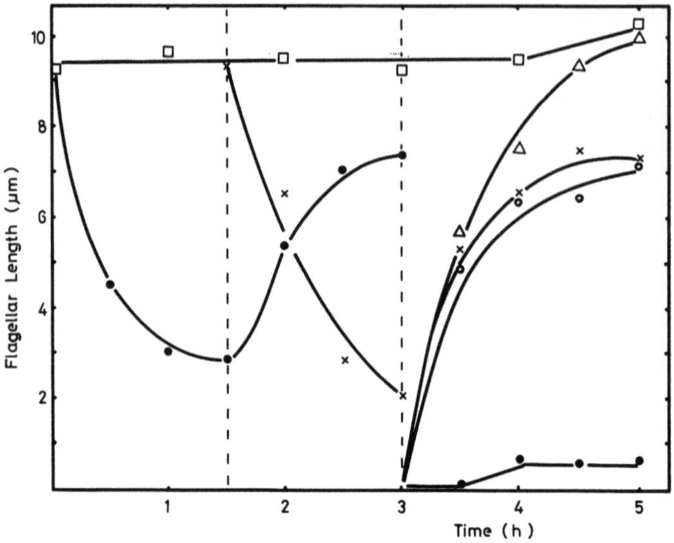

Fig. 2. Attempt to follow the fate of axonemal proteins of *Chlamydomonas reinhardii* cells, e.g., tubulin, through a combined treatment with 75% D_2O and cycloheximide (20 µg/ml). Flagella were induced to shorten by 75% D_2O treatment for 90 min. Cycloheximide was then given (a) during the recovery period of the D_2O treatment to prevent the synthesis of new axonemal proteins and (b) during the regeneration period after untreated full-length or shortened flagella had been detached from the cell bodies by the pH-shock-method. The treatments in detail:
□ —— □ cells which remain untreated throughout the experiment (aliquot D).
● —— ● cells treated with 75% D_2O for 90 min followed by a treatment with Chx during the recovery period, again for 90 min. Thereafter the flagella were detached and regeneration occurred in the presence of Chx (aliquot A).
Flagellar shortening induced by 75% D_2O for 90 min and after deflagellation of the cells flagella were allowed to regenerate:
x —— x in the presence of Chx (aliquot B_1); △ —— △ or without any addition (aliquot B_2).
○ —— ○ At the indicated time (3 h) untreated cells were deflagellated. Regeneration occurred in the presence of Chx (aliquot C).
Further explanations are given in the text

D_2O proceeds gradually throughout the time the experiment has been conducted (Fig. 1). The induction of flagellar shortening by APM is not inhibited by 50% D_2O (Table 2). The rate of the shortening is also not reduced significantly by D_2O when compared to cells only treated with APM.

With the exception of colchicine all other substances shown to prevent flagellar regeneration induce their shortening (see Quader and Filner 1980). What is the fate of the proteins of the axonema, e.g., tubulin, during the process of flagellar regression? Is tubulin resorbed as suggested by Coyne and Rosenbaum (1970), thereby leading to an elevated tubulin pool in the cytoplasm? This question can be partially answered by applying cycloheximide (Chx) which is known to prevent the synthesis of new tubulin molecules but does not block flagellar regeneration (Coyne and Rosenbaum 1970, Quader et al. 1978a).

For suitable time course experiments a cell suspension was divided into four aliquots A, B, C, and D (Fig. 2). Aliquot A was treated with 75% D_2O to induce flagellar shortening. After 90 min the cells were allowed to recover from the D_2O-treatment in the presence of Chx to prevent the synthesis of new tubulin. During the same time period cells of the aliquot B were treated with 75% D_2O. After 3 h the algae of aliquot A, B, and C were deflagellated and aliquot B was subdivided into two aliquots, B_1 and B_2. After flagellar detachment the aliquots A, B_1, and C were allowed to regenerate their flagella in the presence of Chx, whereas aliquot B_2 remained without any addition.

During the recovery period from the 75% D_2O-treatment the cells of aliquot A re-elongated their shortened flagella in the presence of Chx to about 70-80% of their normal length (Fig. 2). After detachment of the recovered flagella these cells cannot regenerate new flagella (Fig. 2), although the potency is still given, because after omitting Chx the cells form new flagella after a lag period (not shown). Cells induced to shorten their flagella with 75% D_2O for a period of 90 min before deflagellation (aliquot B), regenerate their flagella in the presence of Chx to about 70% of that of untreated cells. Cells untreated before their flagella were detached (aliquot C) also recover their flagella to about 70% of their normal length in the presence of Chx.

Assuming that the tubulin of the axoneme is resorbed into the cell body during the period of flagellar shortening and can be re-used, the cells of aliquot A should be able to regenerate at least short flagella after deflagellation. In the case of tubulin re-utilization one would expect that the flagella of cells which have been shortened by D_2O and regenerated in the presence of Chx will become longer than those of cells which did not undergo shortening before flagellar detachment because of the increased cytoplasmic tubulin pool.

Two explanations are, therefore, possible: (1) The axonemal tubulin is not resorbed into the cell body, or (2) the tubulin molecules are resorbed by the cells became, however, modified and are not re-usable without re-modification. The latter suggestion might well be involved in *Chlamydomonas* as it has been shown that the synthesized tubulin needs to be modified in the flagellum before it can be used for the assembly of the axoneme (L'Hernault and Rosenbaum 1983).

In many cases the features of the action of D_2O, at higher concentrations, resembles that of a "temperature-lowering effect" (Siegel and Galun 1978). This is, however, very unlikely in the case of *Chlamydomonas* flagella since the axonemal microtubules are cold stable (Mohri 1976). The results (aliquots B and C) indicate that 75% D_2O causes a cessation of protein synthesis which has been reported for D_2O in *Lemna minor* (Cooke et al. 1979).

Response Toward Sulfhydryl-Group-Blocking Substances

The results obtained with APM in the presence of 50% D_2O indicate that, although cytosolic calcium levels are raised through APM treatment (Quader and Filner 1980, Hertel et al. 1980), calcium alone is not responsible for the disassembly of the axonemal microtubules.

Since both events, flagellar regeneration as well as shortening, are highly dependent on calcium and are possibly based upon the interaction of calcium with protein,

Table 3. Flagella development in the presence of sulfhydryl-group blocking substances. After detachment of the flagella the cells were allowed to regenerate new flagella for 1 h before the inhibitors were added at a concentration of 5×10^{-4} M

Treatment	Flagellar length (μm) at different time intervals (h) after deflagellation				
	0	1	2	3	4
None	0.0	8.1	10.0	10.4	10.3
Diamide	-	-	8.9	7.9	7.1
NEM	-	-	5.7	0.0	0.0
pCMB	-	-	7.4	7.3	7.3
pCMS	-	-	9.7	9.8	10.2

The standard deviations ranged from 0.7 - 1.5 μm

calcium needs either to be removed from or to be moved to the place where the shortening response is sensitized. Are there distinct storage places for calcium? The results obtained through the treatment with EGTA (Quader et al. 1978a) exclude a role for calcium movement from the exterior into the cell. Recent results, however, indicate that EGTA causes a massive flow of calcium from intracellular storage places to the exterior (Quader, unpubl. results). It is suggested, therefore, that calcium is likely to be released from intracellular stores.

Dikstein (1971) discussed the involvement of oxidation and reduction of sulfhydryl groups in the regulation of intracellular calcium movement, e.g., through the control of specific membrane-bound Ca^{2+}-ATPases. To test the possible involvement of sulfhydryls in the regulation of flagellar shortening or regeneration, four substances known to block SH-groups, diamide, N'ethylmaleimide (NEM), p-chloromercuribencoic acid (pCMB), and p-chloromercuriphenylsulfonic acid (pCMS) were applied to cells which had been regenerating their detached flagella for about 1 h (Table 3). Diamide and pCMB prevent the further elongation of the flagella, whereas in the presence of NEM the cells appear to start flagellar shortening but, then, shed off their flagella. p-CMS which is known to penetrate the plasma membrane only very slowly, has no effect on flagellar elongation, indicating an intracellular location of the sulfhydryl-groups in question (Table 3). Although these results are not conclusive, they demonstrate how complex the regulation of flagellar development possibly is.

Ultrastructural Observations on Organelles Possibly Involved in Flagellar Development

The basal bodies are the organizing centers of the axoneme. Organelles in the vicinity of the basal body might be involved in the regulation of the formation of the flagella. Since the importance of calcium on formation and physiology of flagella has been demonstrated these structural investigations focused on organelles believed to be involved in calcium sequestration and storage. Mitochondria and the endoplasmic reti-

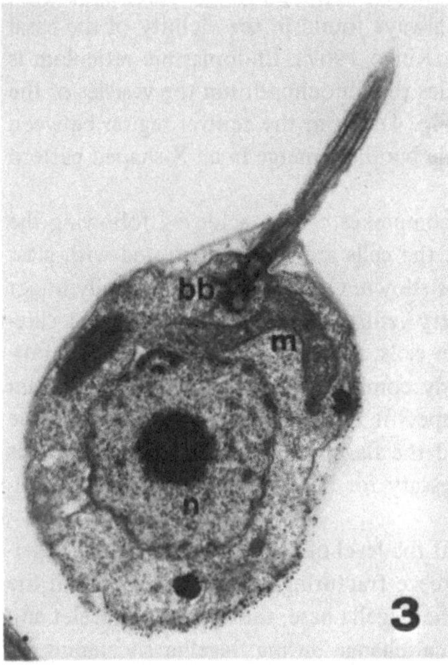

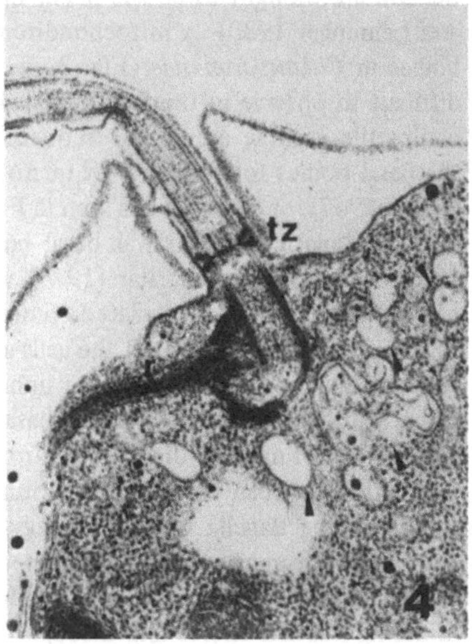

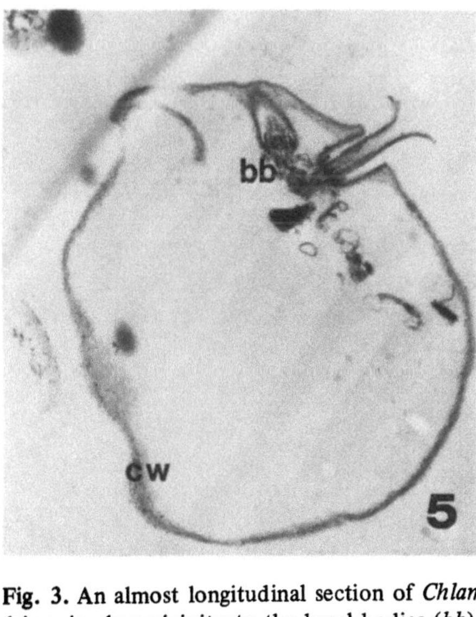

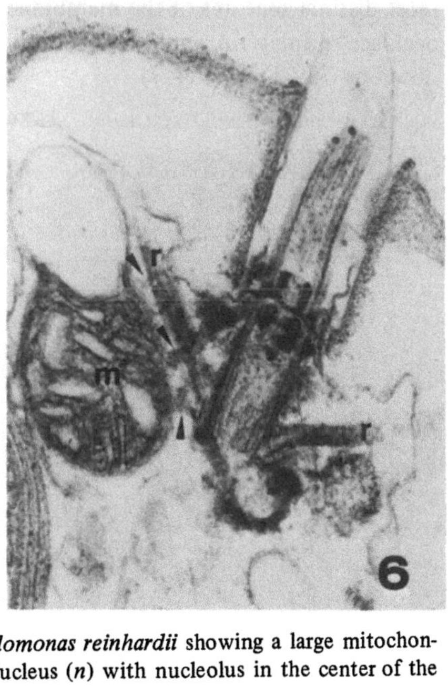

Fig. 3. An almost longitudinal section of *Chlamydomonas reinhardii* showing a large mitochondrium in close vicinity to the basal bodies (*bb*). Nucleus (*n*) with nucleolus in the center of the cell. x 18900

Fig. 4. Longitudinal section through the basal body region. A microtubule rootlet (*r*) emerges from the basal body. Vesicles of the contractile vacuole are visible (*arrows*). Notice the distinct contrast in the transition zone (*tz*). x 45000

Fig. 5. An isolated cell wall-basal body complex of *Chlamydomonas reinhardii.* x 24000

Fig. 6. Isolated cell wall-basal body-complex with intact rootlets (*r*) to one of which a mitochondrium (*m*) is linked. Cross-bridges are indicated by *arrows.* x 64000

ticulum are thought to be the major intracellular storage places for calcium (Becker and Lehninger 1980). A mitochondrium is always found in the vicinity of the basal bodies in *Chlamydomonas* (Fig. 3, see also Ringo 1967). Endoplasmic reticulum is difficult to observe in this cell region. Besides the mitochondrium the vesicles of the contractile vacuole are found frequently (Fig. 4). From the central region between the basal bodies four rootlets of microtubule bundles emerge in an X-shaped pattern (Ringo 1967), one of which is seen in Fig. 2.

The isolation of cell wall-basal body complexes can be achieved following the procedure of Hunt and Ellar (1974) when the cells are gently disrupted with glass beads. In Fig. 5 such an isolated complex is shown. Occasionally, structurally intact flagella are observable. When the cells are very gently broken and followed by a careful centrifugation step to separate unbroken cells, the mitochondrium and the rootlets are still connected to the wall-basal body complex (Fig. 6). The mitochondrium appears to be linked to the rootlet through specific bridges (Fig. 6), indicating a close interaction between the mitochondrium and the flagella. Of course, the main role is to supply the flagella with the energy necessary for flagellar beating. A role in calcium metabolism can only be speculated.

Attempts to study possible alterations at the level of the membrane of the flagella during flagellar shortening through the freeze fracturing method revealed that the most distinct features of the membrane at the flagella base, the so-called bracelet and necklace, display no significant morphological change during flagellar shortening induced by 75% D_2O (Fig. 7).

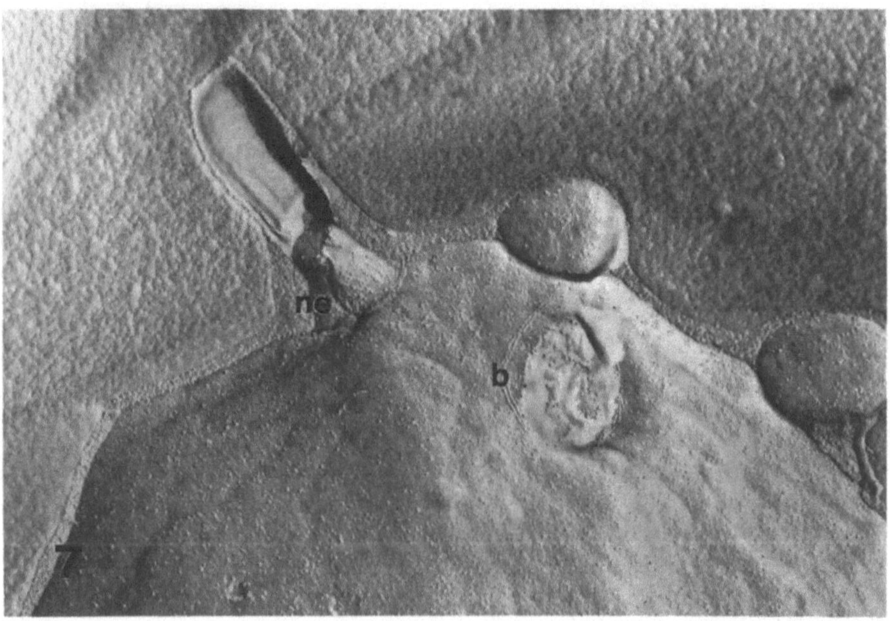

Fig. 7. P-fracture face of a cell treated with 75% D_2O showing the transition zone of the flagella. No difference in the arrangement of the particles of the necklace (*ne*) and the bracelet (*b*) region compared to untreated cells is observable. x 52000

Conclusion

The results indicate that the shortening of the flagella is not simply due to the effect of calcium on the axonemal microtubules. They point to the existence of a distinct calcium-sensitizing system which controls the regeneration and the shortening of the flagella.

Acknowledgements. Frl. C. Angerstein is thanked for excellent technical assistance. Supported by funds from the Deutsche Forschungsgemeinschaft.

References

Becker G, Lehninger AL (1980) J Biol Chem 255:9009
Bloodgood RA (1974) Cytobios 9:143
Cavalier-Smith T (1974) J Cell Sci 16:529
Cooke RJ, Grego S, Oliver J, Davies DD (1979) Planta (Berl) 146:229
Coyne B, Rosenbaum JL (1970) J Cell Biol 47:777
Dikstein S (1971) Naturwissenschaften 58:439
Dustin P (1978) Microtubules. Springer, Berlin Heidelberg New York
Flaumenhaft E, Bose S, Crepi HL, Katz JJ (1965) Int Rev Cytol 18:313
Hertel C, Quader H, Robinson DG, Marmé D (1980) Planta (Berl) 149:336
Hunt R, Ellar DJ (1974) Biophys Biochem Acta 339:173
L'Hernault SW, Rosenbaum JL (1983) J Cell Biol 97:258
Mohri H (1976) Biophys Biochem Acta 456:85
Olmsted JB, Borisy GG (1975) Biochemistry 14:3996
Quader H (1981) In: Robinson DG and Quader H (eds) Cell Walls 81. Wissensch Verlagsgesell, Stuttgart, pp 198–205
Quader H, Filner P (1980) Eur J Cell Biol 21:301
Quader H, Cherniack J, Filner P (1978a) Exp Cell Res 113:295
Quader H, Wagenbreth I, Robinson DG (1978b) Eur J Cell Biol 18:39
Rebhun LI (1976) Am Zool 16:469
Ringo DL (1967) J Cell Biol 33:543
Rosenbaum JL, Moulder JE, Ringo DL (1969) J Cell Biol 41:600
Schliwa M (1976) J Cell Biol 70:527
Siegel SM, Galun E (1978) Plant Cell Physiol 19:851
Solter KM, Gibor A (1978) Nature (Lond) 275:651
Wiberg KB (1955) Chem Rev 55:713
Witman GB, Carlson K, Berliner J, Rosenbaum JL (1972) J Cell Biol 54:507

The Role of Chloroplast in Cell Cycle Control of Algae Dividing by Multiple Fission

I. ŠETLÍK and V. ZACHLEDER[1]

The continuity of life is maintained by the growth and the division of the cells. The sequence of events occurring between two consecutive cell divisions is usually called the cell cycle. In a simple case of a unicellular organism nearly identical cells are reproduced at successive cell divisions. Hence, the basic question is how growth and division processes are coordinated in the cell cycle to secure the constancy of important cell characteristics (Zeuthen 1964, Mitchison 1971, Prescott 1976, Helmstetter et al. 1979).

This question has been first analyzed in eukaryotic cells dividing by binary fission. Although many diverse views were expressed (Padilla et al. 1974, John 1981, Cooper 1982, Nurse and Streiblová 1983), we believe that the following generalization grasps the essence behind most of them: To reproduce viable copies of the cell, all its indispensable functional structures must be duplicated and subsequently carefully divided into two daughter cells. Therefore, after the required amount of growth has been accomplished (after termination of the growth step) the cell becomes competent to divide. It is also said to be committed to cell division or to pass the commitment point. At this point a replication-division sequence (RDS) is started which, as a rule, runs without arrest to its termination. The RDS consists of a DNA replication step, the nuclear division step (mitosis) and the cell division step (cytokinesis). The growth step plus the steps of the RDS form the cell reproductive sequence (CRS). The simplest way to view the temporal pattern of the CRS is that the termination of one step starts the next one. It has been recognized, however, that in many types of cells early events of cytokinesis occur simultaneously with late phases of mitosis so that these two steps overlap partially. Similarly, growth continues beyond the commitment point so that the growth step of the next CRS overlaps with the RDS of the running one (for details see Šetlík and Zachleder 1983, 1984).

A different pattern of cell cycle events emerged from the analysis of reproduction in prokaryotic cells (Helmstetter et al. 1968, Cooper and Helmstetter 1968, Helmstetter et al. 1979). In these, at high growth rates, the time interval between two successive cell divisions can be shorter than the time required for the RDS (or even of one step of it) to run to completion. This was particularly clearly demonstrated in the case of DNA replication, one round of which occupies a certain time interval irrespective of the specific growth rate of the cell. Consecutive replication

1 Department of Autotrophic Microorganisms, Institute of Microbiology, Czechoslovak Academy of Sciences, Třeboň, Czechoslovakia

Compartments in Algal Cells and Their Interaction
ed. by W. Wiessner, D. Robinson, and R.C. Starr

rounds are started at intervals corresponding to cell doubling times. At higher specific growth rates cell doubling time becomes shorter, and at maximum growth rates several times shorter, than the time required for one replication of the bacterial chromosome. Concomitantly, the DNA content of one cell increases in the proportion given by the ratio of the two times. It follows that several replication rounds occur in the cell simultaneously. The overall DNA content of the cell doubles within the cell doubling time, but one actual duplication of the chromosome extends over two to several cell division cycles. The same may apply to other steps of the RDS. Thus, at high growth rates, two to several CRS occur simultaneously within one cell with a phase shift dictated by the growth doubling time. By each closure of the cell wall at cell division two entities become separated, which actually represent more than one cell. We say that the CRS overlap in the cell division cycles.

This situation has a counterpart in eukaryotic cells reproducing by multiple fission (Šetlík and Zachleder 1983). This division pattern became popular since it characterizes, among others, the cell cycles of fast-growing algae of the orders Chlorococcales and Volvocales which have long been used for research of algal physiology, biochemistry and genetics. In these 2nd daughter cells are released by one mother cell at the end of the division cycle. Evidently then n CRS take place in the course of one cell division cycle. The similarity with the prokaryotes is reflected by the fact that the value of n (and, consequently, the number of daughter cells produced at cell division) varies in proportion to the specific growth rate of the cells (Molloy and Schmidt 1968, Kates et al. 1968, Mihara and Hase 1971, Poynton and Branton 1972, Poynton 1973, Donnan and John 1983). The higher the growth rate, the more RDS are consecutively triggered in one cell cycle. The basic difference as compared to prokaryotic cells, that divide always by binary fision, is the periodic release of 2^n daughters in eukaryotic cell division. This means that after a certain number of the RDS have been started in one cell, further sequences are not initiated and the triggered ones are allowed to run to completion. The primary cause of this periodicity lies in the complicated pattern of chromosome segregation during mitosis, which rules out its simultaneous course with DNA replication (Cooper 1979, Šetlík and Zachleder 1983).

The concept of cell cycle in algae dividing by multiple fission here outlined is not a generally accepted one. In two recent publications we have attempted to elaborate it in some detail and to analyze the possible control mechanisms involved (Šetlík and Zachleder 1983, 1984). Essential points of our approach are as follows: The key to the analysis of multiple fission in algae is provided by the observation that commitments to divide into the increasing number of daughter cells (i.e., 2nd with n = 1,2,3 etc.) are attained stepwise after a definite amount of growth (the growth step) has been realized. Growth is a cumulative designation for synthetic and constructional work required for the generation and assembly of essential macromolecules as well as for accumulation of reserves. The critical amount of growth required to attain the critical size for commitment corresponds approximately to doubling of various cell parameters such as volume, content of RNA and protein etc. In the above quoted publications we detail the reasons for considering the number of ribosomes in the cell to be the most important single parameter that reflects the complex quantity of growth.

In phototrophically grown algae growth markedly depends on material and energy supply from photosynthesis. Darkening of the cells during their growth phase considerably slows down and after a short time completely stops all synthetic processes that form the backbone of growth, particularly the increase of rRNA (Zachleder and Šetlík 1982). On the other hand, events in the replication-division sequence can proceed without supply of external energy at the expense of material and energy reserves accumulated during the growth step leading to commitment. This can best be demonstrated by taking samples from a synchronously growing population at regular intervals and incubating them in the dark with aeration Šetlík et al. 1972). After a sufficiently long dark period, the samples are then examined to assess the percentage of cells showing various division patterns. In the samples withdrawn at increasing length of the light period, cells dividing into two, four, eight etc. daughters will be found at the end of the dark period. Therefore, the commitments to the consecutive steps of division are attained and, herewith, the number of daughter cells produced is determined, earlier than division starts.

An important question is what determines the number of effective commitments in the given cycle, i.e., at what point the continuously illuminated cell will start to divide. The formal answer is, that the specific growth rate of the cell is not constant throughout the cycle but gradually declines. Therefore, at the end of the cycle the doubling time and, herewith, the interval between two subsequent commitments become so long that all the replication-division sequences already triggered may run to completion before another commitment becomes effective.

Looking for a possible mechanism of this interaction between the changing growth rate of the cell and the rate of the RDS's, it is necessary to know what happens between the commitments and the cell division. Surprisingly enough, rather different patterns of events are found with various algae. We have examined this question with *Scenedesmus quadricauda* by following the course of DNA replications, nuclear divisions and protoplast fissions. Our conclusion was that the RDS's triggered at successive commitments actually run in continuity and in parallel in the cell. This was suggested by the fact that consecutive rounds of DNA replication and nuclear division are well separated in time and are spaced more or less evenly throughout the second half of the growth phase of the cycle.[2] The DNA replication steps appear soon after the corresponding commitment is recorded. The time intervals between commitments and the corresponding nuclear divisions are approximately equal in the consecutive RDS's, showing a noticeable tendency to become shorter with later commitments, which is variably expressed in various cycles. The protoplast fissions start, as a rule, only after the last nuclear division is terminated and follow then in close sequence.

Material continuity of the RDS's triggered at the successive commitments can hardly be proposed in another extreme type of multiple fission in algae, which is characteristics, among others, of some strains of *Chlorella* and *Chlamydomonas.* Unfortunately the only detailed data on timing of commitments and of the steps in the RDS's under various growth rates are those by Donnan and John (1984). Combining

2 We designate as growth phase of a synchronous population the period from the start of the cycle to the appearance of first dividing cells in light; the triggered RDS's are running in parallel with growth during this period

their results with pieces of evidence scattered in numerous papers by various authors, one obtains the following picture:

The first DNA replication starts only after the last growth step in the given cell cycle is terminated. It is thus separated from its commitment by a relatively long period, during which no process maintaining continuity can be traced. The first replication is closely followed by its corresponding nuclear division and protoplast fission. Moreover, the DNA replications, nuclear divisions and protoplast fissions belonging to consecutive commitments then take place successively with the minimum possible delay. The result seen at the population level is one big wave of DNA synthesis followed closely by the waves of nuclear division and protoplast fission. We call this pattern the clustered type of multiple fission in contrast to the consecutive type described for *Scenedesmus quadricauda.*

We believe that there is no fundamental difference in the control mechanisms governing these two types of cell cycles. As detailed in our earlier paper (Šetlík and Zachleder 1983) different species of the same genus are often found to follow either one of the two different patterns and transitions from one to the other can also be traced in one strain under different growth conditions. We have proposed that different strength of competitive interaction between growth and the steps in the replication-division sequence can decide the type of multiple fission. In the consecutive type the steps of the RDS's can run concomitantly with the subsequent steps; early in the cycle when the growth rate is high, the RDS's proceed slowly and they become accelerated towards the end of the cycle when the growth rate declines. On the other hand, in the clustered type, the first step of the RDS's, i.e., DNA replication, is prevented as long as the specific growth rate of the cell remains above a critical value.

This general statement gains a very attractive specific content if one considers the role that interaction between the chloroplast and the cytoplasm-nucleus compartments can play in the control of the cell cycle events. As yet, in most cell cycle studies on algae, including those in our laboratory, this aspect has been largely neglected. One notable exception was made by Chiang (1975), who noted that in *Chlamydomonas reinhardii* the number of chloroplast DNA replications that occurred in the middle of the growth phase of the cycle was always equal to the number of nuclear DNA replications (and the corresponding nuclear and cell divisions) which occurred at the end of the cycle. He reasoned that either a regulatory mechanism must have coordinated in anticipation the chloroplast DNA replication with the cell division that occurred many hours later or, conversely, that the number of eventual cell divisions is dictated by the extent of chloroplast DNA replications that occurred earlier in the cycle. In the rest of the paper we shall try to show that the latter variant fits well the known facts on multiple fission cell cycles of algae, and that it provides a convenient starting point for a working hypothesis on control mechanisms operating in these cycles. Although many links in this hypothesis are speculative in nature, we feel it worth while to elaborate as a challenge for experiments that may either confirm or disprove it.

First data on the replication of chloroplast DNA (cDNA) in *Chlamydomonas reinhardii* have been provided by Chiang and Sueoka (1967). Chiang remained involved in DNA studies in this species for many years, so that his conclusion quoted above is based on extensive experience (Kates et al. 1968, Chiang 1971, Chiang et al. 1975,

Grant et al. 1978). Moreover very similar results on cDNA replication and its relation to nDNA replication in *Chlamydomonas reinhardii* were recently reported by Keller and Ho (1981). These authors complemented the in vivo measurements at three growth rates by an experiment in which the cDNA-synthesizing capacity was assessed in chloroplasts isolated at various stages of a synchronous culture. Variation in rates of cDNA synthesis during the cell cycle reported by Dalmon et al. (1975) for two strains of *Chlorella pyrenoidosa* and by Iwamura et al. (1982) for *Chlorella ellipsoidea* are in good agreement with the observations on *Chlamydomonas.* Two other papers confirm that cDNA replication is continuous throughout the cell cycle: Wanka et al. (1970) for *Chlorella,* giving no details on rate variation and Turmel et al. (1980) reporting for *Chlamydomonas* the rather exceptional case when maximum rate of cDNA synthesis follows closely the nDNA replication in the dark period.

Unfortunately all these investigations concern algae with the clustered type of multiple fission. For these we may draw the following general picture: cDNA replication starts early in the cell cycle and extends over a large part or throughout its growth phase. At higher growth rates, the rate of cDNA synthesis traces sometimes perhaps a bell shaped curve with the maximum attained around the middle of the growth phase but more often a strongly skewed curve whose maximum is found early in the cycle. Thus the rate of cDNA synthesis follows what can be assumed to be the course of chloroplast growth rate (John et al. 1973, Atkinson et al. 1974). The variation of photosynthetic activity in cell cycles of chlorococcal and volvocal algae also yields curves of comparable shape (Senger 1975, Berková et al. 1972, Šetlík et al. 1981). Since the general character of growth and of changes in photosynthetic activity is similar in algae characterized by the consecutive and clustered type of multiple fission, it can be expected that also the courses of cDNA synthesis rates will agree in the two types of algae.

These results indicate that in multiple fission cell cycles cDNA replication proceeds uncoupled from nDNA replication. They further suggest that cDNA replication is initiated at each doubling of the chloroplast's critical size and that it is linked with chloroplast growth in the manner characteristic of prokaryotic cells. At low growth rates distinct cDNA replication rounds may occur separated by intervals without any cDNA synthesis as has been found in the famous first experiment of Chiang and Sueoka (1967). At high growth rates cDNA synthesis is continuous and its rate follows the bell-shaped curve. An important piece of circumstantial evidence pointing to the same conclusion is provided by the observations of Hesse (1974) on *Bumilleriopsis.* This xantophycean alga, which displays the consecutive type of multiple fission, has numerous chloroplasts which divide during the growth phase of the cell cycle. Their rate of division traces a similar curve as the rate of cDNA synthesis in the above-mentioned algae.

In this context the prokaryotic feature of multiple fission cell reproductive sequences in algae results from the behaviour of their chloroplast. The prokaryotic character of its growth-replication control renders the chloroplast capable of high specific growth rate since at each cDNA replication its gene dosage increases. A high growth rate of the chloroplast requires a corresponding activity in the cytoplasm. The assembly of functional chloroplast ribosomes depends on supply of protein made in the cytoplasm (cf. Ledoigt and Freyssinet 1982, Schmidt et al. 1983); chloroplast

DNA polymerase and DNA-dependent RNA polymerase are coded in the nucleus and many essential chloroplast structures consist of supramolecular complexes involving both chloroplast and cytoplasmic translates (cf. Ohad and Drews 1982, Parthier 1982). The selfish chloroplast has, therefore, to keep the cytoplasmic protein-synthesizing machinery running at a rate satisfying its demands. This is, in fact, reflected by the strict proportionality in the increase of chloroplast and cytoplasmic ribosomes observed by Wilson and Chiang (1977) in the course of the *Chlamydomonas* cell cycle. Cattolico et al. (1973) with *Chlamydomonas* came to a somewhat different conclusion, namely that the increase of chloroplast rRNA levels off somewhat earlier in the light phase of the cycle while the cytoplasmic species continues to increase to the moment of darkening. The results by Hirai et al. (1978) on *Chlorella* are questioned because of the possible preferential incorporation of tritium labeled uridine into the chloroplast rRNA but their general trend agrees with the results of Cattolico et al. (1973).

The proportionality of the rates of ribosome assembly in the chloroplast and in the cytoplasm can be maintained on the basis of a feedback control. It may be assumed that both in the chloroplast and in the cytoplasm rRNA transcription is controlled by the availability of amino acids. The rate of amino acid synthesis depends on light since photochemical reduction of nitrite to amino groups is the only source of the latter for the whole cell. This is the cause for the strong control of rRNA synthesis by light which resembles the behaviour of stringent bacterial strains towards amino-acids. In *Chlamydomonas* it was even shown that restriction in rRNA synthesis is accompanied by increase in ppGpp concentration (Heizmann and Howell 1978). It seems logical to assume that the chloroplast will preferentially keep running its synthesis of amino acids, RNA's and proteins. As soon as it becomes limited by the supply of proteins from the cytoplasm, excess of amino groups will be exported and thus the balance restored.

The above-quoted results indicate that the proportionality in the rates of chloroplast and cytoplasm ribosome assembly is preserved for a large part of the cell cycle, that means even after the cDNA passes the first or even the second round of replication. Herewith the gene dosage in the chloroplast is increased and its rRNA transcription can proceed at a twofold or fourfold rate. The simplest way in which a corresponding increase in the rate of nucleolar rRNA synthesis can be secured is to increase proportionally the number of nuclear rDNA cistrons, i.e., to replicate the rDNA. We propose that this actually occurs. This proposal has a largely hypothetical character although Howell (1972) has observed an amplification of nucleolar rDNA to take place during the cell cycle of *Chlamydomonas reinhardii.* An increase in rDNA proportional to the growth accomplished was also clearly demonstrated in *Acetabularia* (Berger and Schweiger 1975), whose cell cycle has many features in common with the multiple fission in algae considered here. In which way the nucleolar rDNA replication could be linked to the chloroplast DNA replication is not clear.

It seems justified to propose that the consecutive cDNA replications mark the consecutive commitments to divide into increasing numbers of daughter cells in multiple fission cell cycles. In the earlier paper we have arrived at the conclusion that growth responsible for commitments must be articulated into definite growth steps. This formal postulate seems to gain in the present context a material content:

cDNA replications delimit distinctly the chloroplasts growth steps. It can be also understood why increase in various growth parameters during the cell cycle may in some cases have a well-expressed stepwise character. During periods of cDNA replication, the rate of chloroplast transcription may be slowed down. If the transcriptional and translational activity of the chloroplast controls the specific growth rate of the cell, it also becomes easy to explain why the latter gradually decreases and why this decline is sometimes particularly fast towards the end of the cycle. In contrast to the chloroplast, the cytoplasmic protein synthesis operates throughout the cell cycle with mRNA transcribed from only one genome present at the start of the cycle. Thus at a certain point in the cycle the supply of cytoplasmic translation products may become limiting for further assembly of various essential chloroplast structures and of its ribosomes in particular. This disparity may be further emphasized by the fact that transcription of nuclear genes may not be continuous, as suggested by the results from the laboratories of John (John et al. 1982, Rollins et al. 1983) and of Howell (Howell 1975, Howell and Walker 1977, Ares and Howell 1982). If some message released into the cytoplasm early in the cycle, whose translation product is required for the growth of the chloroplast, has a lifetime comparable with the duration of the cell cycle its translation rate will noticeably fade during the cycle and the limitation imposed on chloroplast growth will be particularly strong.

The molecular nature of the signal mediating between some characteristic reflecting growth and the initiation of DNA replication remains as yet enigmatic in cell cycle studies concerning both prokaryotes and eukaryotes. As here proposed, in multiple fission cell cycles of algae we have to look for still another mediator linking cDNA replication rounds with nDNA replication. This can occur either by a specific signal released at cDNA replication, or nDNA replication can be triggered by growth in the cytoplasm, which is proportional to that of the chloroplast. In either case, with the clustered type of multiple fission a record of the consecutive commitments must be stored in some way up to the moment when the cascade of nDNA replications is released, so that an appropriate number of replication rounds occur. How this is achieved is obscure. If it were proved that nuclear rDNA amplification actually follows each round of cDNA replication, as proposed above, the number of the replicas could provide the record.

Another question is what in the clustered type of multiple fission precludes the initiation of chromosomal nDNA replication for the most part of the cycle, and what then releases the cascade of replications at the end of the cycle. The control must involve a quantitative factor responding to chloroplast growth rate, since in *Scenedesmus quadricauda,* at high growth rates, indications of a transition to the clustered type of multiple fission can be observed. These requirements might be met by the regulation of the deoxyribonucleotide pool sizes. During the chloroplast-dominated period of rapid growth, ribonucleotides may be preferentially consumed for RNA synthesis. Deoxyribonucleotide synthesis may be hampered either by limited availability of substrate or by low activity of the corresponding enzymes if the latter were regulated in some way by substrate pool sizes. Knutsen et al. (1974) have observed that an abrupt increase in the pool sizes of deoxyribonucleotides precedes the clustered DNA replication waves at the end of *Chlamydomonas reinhardii* cell cycle, while Bachmann et al. (1983) found that activities of ribonucleoside diphosphate reductase and thymidylate synthase display a rapid rise to a maximum just before nDNA replication starts in *Scenedesmus obliquus.* Earlier, Johnson and Schmidt (1966) and Shen and Schmidt (1966) observed periodic changes of thymidylate kinase and of deoxycytidylate deaminase activity in synchronous cultures of *Chlorella sorokiniana.*

In spite of the piece of justified scepticism by Wanka and Poels (1969), based on their results obtained with *Chlorella pyrenoidosa,* general evidence seems to indicate that some causal connection exists between markedly enhanced deoxyribonucleotide synthesis and the burst of nDNA replications. Other data relevant to this question concern the compartmentation of enzymes involved in deoxyribonucleotide metabolism. It has long been known that thymidine kinase activity is confined to the chloroplast in *Chlamydomonas reinhardii* (Swinton and Hanawalt 1972, Chiang et al. 1975) and in some other algae. As to the other enzymes, we can so far rely only upon the most interesting results obtained by Schweiger and his collaborators with *Acetabularia.* They found that thymidine kinase (Bannwarth et al. 1977), dCMP deaminase (Bannwarth et al. 1982, Bannwarth and Schweiger 1983) and ribonucleoside reductase (de Groot and Schweiger 1981) are encoded and translated in the chloroplast. The only exception seems to be the thymidylate kinase (de Groot and Schweiger 1983), which appears to be encoded in the nucleus. Schweiger and his colleagues have also shown that the activity of the above-mentioned enzymes increases dramatically by their de novo synthesis just before the train of nuclear divisions starts in *Acetabularia* (cf. Bannwarth et al. 1977, Schweiger 1982).

If it were confirmed that some of the essential enzymes required for deoxyribonucleotide synthesis are the product of chloroplast genome expression also in chlorococcal and volvocal algae, the following speculation would deserve scrutiny by further experiments: During the period of vigorous chloroplast growth enzymes involved in dNTP synthesis are produced in the chloroplast in amounts sufficient to supply the consecutive cDNA replications with precursors. At the time when the lack of cytoplasmic translates restricts the growth of the chloroplast, its high gene dosage and translational capacity become involved, among others, in high rate of synthesis of the respective enzymes. These could then secure the high rate of dNTP supply for the consecutive waves of nuclear DNA replication in the clustered type of multiple fission. An evident objection is that export of deoxynucleotide phosphates from the chloroplast may be difficult or even impossible. On the other hand, ultrastructural studies offer many examples of a very close contact between the chloroplast and the nucleus in algae.

In conclusion, we believe that the facts surveyed above put beyond doubt the essential role of the chloroplast in determining the character of the algal multiple fission cycles. Some of our proposals may be found wrong and many details of the control mechanisms involved have yet to be explored by further experimental work. We are aware of only one fundamental objection. The colorless alga, *Prototheca zopfii,* divides by multiple fission and, when grown heterotrophically at various growth rates, displays the same variation of daughter cell number on specific growth rate as do its phototrophically growing relatives (Poynton and Branton 1972, Poynton 1973). Unfortunately no details are known as to the extent of deficiency in the chloroplast of this alga. If only the thylakoid membranes are affected and the genetic, transcriptional and translational capacities of the chloroplast are preserved, the chloroplast still may play a crucial role in processes determining the character of growth. Such a situation was found with *Polytoma obtusum*, the colorless counterpart of *Chlamydomonas* (Siu et al. 1976a,b), as well as with several mutants of *Chlamydomonas* whose photosynthetic function is affected by the mutation (cf. Wilson et al. 1980). The crucial question is how far and in what manner the role of chloroplast as a main source of energy for growth is preserved in these nonphotosynthetic algae. If the mitochondrion takes over a substantial part of this function, it might become involved also in the control of the cell cycle events in a manner similar to that proposed here for the chloroplast. In any case a detailed investigation of the apparent striking inconsistency of the hypothesis advanced above with the behaviour of colorless algae dividing by multiple fission will bring new fundamentally important pieces of knowledge on the role of energy-converting organelles in the control of cell cycles.

References

Ares M Jr, Howell SH (1982) Proc Natl Acad Sci USA 79:557–5581
Atkinson AW Jr, John PCL, Gunning BES (1974) Protoplasma 81:77–109
Bachmann B, Hofmann R, Follmann H (1983) FEBS Lett 152:247–250
Bannwarth H, Ikehara N, Schweiger HG (1977) Proc R Soc Lond B 198:177–190
Bannwarth H, Ikehara N, Schweiger HG (1982) Eur J Cell Biol 27:200–205
Bannwarth H, Schweiger HG (1983) Cell Biol - Internat Rep, in press
Berger S, Schweiger HG (1975) Protoplasma 83:41–50
Berková E, Doucha J, Kubín Š, Zachleder V, Šetlík I (1972) In: Forti G, Avron M, Melandri A (eds) Photosynthesis. Two centuries after its discovery by Joseph Priestley, vol III. Dr W Junk N V Publ, The Hague, pp 2619–2632
Cattolico RA, Senner JW, Jones RF (1973) Arch Biochem Biopyhs 156:58–65
Chiang KS, Sueoka N (1967) Proc Nat Acad Sci USA 57:1506–1513
Chiang KS (1971) In: Boardman NK, Linnane AW, Smillie RM (eds) Autonomy and biogenesis of mitochondria and chloroplasts. North-Holland Publ Co, Amsterdam, pp 235–249
Chiang KS, Eves E, Swinton D (1975) Develop Biol 42:53–63
Chiang KS (1975) In: Colloques Internat CNRS 240:147–158
Clay WF, Matsuda K, Hoshaw RW, Rhodes PR (1975) Ann Bot 39:525–533
Cooper S, Helmstetter CE (1968) J Mol Biol 31:519–540
Cooper S (1979) Nature 280:17–19
Cooper S (1982) In: Nicolini C (ed) Cell Growth. Plenum Publ Co, New York, pp 315–335
Dalmon J, Bayen M, Gillet R (1975) In: Colloques Internat CNRS 240:179–183
DeGroot EJ, Schweiger HG (1981) Eur J Cell Biol 24:5
DeGroot EJ, Schweiger HG (1983) J Cell Sci, in press
Donnan L, John PCL (1983) In: Nurse P, Streiblová E (eds) The Microbial Cell Cycle. CRC Press, Boca Raton, in press
Grant D, Swinton DC, Chiang KS (1978) Planta 141:259–267
Heizmann P, Howell SH (1978) Biochim Biophys Acta 517:115–124
Helmstetter CE, Cooper S, Pierucci O, Revelas E (1968) Cold Spring Harb Symp Quant Biol 33:809–822
Helmstetter CE, Pierucci O, Weinberger M, Holmes M, Tang MS (1979) In: Ornston LN, Sokatch JR (eds) The Bacteria, vol VII. Academic Press, New York, pp 517–579
Hesse M (1974) Ber Deutsch Bot Ges 87:373–381
Hirai A, Nishimura T, Iwamura T (1979) Plant Cell Physiol 20:93–102
Howell SH (1972) Nature New Biol 240:264–267
Howell SH (1975) In: Colloques Internat CNRS 240:159–164

Howell SH, Walker LL (1977) Develop Biol 56:11–23
Iwamura T, Katoh K. Nishimura T (1982) Cell Struct Funct 7:71–86
John PCL (ed) (1981) The cell cycle. Soc Exp Biol Seminar Ser 10. Cambridge University Press
John PCL, McCullogh W, Atkinson AW Jr, Forde BG, Gunning BES (1973) In: Balls M, Billett FS (eds) The cell cycle in development and differentiation. Symp Soc Dev Biol. Cambridge University Press, pp 61–86
John PCL, Lambe CA, McGookin R, Orr B, Rollins MJ (1982) J Cell Sci 55:51–67
John PCL, Donnan L, Harper JDI, Rollins MJ, Keenan CA (1984) In: Chaloupka J, Kotyk A, Streiblová E (eds) Principles of cell cycle controls. Inst of Microbiology, ČSAV, Prague
Johnston RA, Schmidt RR (1966) Biochim Biophys Acta 129:140–144
Kates JR, Chiang KS, Jones RF (1968) Exp Cell Res 57:121–135
Keller SJ, Ho Ch (1981) Int Rev Cytol 69:157–190
Knutsen G, Lien T, Skoog L (1974) Exp Cell Res 83:442–444
Ledoigt G, Freyssinet G (1982) Biol Cell 46:215–236
Mihara S, Hase E (1971) Plant Cell Physiol 12:225–236
Mitchison JM (1971) The biology of the cell cycle. Cambridge University Press
Molloy GR, Schmidt RR (1968) Biochem Biophys Res Commun 40:1125–1135
Nurse P, Streiblová E (eds) (1983) The microbial cell cycle. CRC Press, Boca Raton
Ohad I, Drews G (1982) In: Govindjee (ed) Photosynthesis, vol II. Development, carbon metabolism and plant productivity. Academic Press, New York London, pp 90–140
Padilla GM, Cameron IL, Zimmerman A (eds) (1974) Cell cycle controls. Academic Press, New York
Parthier B (1982) Biochem Physiol Pflanzen 177:283–317
Poynton RO, Branton D (1972) Proc Natl Acad Sci USA 69:2346–2350
Poynton RO (1973) J Bacteriol 113:203–211
Prescott DM (1976) Reproduction of eukaryotic cells. Academic Press, New York
Rollins MJ, Harper JDI, John PCL (1983) J Gen Microbiol 129:1–21
Schmidt RJ, Richardson CB, Gillham NW, Boynton JE (1983) J Cell Biol 96:1451–1463
Schweiger HG (1982) In: Parthier B, Boulter D (eds) Nucleic acids and Proteins in Plant II, Encyclopedia of plant physiology, new ser 14, part B. Springer, Berlin, pp 645–662
Şenger H (1975) In: Colloques Internat CNRS 240:101–108
Šetlík I, Berková E, Doucha J, Kubín Š, Vendlová J, Zachleder V (1972) Arch Hydrobiol/Suppl 41 Algolog Studies 7:172–213
Šetlík I, Zachleder V, Doucha J, Berková E, Bartoš J (1975) Arch Hydrobiol/Suppl 49 Algolog Studies 14:70–104
Šetlík I, Zachleder V (1983) In: Nurse P, Streiblová E (eds) The Microbial Cell Cycle. CRC Press, Boca Raton
Šetlík I, Zachleder V (1984) In: Chaloupka J, Kotyk A, Streiblová E (eds) Principles of cell cycle controls. Inst of Microbiology, ČSAV, Prague
Šetlík I, Šetlíková E, Masojídek J, Kalina T, Mader P (1981) In: Akoyunoglou G (ed) Photosynthesis, Proc Fifth Internat Congr Photosynthesis, vol V. Balaban Internat Sci Service, Philadelphia, pp 481–490
Shen SRC, Schmidt RR (1966) Arch Biochem Biophys 115:13–20
Siu CH, Swift H, Chiang KS (1976a) J Cell Biol 69:352–370
Siu CH, Swift H, Chiang KS (1976b) J Cell Biol 69:371–382
Swinton DC, Hanawalt PC (1972) J Cell Biol 54:592–597
Turmel M, Lemieux C, Lee RW (1980) Cur Genet 1:113–125
Wanka F, Poels CLM (1969) Eur J Biochem 9:478–482
Wanka F, Joosten HFP, DeGrip WJ (1970) Arch Mikrobiol 75:25–36
Wilson R, Chiang KS (1977) J Cell Biol 72:470–481
Wilson RE, Swift H, Chiang KS (1980) J Cell Sci 42:127–132
Zachleder V, Šetlík I (1982) Biol Plant 24:341–353
Zeuthen E (ed) (1964) Synchrony in cell division and growth. Interscience Publ Inc, New York

Metabolite Regulation of the Chloroplast Genome Expression and the Chloroplast-Cytoplasm Regulatory Relationships

V. E. SEMENENKO[1], M. G. ZVEREVA[2], E. S. KUPTSOVA[1], L. A. KLIMOVA[2] and M. G. VLADIMIROVA[1]

Introduction

It has been shown that stereochemical analogs of glucose (2-deoxy-D-glucose) completely suppress RNA synthesis, synthesis of the main enzymes of the Calvin cycle and components of the chloroplast photochemical apparatus and inhibit photosynthetic evolution of oxygen in *Chlorella* cells. Chlorophyll is deaggregated and its synthesis inhibited. The activity of the reaction centers of photosystems I and II is impaired and the concentration of P700 reduced. The repression process is accompanied with a destruction of chloroplast ultrastructure: the degradation and disappearance of the pyrenoid occurs. The glucose-induced effect of repression is light-dependent and made completely reversible by washing out the analog. The synthesis of RNA is one of the first to be suppressed and restored when the analog is introduced and washed out. Phenomena similar to those evoked by the glucose analogs could be observed in natural models in which hypertrophic accumulation of assimilates occurs in the *Chlorella* chloroplast. The data show that a molecular mechanism of metabolite regulation of chloroplast genome expression exists which underlies the regulation of the functional activity of the chloroplast and may play the role of a "terminal actuating mechanism" in the sink-source relationship between chloroplast and cytoplasm within the whole cell.

The problem of the regulatory interactions between different compartments of photosynthesizing cells is a very important in modern physiology and molecular biology of plant cells, and requires the investigation of the intracompartmentary regulatory mechanisms.

The present work is devoted to studying the problem of the existence of a chloroplast system of metabolite regulation of photosynthesis at a protein synthesis level with carbon reduction products as effectors and the role of that mechanism in sink-source interactions between chloroplast and cytoplasm.

The clarification of the molecular mechanisms of endogenous regulation of photosynthesis is a major problem which determines not only the search for new ways to optimize the photosynthetic productivity of plants, but also further studies of the elementary mechanisms underlying transformation of light energy in this process.

1 Institute of Plant Physiology, USSR Academy of Sciences, Moscow

2 Soil Science and Photosynthesis Institute, USSR Academy of Sciences, Pushchino

The chloroplast is known to be not only the site of processes and reactions directly involved in the transformation of light energy and carbon reduction; it also possesses its own DNA and protein-synthesizing system which ensures synthesis of many important components of the photosynthetic apparatuses. The integration and interaction of these two systems seem to underlie the action of the mechanisms of endogenous regulation of photosynthesis and determine the adaptive properties of the chloroplast [1].

It is known that the physicochemical properties of the DNA and protein-synthesizing systems of the chloroplast are similar in many respects to those of bacterial cells. It may be thought, therefore, that the system regulating chloroplast genome expression also employs principles characteristic of prokaryotic organisms and based on the mechanisms of metabolite induction-repression [1,2].

Nature of the Effector Metabolite

Logical analysis leads to the conclusion [3,4] that the metabolite which performs functions of a corepressor must be found in the metabolic fork between synthesis of products transported from the chloroplast and of products deposited in storage. In this case, causing the repression of chloroplast protein synthesis, such a metabolite could accomplish this repression only after substrates are provided for cytoplasmic processes, and polysaccharides are stored in the chloroplast. These requirements are satisfied by ADPG, UDPG, glucose-1-phosphate, glucose-6-phosphate, and glucose which can appear in the chloroplast after the replenishment of the storage starch pool. In this connection, a high regulatory activity of glucose in coordinating the metabolism of microorganisms is noteworthy.

At the same time numerous studies of the effect of exogenous glucose on photosynthesis fail to reveal depression of photosynthesis. This, however, may be due to

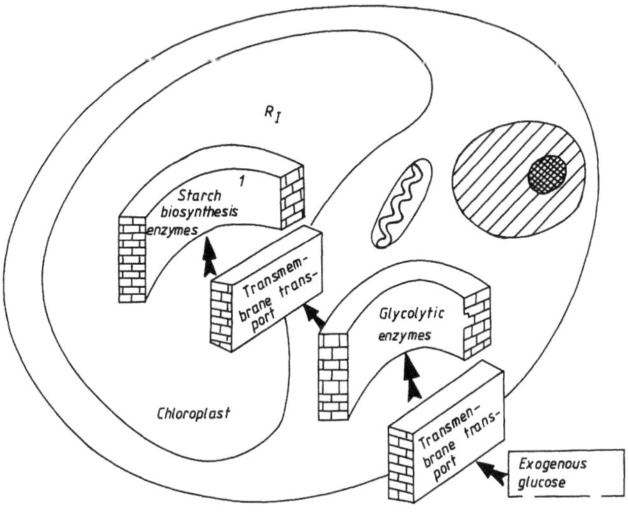

Fig. 1. Metabolic barriers on the path of exogenous glucose to the site (R_X) of its possible regulatory action in the chloroplast

the compartmentation of a cellular structure and the presence of many metabolite barriers (Fig. 1) on the path of exogenous glucose to the centers of its possible regulatory action in chloroplast [1,3].

In the research on the mechanisms of endogenous regulation of photosynthesis conducted in our laboratory to study the regulatory role of glucose in the chloroplast, new approaches [3] have been employed based on utilization of stereochemical analogs of glucose (2-deoxy-D-glucose, 3-O-methylglucose, xylose) that are poorly metabolizable (or nonmetabolizable), but are readily transported into the cell and are "biologically full-valued".

Effect of 2-Deoxy-D-Glucose (2dDG) on Photosynthetic Oxygen Evolution, Chlorophyll Synthesis, and Synthesis of Fraction I Protein and its Reversibility

Adding 2dDG to algal cultures under optimal cultivation conditions (on complete nutrient media with illumination) leads (Fig. 2) to rapid (setting in after 2-3 h) depression of photosynthesis, and inhibition of chlorophyll synthesis with the appearance of colorless (white) cells after 24-48 h. This effect is observed in various green and blue-green algae and is completely reversible after washing out of 2dDG from the cells [1]. 3-O-methylglucose and xylose exert a similar action, in particular on chlorophyll synthesis in *Chlorella*. 2dDG causes also specific and reversible depression of the synthesis of fraction I protein, contrary to the sum of other cellular proteins, whose content not only fails to decline under the influence of 2dDG, but even rises slightly [5].

Repression of the Synthesis of RNA and Certain Enzymes of the Calvin Cycle under the Effect of 2dDG

The stereochemical analog of glucose not only depresses the synthesis of fraction I protein in photosynthesizing cells, but also inhibits incorporation of ^{14}C-uracyl into RNA and the synthesis of ribose-5-phosphate isomerase, phosphoribulokinase, ribulose-1, 5-bisphosphate carboxylase and carbonic anhydrase (Fig. 3). 2dDG did not af-

Fig.2A-C. Influence of 2-deoxy-D-glucose on the rate of photosynthesis (*A*), configuration of the light curves of oxygen evolution (*B*) and chlorophyll synthesis (*C*) in *Chlorella sp.K*, and reversibility of its action. *A* 1 control; 2 experiment; *B* 1 control; 2,3 after 1 and 2 h of 2dDG action, respectively; *C* 1 control; 2 experiment. Final concentration of 2dDG, 0.5%

Fig. 3A-C. Kinetics of depression of RNA (*A, B*) and enzymes of Calvin cycle (*C*) synthesis in cells of *Chlorella sp.K* under the effect of 2-deoxy-D-glucose. *A* Incorporation of 2-^{14}C-uracyl into total nucleic acids. *1* control; *2* with 2dDG; *B* Fragment of *A; C* 1 rate of RNA synthesis; *2* activity of ribose-5-phosphate isomerase; *3* phosphoribulokinase; *4* ribulosebisphosphate carboxylase; *5* carbonic anhydrase. Values are given in percent of the starting value per unit of culture volume

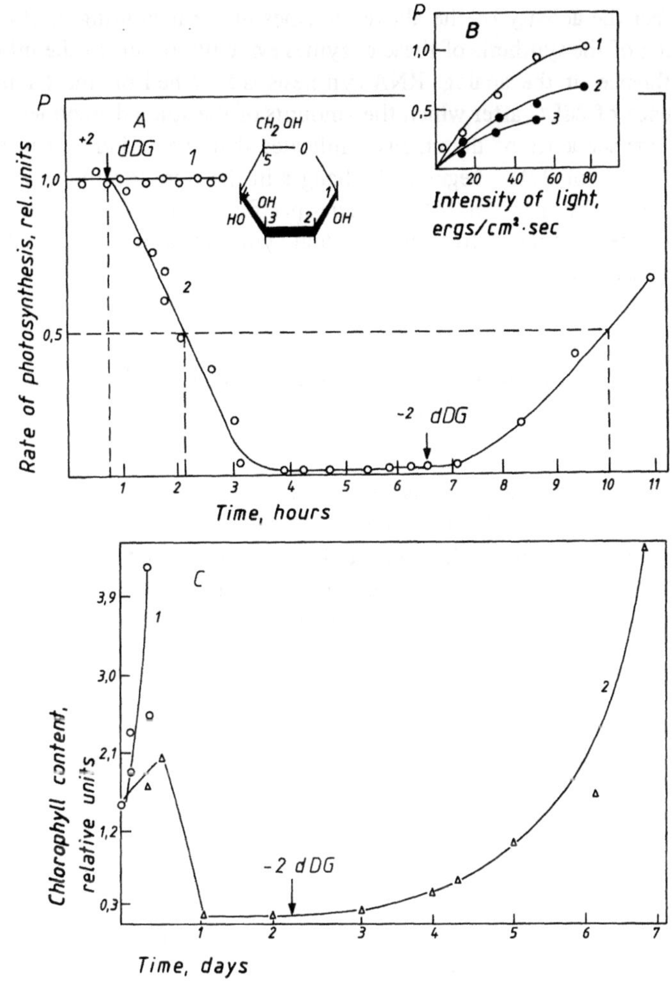

Fig. 2A-C. (Legend see p. 130)

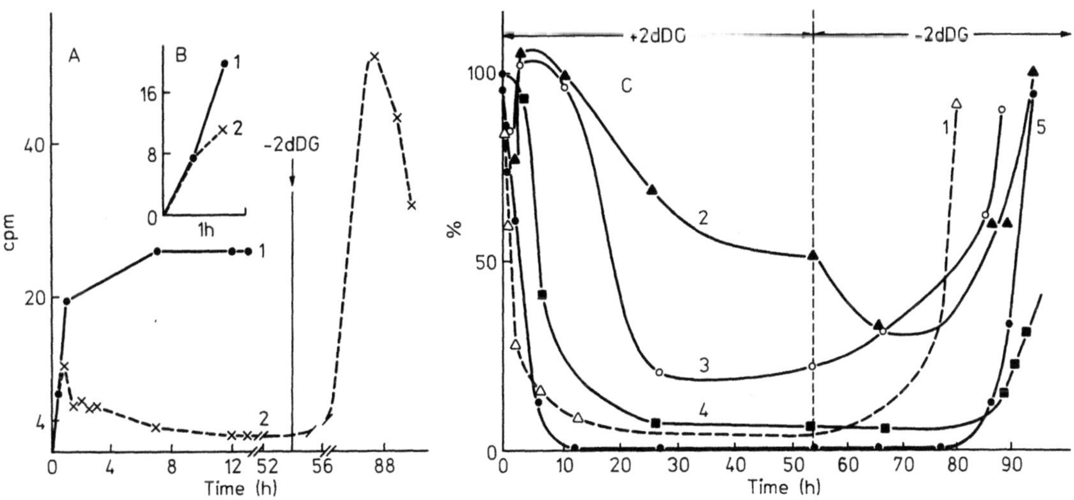

Fig. 3A-C. (Legend see p. 130)

fect the activity of the above enzymes in the measuring mixtures. Therefore, blocking of the synthesis of these enzymes evidently occurs in the intact cells under the influence of the analog. RNA synthesis is switched off most rapidly under the influence of 2dDG after which the amounts of the studied enzymes start to decrease. This time sequence of the changes indicates that events induced in the cell under the influence of 2dDG begin with changes in the transcription processes. This is also evidenced by the effect of organelle-specific transcription and translation inhibitors on the restoration of fraction I protein synthesis after washing 2dDG off the *Chlorella* cells [1,5].

Changes in Ultrastructural Organization of Chloroplast in *Chlorella* Cells under the Influence of 2dDG

The glucose analog seems to depress the synthesis not only of enzymes of the Calvin cycle, but also of structural proteins of the chloroplast, as is evidenced by significant disturbances in its ultrastructural organization [6]. Electron microscope photographs show a peculiar "dismantling" of the thylakoid system caused by the effect of 2dDG: first exfoliation of the system, then divergence and fragmentation of the thylakoids

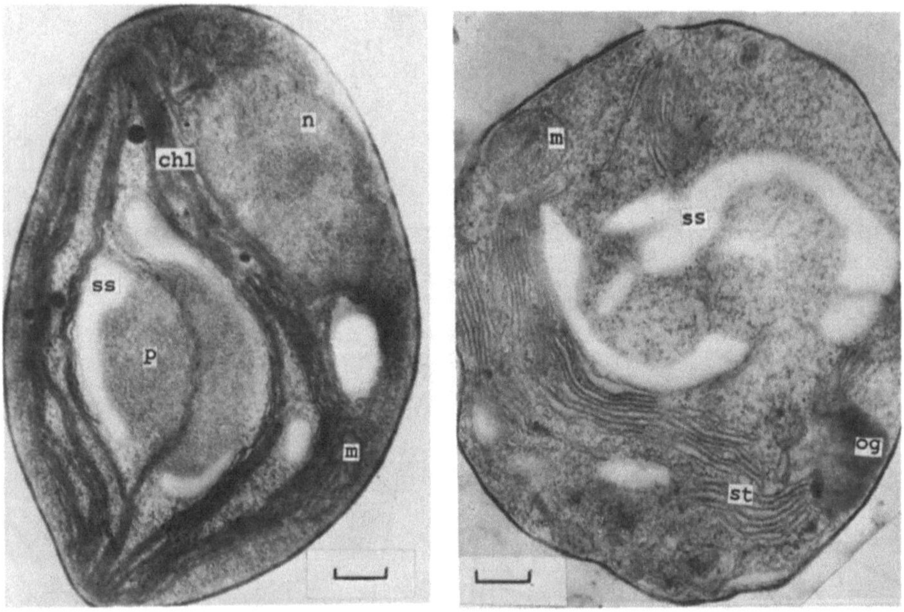

Fig. 4A,B. Changes in ultrastructural organization of the cell of *Chlorella sp.K* under the effect of 2-deoxy-D-glucose, **A** control cell; **B** cell after 48 h of 2dDG action. *chl* chloroplast; *m* mitochondrion; *n* nucleus; *p* pyrenoid; *ss* grains of pyrenoid starch sheath; *st* single thylakoids; *og* osmiophilic globulin; bar = 0.2 μm

to the point of forming free-lying single disks. Another characteristic change is the complete disappearance of the pyrenoid whose localization in the chloroplast can be judged by the remaining grains of the starch sheath. In addition, multivesicular bodies (globules of unidentified substances of a varying electron density) appear in the cell (Fig. 4).

Changes in the State (Deaggregation) of Photosynthetic Apparatus Pigments in Intact *Chlorella* Cells under the Effect of 2dDG

The disturbance of the structural organization of the chloroplast membrane systems caused by 2dDG is also evidenced by changes in the characteristic structure of the red absorption maximum for chlorophyll a and the spectra of low-temperature fluorescence of *Chlorella* cells. A peculiar feature of these changes is the disappearance of long wavelength peaks at 687,692,697 and 702-703 nm and gradual extinction of fluorescence at 726 nm characteristic of photosystem I [7] which seems to point to the suppression of the synthesis of pigment-protein components in this photosystem by 2dDG. After washing out the analog, highly aggregated chlorophyll forms appear and a characteristic structure of the low-temperature fluorescence spectrum is restored.

Effect of 2dDG on the Activity of Reaction Centers in Photosystem I (P700 Concentration)

Another indication of the upset functional activity of the reaction centers in photosystem I caused by 2dDG is a gradual decrease in light-induced absorption at 700 nm (Fig. 5A) which is associated with photooxidation of P700 [8]. Interestingly, the magnitude of the signal associated with the photooxidation of P700 is reduced, whereas its kinetics undergoes no significant changes (P700 is rapidly oxidized and then quickly reduced in darkness). In this connection, it would seem that a gradual decrease in Δ D is not due to a lower activity of the reaction centers, but due to a decrease in the concentration of P700 and, hence, in the number of photochemically active centers of photosystem I.

Indeed, the calculation of the number of P700 molecules and the total number of chlorophyll molecules has shown [9] that the starting culture (control) contains 600 chlorophyll molecules per P700 molecule. Under the effect of 2dDG these ratios change quickly: there are 900 chlorophyll molecules per P700 molecule after 3.5 h, 1800 after 8 h, and as many as 20,000 chlorophyll molecules after 24 h. After the analog is washed out the initial concentration of P700 capable of reversible phototransformation is restored.

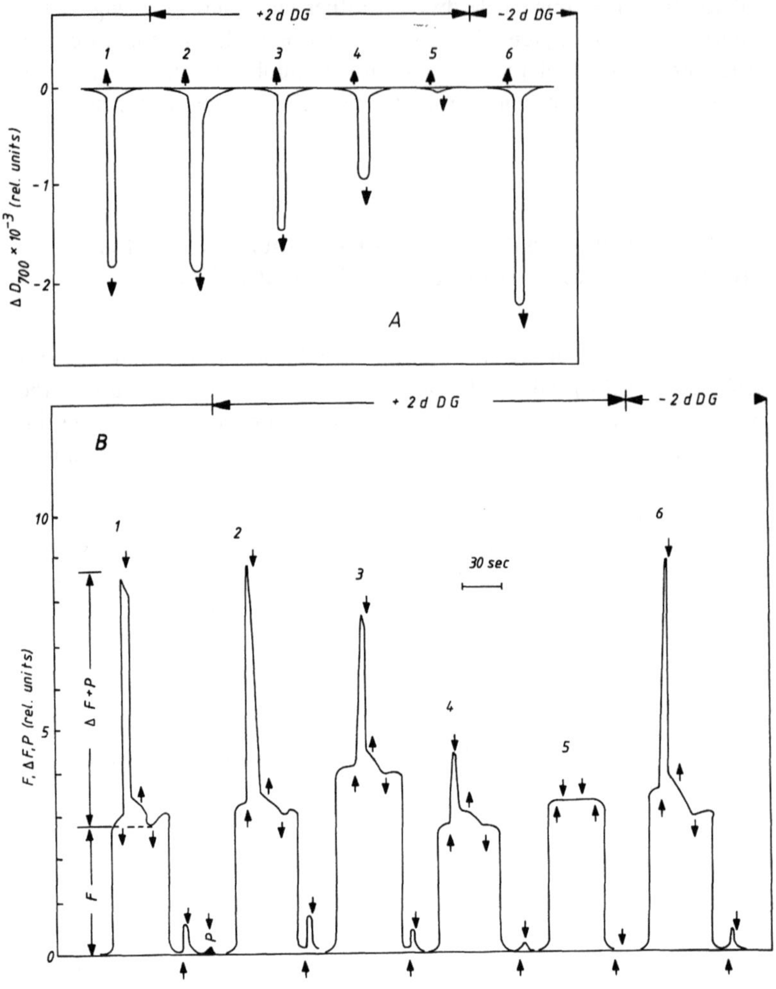

Fig. 5A,B. Effect of 2dDG on the activity of photosystem I (**A**) and photosystem II (**B**) and on their interaction in intact cells of *Chlorella sp.K.* **A** Kinetics of light-induced decrease of absorption at 700 nm ($\triangle D_{700}$); **B** Kinetics of light-induced changes in chlorophyll fluorescence yield of photosystem II (△F); "dark" fluorescences (F). **A** and **B** *1* control; *2,3,4,5* after 1, 2, 6, 30 h of 2dDG action, respectively, 6-71 h after washing out of 2dDG. Upward arrows switching on of effective light absorbed by photosystem II (↑) or photosystem I (↑); *downward arrows* switching off of light

Reversible Repression of Reaction Centers of Photosystem II under the Effect of 2dDG

The glucose analog causes also a reversible repression of photosystem II as was demonstrated in the studies of photo-induced changes in the chlorophyll fluorescence yield (△F) of this photosystem in intact *Chlorella* cells. It can be seen from Fig. 5B that under the action of 2dDG in light the magnitude of positive $\triangle$F declines rapidly to the level of about 30% of control $\triangle$F after 4-5 h. After 20-22 h of 2dDG action, $\triangle$F practically vanishes which points to upset functioning of the reaction centers of photosystem II [9].

The above data on disturbance in the chloroplast photochemical reactions under the effect of 2dDG seem to show that the glucose analog represses the synthesis of photosynthetic membrane components, in particular, proteins of the pigment-protein complexes which are specifically responsible for the native forms of chlorophyll in the photochemical systems of the chloroplast. It is important to note that in all the experiments, repression of the photochemical apparatus activity was preceded by blocking of RNA synthesis similar to the case of inhibiting the enzymes of the Calvin cycle.

Light Dependence of 2dDG Effect on Repression of Biosynthesis and Functional Activity of Chloroplast

Glucose-induced effect of repression of biosynthesis and functional activity of chloroplasts is light-dependent; it is not observed in the dark but is saturated by low light intensities. An important point is that the light dependence of the 2 dDG effect obeys the dose law [4,10]. Suppression of the functional activity of the photosynthetic apparatus by glucose analogs occurs only in a narrow spectral range (650-665 nm), and does not occur in blue and far-red light (Fig. 6).

Role of Feed-Back Mechanism of Metabolite Regulation of Chloroplast Genome Expression in Chloroplast-Cytoplasm Interaction

The data obtained using glucose analogs demonstrate that the chloroplast (Fig. 7) possesses feed-back mechanism of metabolite regulation of photosynthesis at a level of the synthesis of photosynthetic apparatus components involving glucose as an effector. This mechanism underlies the adaptive properties of the chloroplast and may play the role of a "terminal actuating mechanism" in the sink-source relationship between chloroplast and cytoplasm within the whole cell [11].

Indeed the phenomena evoked by glucose analogs, and in particular, inhibition of RNA synthesis and activity of the chloroplast photochemical systems [9,12] has been observed in natural models consisting of nondividing *Chlorella* cells in which hypertrophic accumulation of assimilates in the chloroplast occurs [13]. It would be interesting to conduct further studies of the phenomenon in in vitro systems using the methods of molecular biology and genetics.

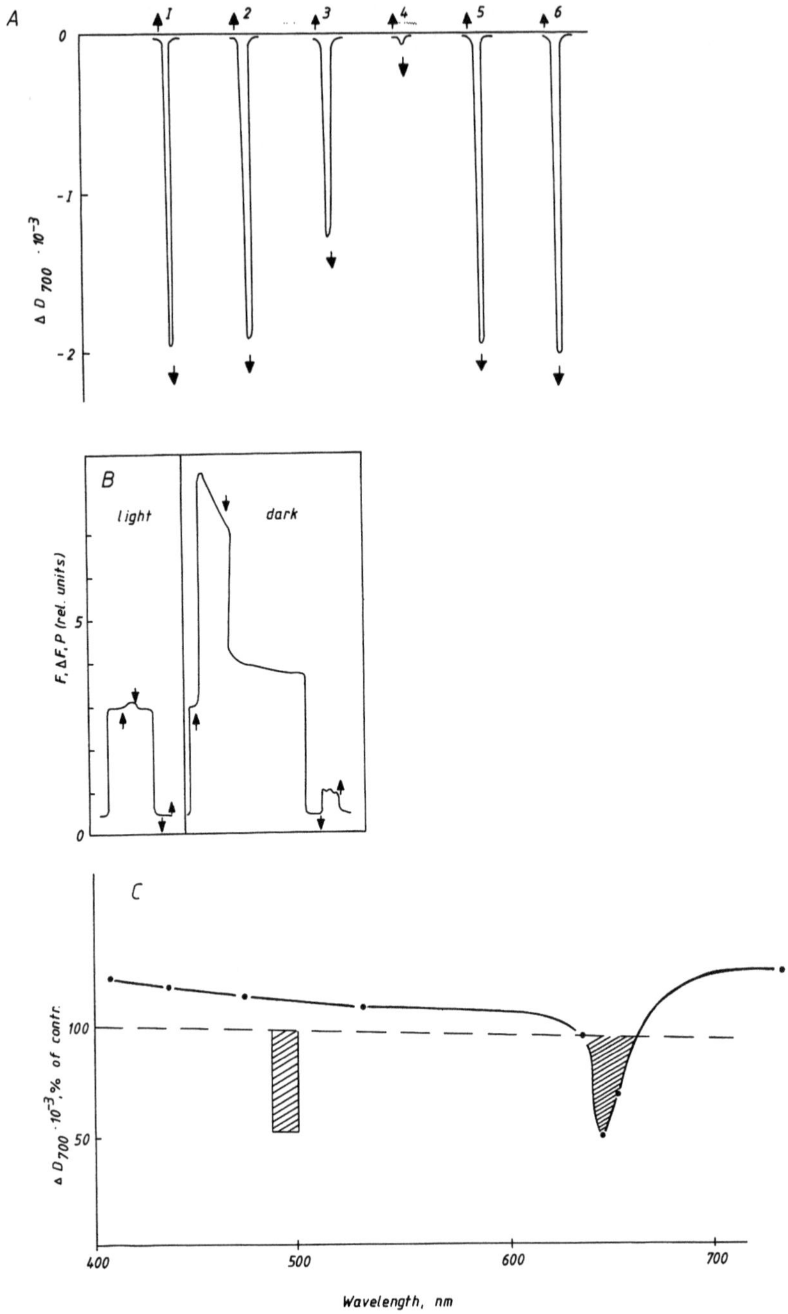

Fig. 6A-C. Light dependence of 2dDG-repression effect on the photosynthetic apparatus of *Chlorella* cells. **A** concentration of P_{700}. *1* Light without 2dDG; *2* dark without 2dDG; *3* with 2dDG on light, 6h; *4* with 2dDG on light, 50 h; *5* with 2dDG in dark, 6 h, *6* with 2dDG in dark, 50 h. **B** ΔF 22 h after 2dDG addition under light and dark. **C** 2dDG-effect on P700 at different spectral ranges

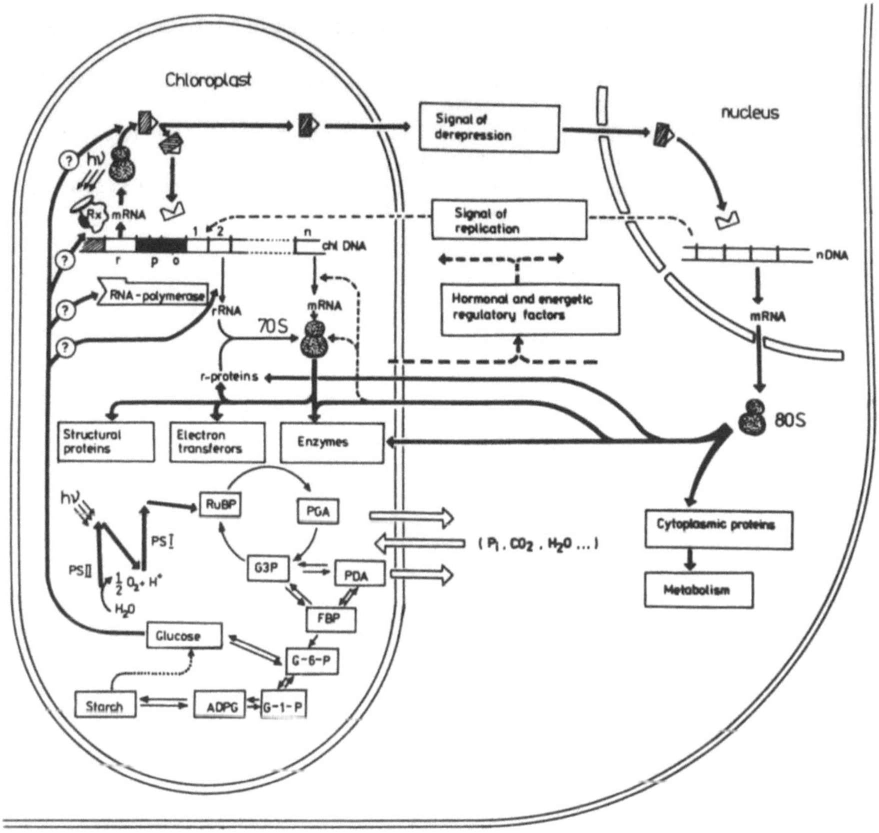

Fig. 7. A scheme of metabolite regulation of chloroplast genome expression and chloroplast-cytoplasm interrelation

References

1 Semenenko VE (1978) Sov Plant Physiol (Engl Transl Fiziol Rast) 25:715–731
2 Kung S (1977) Annu Rev Plant Physiol 28:401–437
3 Semenenko VE, Afanasyeva VP (1972) Soviet Plant Physiol (Engl Transl Fiziol Rast) 19:918–923
4 Semenenko VE (1975) In: Kursanov AL and Voskresenskaya NP (eds) Photoregulation of Morphogenesis of Plants, Nauka Publ, Moscow, pp 135–157
5 Semenenko VE, Kasatkina TI, Rudova TS (1976) Soviet Plant Physiol (Engl Transl Fiziol Rast) 23:1036–1041
6 Vladimirova MG (1976) Soviet Plant Physiol 23:847–852
7 Zvereva MG, Shubin LM, Klimova LA, Semenenko VE (1979) Dokl Akad Nauk SSSR 244:1244–1247
8 Kok B (1956) Biochem Biophys Acta 22:399–401
9 Zvereva MG, Klimova LA, Semenenko VE (1980) Soviet Plant Physiol (Engl Transl Fiziol Rast) 27:1218–1228

10 Kuptsova ES, Semenenko VE (1981) Soviet Plant Physiol (Engl Transl Fiziol Rast) 28: 743–748
11 Semenenko VE (1982) In: Nichiporovich AA (ed) Physiology of Photosynthesis. Nauka Publ, Moscow, pp 164–187
12 Semenenko VE (1981) In: Akoyunoglou G (ed) Photosynthesis. V. Chloroplast Development, Balaban Intern Sci Services, Philadelphia, pp 767–776
13 Semenenko VE, Vladimirova MG, Orleanskaya OB, Raikov NI, Kovanova ES (1969) Soviet Plant Physiol (Engl Transl Fiziol Rast) 16:172–181

Contractile Vacuole Complexes in Algae

K. HAUSMANN[1] and D. J. PATTERSON[2]

Definition

Contractile vacuoles are organelles which occur in many protists and some freshwater sponges. They are defined by their behaviour of slowly filling with fluid and periodically expelling this fluid from the cell (for review see Patterson 1980). Contractile vacuoles form one part of a more extensive organelle, the contractile vacuole complex. The organization of the contractile vacuole complex (CVC) can best be described from ultrastructural studies. The CVC typically includes (1) the contractile vacuole, (2) a system of membranous elements called the spongiome and possibly (3) a permanent pore supported by cytoskeletal structures which ensure the expulsion of the fluid at a fixed site on the surface of the cell. Contractile vacuole complexes have usually been implicated in the osmoregulatory competence of the cells.

Distribution

CVC's are found in most types of algae which have species living in freshwater environments and which lack complete cell walls. CVC's are consequently found in many green volvocalean algae, prasinophytes, euglenoids, cryptophytes, chrysophytes, prymnesiophytes (haptophytes), chloromonadophytes (raphidiophytes) and some heterochlorid flagellates. CVC's have not been identified with certainty in the eustigmatophytes. They are typically absent from diatoms, but may occur in some naked stages of their life-cycle (Pascher 1932). CVC's are absent from the non-volvocalean green algae, but in groups such as the Conjugales, they may appear in certain wall-less stages of their life-cycle (Lloyd 1928). CVC's do not appear to occur in the exclusively marine silicoflagellates (van Valkenburg 1980). They are absent from dinoflagellates, but these organisms have an organelle called the pusule, which has been argued to be analogous to contractile vacuoles (Dodge 1972, Mignot 1970, Schnepf and Deichgräber 1972). It has recently been suggested that the fluid, and sometimes particulate, contents of the pusule can be expelled by the contraction of an adjacent flagellar root (Cachon et al. 1983).

1 Lehrstuhl für Zellenlehre, Universität Heidelberg, Im Neuenheimer Feld 230, 6900 Heidelberg, FRG

2 Department of Zoology, University of Bristol, Bristol BS8 1UG, England

Compartments in Algal Cells and Their Interaction
ed. by W. Wiessner, D. Robinson, and R.C. Starr

Behaviour

Algal flagellate cells typically contain one or two CVC's. Most frequently they lie in the vicinity of the flagellar bases, and discharge their contents at fixed points in the surface of the cell. Discharge of vacuolar contents forms one part of a recurring cycle of behaviour. This incorporates two major phases, filling or diastole, and expulsion (contraction) or systole. A typical pattern of behaviour is illustrated by Fig. 1. At the light-microscopical level filling appears to occur by the fusion of small accessory or contributory vesicles. The term "contractile vacuole" is arbitrarily applied to the largest of the vesicles. Diastole usually terminates with the rounding up of the vacuole. Simultaneously the contractile vacuole becomes slightly smaller in diameter (Leedale 1967, Patterson 1980, Patterson and Hausmann 1981). These events have been interpreted as indicating the contraction of elements intimately associated with the vacuole. During the systole the vacuole appears to fuse with the plasma membrane, and disappears from view as its contents are lost, when observed with the light microscope.

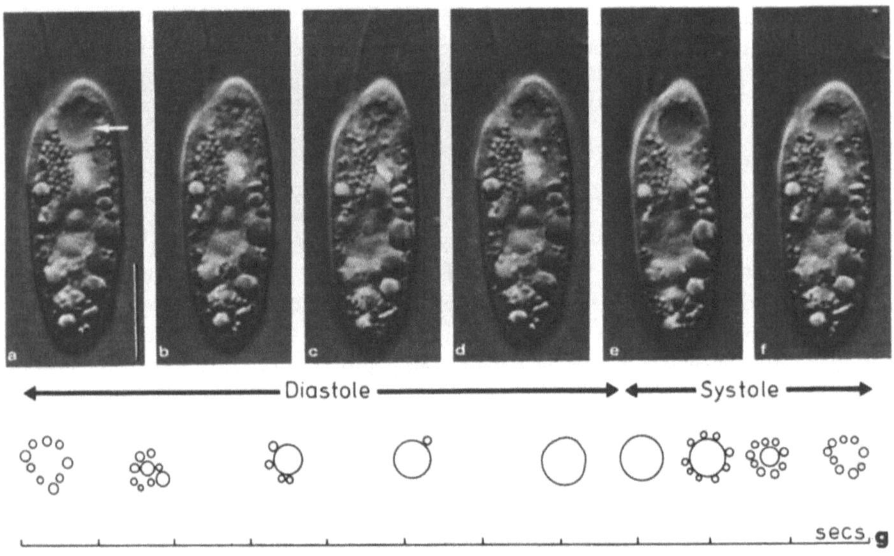

Fig. 1a-g. Light-microscopical appearance of the contractile vacuole cycle in *Chilomonas paramecium.* **a** The vacuole (*arrow*) in round-up just prior to expulsion of its contents; **b,c** The complex in diastole, the vacuole fills by the fusion of contributory vesicles; **d** Nearing the end of diastole, a single vacuole can be seen; **e** The end of diastole is marked by the rounding up of the vacuole; **f** Expulsion follows shortly afterwards, at which time more contributory vesicles appear. Scale bar = 5 μm; **g** Diagrammatic representation of a typical cycle of contractile vacuole behaviour. (**a-g**) from Patterson and Hausmann 1981)

Ultrastructure

In thin sections the contractile vacuole typically appears as a fluid-filled membrane-bounded vacuole. In some species, e.g., *Vacuolaria virescens,* the vacuole appears to contain organic matter (Heywood 1978). Some regions of the vacuolar membrane may have a membrane coat on the cytoplasmic surface. The possible function of this coat will be discussed below.

The spongiome appears to be an ubiquitous component of CVC's (Patterson 1980). It is thought to be involved in the extraction of fluid from the cytoplasm. The spongiome cannot be seen in living cells. In most types of algal flagellates, ultrastructural studies indicate that the spongiome is comprised of a bed of fine vesicles which cluster around the vacuole (Fig. 2a). This is the situation in volvocalean green algae and the zoospores of other green algae (Gruber and Rosario 1979, Hobbs 1971, Hoffman 1976, Manton 1964, Melkonian 1975, Moestrup 1974, 1975, Weis et al. 1977), prasinophytes (Maiwald 1971), euglenoids (Leedale 1967, Mignot 1966), cryptophytes (Hausmann and Patterson 1981, Patterson and Hausmann 1981, 1982), xantophytes (Hibberd 1970, Hibberd and Leedale 1971, Massalski and Leedale 1969), prymnesiophytes (Green and Hibberd 1977) and some chrysophytes (Belcher 1969, Hibberd 1971). A number of ochromonadine chrysophytes have a spongiome of flattened tubules with distinctive external and internal membrane coats (Fig. 2b) (Aaronson and Behrens 1974, Belcher and Swale 1971, 1972, Cole and Wynne 1973, Tsekos and Schnepf 1972, Wessel and Robinson 1979). One reference to tubular elements in the spongiome of a species of *Spumella* (Belcher and Swale 1976) is not supported by an illustration. In a number of those species with a vesicular spongiome, the spongiome contains vesicles with prominent coats on the cytoplasmic face of the membrane (Fig. 2a) (Green and Hibberd 1977, Hausmann and Patterson 1981, Hibberd and Leedale 1971, Leedale 1967).

Many algal contractile vacuoles repeatedly discharge their contents at a discrete point of the cell surface. In ciliates, where contractile vacuoles also discharge through discrete sites in the cell, there is a differentiated pore which holds the vacuole in position by means of microtubules and provides a site for discharge (Hausmann and Allen 1977). Similar structures might be expected to occur in many flagellates. The occurrence of the complex in the vicinity of the flagellar basal bodies with their extensive cytoskeletal (rootlet) structures may be more than fortuitous. In a few cases, cytoskeletal elements in the vicinity of the discharge site in flagellates have been described (Patterson and Hausmann 1981). However, there is much evidence for a local differentiation of the plasma membrane at the site of fluid discharge. This may take the form of a close association between the plasma membrane and the membrane of the contractile vacuole (Aaronson and Behrens 1974, Belcher and Swale 1972, Cole and Wynne 1973, Hibberd 1970,1971, Maiwald 1971, Massalski and Leedale 1969, Moestrup 1975, Wessel and Robinson 1979). Alternative evidence for local differentiation of the plasma membrane is provided from freeze-fracture studies (Weiss et al. 1977). Green and Hibberd (1977) have described a highly differentiated expulsion tube in *Diacronema vulkianum.*

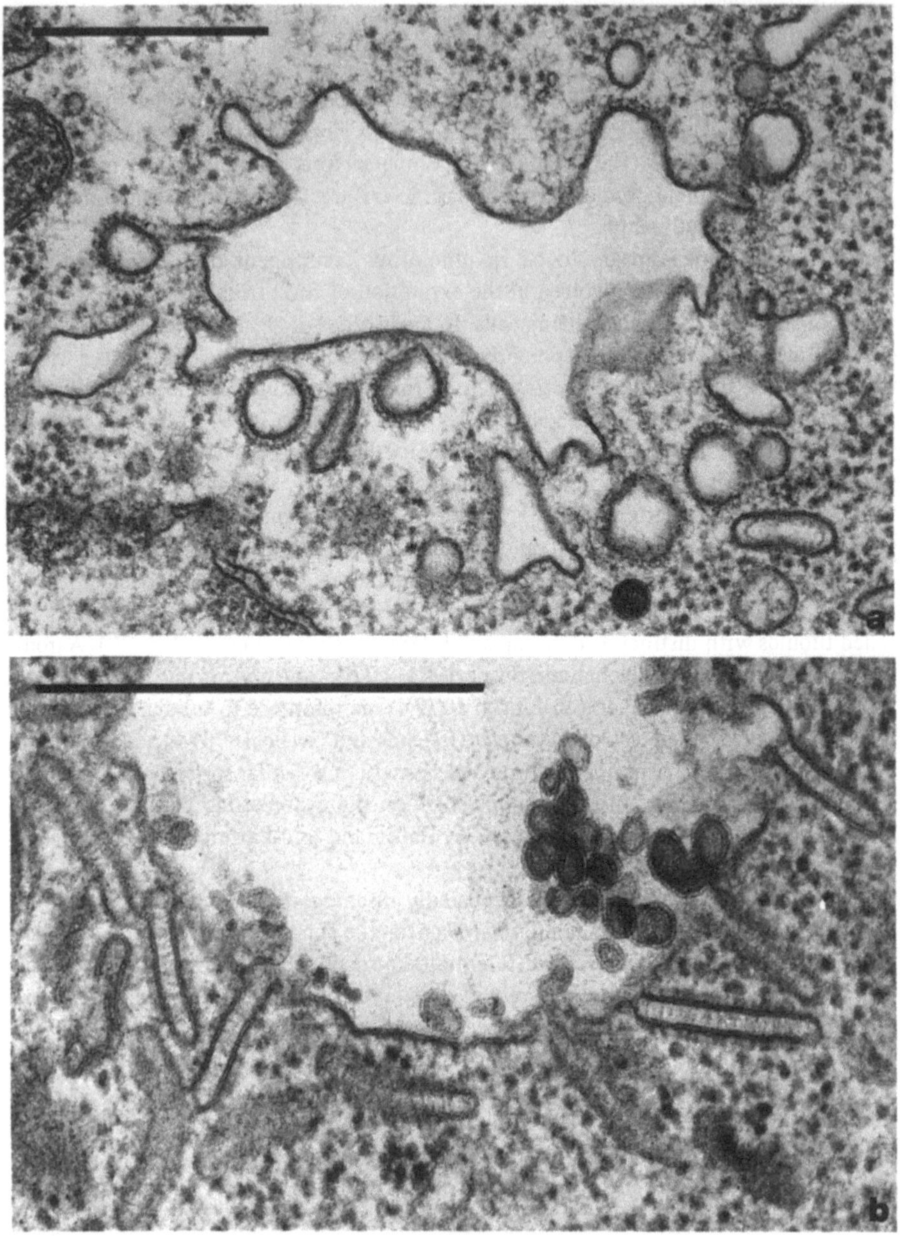

Fig. 2a. Electron micrograph of the contractile vacuole complex of *Cyathomonas truncata*. The spongiome contains coated and uncoated vesicles. Regions of the membrane of the contractile vacuole are also coated. Scale bar = 0.5 μm. **b** Electron micrograph of the contractile vacuole complex of *Poterioochromonas stipitata*. The spongiome is comprised of flattened coated vesicles with a distinct internal structure. Scale bar = 0.5 μm. (Original by W. Herth, Heidelberg)

The ultrastructural organization of these organelles is clearly only sparsely documented. Nonetheless, there are indications that there is a variety of patterns of organization among algal flagellates as well as among other protists (Patterson 1980). It is likely that this diversity is due to the polyphyletic origin of these organelles, rather than to diversification from a common ancestry. This suggestion derives from the supposed high adaptive pressure among protists to become mobile and/or heterotrophic. In both instances, the cells would need to relinquish a cell wall and would need to develop some mechanism to offset the osmotic influx of water. If these organelles are polyphyletic, then any generalising statements should be interpreted with caution.

Functioning of the Complex

The spongiome is widely believed to be the site at which fluid is isolated from the cytoplasm (Patterson 1980). In most flagellates, the vacuole forms and fills by the fusion of the fluid-containing vesicles of the spongiome (see Patterson and Hausmann 1981, for a typical account of this process). Some of the vacuolar fluid, if not all of it, must accumulate via this route. The mechanism by which fluid is segregated in CVC's in general is not known. But one possibility is that ion (or other solute) pumps, located in the membranes of the vesicles, create osmotic gradients directed into the lumen of the vacuole. Such gradients could draw water across the membranes of the vesicles. Circumstantial evidence for an active role of the spongiome comes from evidence of high phosphatase activity within the vesicles of the spongiome of some amoebae (Bowers and Korn 1973, Quiviger, de Chastellier and Ryter 1978). Comparable evidence has now been obtained for *Poterioochromonas* (Quader et al. 1983).

Studies earlier in this century suggested that the vesicles which contribute to the filling vacuole are derived from dictyosomes (see below). This appears certainly to be the case in the Chloromonadophyceae (Heywood 1978, Mignot 1967, Schnepf and Koch 1966). In the other flagellates in which this has been investigated, there is no evidence for a dictyosomal origin of the vesicles of the spongiome (Patterson and Hausmann 1982). Consequently, the vesicles of the spongiome must be derived by a process of recycling, involving the fragmentation of the vacuolar membrane as suggested for amoebae (McKanna 1973). This appears to be the situation in cryptophytes (Patterson and Hausmann 1981).

What is the significance of the coat associated with some vesicles and some regions of the contractile vacuole? In reference to other protists, McKanna (1974, 1976) has argued that the coat contributes to the process of fluid segregation. Hausmann and Patterson (1981) have drawn attention to the similarity between the coat of spongiome vesicles and clathrin coats. They have suggested that the coated vesicles are involved in the recycling of membranes or membrane components. In investigating this phenomenon, care must be taken to ensure that only one functional system is being investigated. Leedale (1967) has interpreted coated vesicles as integral components of the euglenid CVC. Later, however, Kivic and Vesk (1974) argued that these vesicles may in fact be involved in the pinocytotic uptake of materials from the reservoir. These vesicles may therefore not be a component of the CVC.

The cycle of behaviour may be envisaged as involving the segregation of fluid into vesicles by the action of energy-dependent membrane pumps. As a result the vesicles enlarge and fuse together. This process of filling creates an excess of membrane in the vacuole and this may be retrieved by selective vesiculation. It has been assumed that the retrieval occurs via the coated regions of the vacuole (Hausmann and Patterson 1981). After a certain interval, a timing mechanism induces contraction of elements around the vacuole, which cause it to round up and fuse with the plasma membrane. The cellular hydrostatic pressure then causes the vacuolar contents to be expelled from the cell. As fluid is lost from the vacuole, the membrane collapses and fragments into the vesicles of the spongiome. The cycle is completed.

This account presumes that expulsion occurs after complete fusion of the vacuolar membrane with the plasma membrane. This is suggested by some electron micrographs which show continuity of the two membranes (Hoffman 1976, Patterson and Hausmann 1981). However, Weiss et al. (1977), on the basis of some detailed ultrastructural studies on *Chlamydomonas reinhardii,* have suggested that fluid is discharged without membrane fusion. They suggested the fluid leaves the cell via hydrophilic transmembrane channels. Linder and Staehelin (1979) have suggested an intermediate model for a trypanosomatid in which components of adjacent membranes fuse to form microchannels. If such mechanisms are proven to operate in these species, they are unlikely to be universal, as direct observation on living protists have clearly shown membrane fusion (Organ et al. 1968).

Osmoregulation

It has, for a long time, been thought that contractile vacuoles satisfy the osmoregulatory requirements of the cell (see Patterson 1980 for review). Although there is no experimental evidence for this function in algal cells, their occurrence in cells from media of low or changeable osmotic pressure, and their occurrence in cells which lack continuous cell walls supports this function. This is because it is in these cells that one would expect the greatest osmotic influx of water, and it is these cells which lack a cell wall which could limit volume changes. The ability of the contractile vacuoleless mutants of *Chlamydomonas moewusii* (Guillard 1960) to live only in media of high osmotic pressure supports this interpretation. However, the growing awareness of the presence of a volume regulatory mechanism in algal and other protist cells (Brown 1976, Kauss 1979, Kauss and Rausch, this vol., Patterson 1980), reveals that the CVC has a secondary role in the establishing of osmoregulatory competence. The volume regulatory mechanism has the primary role in establishing the osmotic gradient across the cell surface, the CVC only serves to eliminate fluid to compensate for the osmotic influx of water.

CVC: An Independent Organelle?

A protist cell functions in its entirety, and attempts to delineate components of the cell are arbitrary. Nonetheless, it is still possible to identify cell "organelles", such as nuclei, dictyosomes, endoplasmic reticulum, mitochondria, chloroplasts etc. Is the CVC a discrete organelle in the sense that these other structures are discrete organelles, or is it merely a part of some other organelles? Is the complex a transient or stable structure?

In the earlier days, there was a common view that contractile vacuoles are not stable structures and that they are an extension of the Golgi complex. This concept envisages the vacuole loosing membrane to the plasma membrane at systole, and that this is replenished by fusion of small vesicles derived from the dictyosomes. This view derives from studies earlier in this century which revealed that CVC's and dictyosomes shared some staining properties (for literature see Grasse and Hollande 1941, Patterson 1980). There is little doubt that the CVC's of chloromonadophyceans correspond in several respects to this concept (Heywood 1978, Mignot 1976, Schnepf and Koch 1966). The same view persists for CVC's of some other taxa (Gruber and Rosario 1979, Leedale 1967, Manton 1964, Schnepf 1969). This is probable because both organelles frequently include small membranous vesicles and such vesicles could be implicated in the transfer of material from one organelle to the other. In a recent study of cryptomonads Patterson and Hausmann (1982) showed that the two populations of vesicles were distinct and that there is no evidence for direct transfer from one organelle. If there is no marked recruitment of membrane, it seems unlikely that there is any great loss of membrane from the CVC. The CVC would therefore appear to be a discrete, functionally independent and stable structure in most types of algal cells, the exception being the chloromonadophyceae.

Acknowledgements. This work was supported by the Deutsche Forschunggemeinschaft. DJP thanks the Royal Society of London and the British Council for financial support.

References

Aaronson S, Behrens U (1974) J Cell Sci 14:1–9
Belcher JH (1969) Br Phycol J 4:105–117
Belcher JH, Swale EMF (1971) Br Phycol J 6:157–169
Belcher JH, Swale EMF (1972) Br Phycol J 7:53–59
Belcher JH, Swale EMF (1976) Arch Protistenkd 118:215–220
Bowers B, Korn ED (1973) J Cell Biol 59:784–791
Brown AD (1976) Bacteriol Rev 40:803–846
Cachon J, Cachon M, Boillot A (1983) Cell Motil 3:61–67
Cole GT, Wynne MJ (1973) Cytobios 8:161–173
Dodge JD (1972) Protoplasma 75:285–302
Grassé PP, Hollande A (1941) Arch Zool Exp Gen 82:301–319
Green JC, Hibberd DJ (1977) J Mar Biol Assoc UK 57:1125–1136
Gruber HE, Rosario B (1979) Cytologia (Tokyo) 44:505–526
Guillard RRL (1960) J Protozool 7:262–268

Hausmann K. Allen RD (1977) Cytobiologie 15:303–320
Hausmann K, Patterson DJ (1981) Exp Cell Res 135:449–453
Heywood P (1978) J Cell Sci 31:213–224
Hibberd DJ (1970) Br Phycol J 5:119–143
Hibberd DJ (1971) Br Phycol J 6:207–223
Hibberd DJ, Leedale GF (1971) Br Phycol J 6:1–23
Hobbs MJ (1971) Br Phycol J 6:81–103
Hoffman LR (1976) Protoplasma 87:191–219
Kauss H (1979) Prog Phytochem 5:1–27
Kivic PA, Vesk M (1974) Arch Mikrobiol 96:155–159
Leedale GF (1967) Euglenoid Flagellates. Prentice-Hall, NJ
Linder JC, Staehelin LA (1979) J Cell Biol 85:371–382
Lloyd FE (1928) Protoplasma 3:45–66
Maiwald M (1971) Arch Protistenk 113:334–344
Manton I (1964) J Exp Bot 15:399–411
Massalski A, Leedale GF (1969) Br Phycol J 4:159–180
McKanna JA (1973) Science (Wash DC) 179:88–90
McKanna JA (1974) J Cell Biol 63:317–322
McKanna JA (1976) J Ultrastruct Res 54:1–10
Melkonian M (1975) Protoplasma 86:391–404
Mignot JP (1966) Protistologica 2:51–117
Mignot JP (1967) Protistologica 3:25–60
Mignot JP (1970) Protistologica 6:267–281
Mignot JP (1976) Protistologica 12:279–293
Moestrup O (1974) Biol J Linn Soc 6:11–125
Moestrup O (1975) Biol J Linn Soc 7:23–34
Organ AE, Bovee EC, Jahn TL, Wigg D, Fonseca JR (1968) J Cell Biol 37:139–145
Pascher A (1932) Beih Bot Centrbl 49:703–709
Patterson DJ (1980) Biol Rev 55:1–46
Patterson DJ, Hausmann K (1981) Br Phycol J 16:429–439
Patterson DJ, Hausmann K (1982) Cell Biol Int Repts 6:513–521
Quader H, van Kempen R, Stelzer R, Robinson DG (1983) Eur J Cell Biol 30:283–287
Quiviger B, de Chastellier C, Ryter A (1978) J Ultrastruct Res 62:228–236
Robinson DG, Quader H (1980) Planta (Berl) 148:84–88
Schnepf E (1969) Protoplasmatologia VIII, 8:1–181
Schnepf E, Deichgräber G (1972) Protoplasma 74:411–425
Schnepf E, Koch W (1966) Arch Mikrobiol 54:229–236
Tsekos I, Schnepf E (1972) Naturwissenschaften 6:272–273
Van Valkenburg SD (1980) In: Cox ER (ed) Phytoflagellates. Elsevier/North-Holland, New York, pp 335–350
Weiss RL, Goodenough DA, Goodenough UW (1977) J Cell Biol 72:133–145
Wessel D, Robinson DG (1979) Eur J Cell Biol 19:60–66

Compartmentation of Ca^{2+} and its Possible Role in Volume Regulation of *Poterioochromonas*

H. Kauss and U. Rausch[1]

Introduction

There is growing evidence from animal systems that the concentration of free Ca^{2+} in the extra-organelle cytoplasm of resting cells is in the range of 10^{-7} M. External stimuli (e.g., antibody binding, light, hormones, nerve pulses) can induce an increase in the $[Ca^{2+}]_{cyt}$ which in turn regulates many metabolic and developmental steps, rendering Ca^{2+} an important "second messenger". Indirect evidence has often led to the suggestion, that Ca^{2+} may be equally important for plant cells, although in most cases measurements of $[Ca^{2+}]_{cyt}$ and its changes have not been performed (Williamson 1981). There appears to be only one plant example where direct measurements of $[Ca^{2+}]_{cyt}$ are available, namely internodal cells of *Chara* (Williamson and Ashley 1982). Using injection of the photoprotein aequorin the $[Ca^{2+}]_{cyt}$ was found to be about 0.2 μM and was raised by electrical stimulation to 7 μM with a concommitant cessation of cytoplasmic streaming. Total Ca^{2+} in the cytoplasm of *Chara* is 2-8 mM, indicating that great amounts must be sequestered in organelles (as far as known mainly in mitochondria and endoplasmatic reticulumn, and vesicles derived from it). According to the above reference, short-term regulation of $[Ca^{2+}]_{cyt}$ appears to occur in *Chara* at transport processes into and out of these organelles. Long-term regulation additionally involves transport over the tonoplast and plasma membrane, as the vacuolar and external concentrations of Ca^{2+} are in the mM range.

Most techniques which are presently available for direct measurements of $[Ca^{2+}]_{cyt}$ require microinjection or insertion of microelectrodes and thus can only be applied on certain cell-types. There is, therefore, little chance to employ these techniques on *Poterioochromonas* which has motile, small and wall-less cells. Conclusions as to the role of Ca^{2+} in this microalga thus can only be drawn from indirect observations.

An increase in external solutes in a suspension of *Poterioochromonas* causes the cells to shrink within one to two minutes; the water lost is regained in the following 1 to 2 h due to an internal accumulation of solutes, mainly isofloridoside (Kauss 1983, Kauss and Thomson 1982). Some physiological and biochemical details of IF metabolism have been clarified and are summarized in Fig. 1. The substance is under turnover even when present at constant concentrations, the turnover rates being higher as higher the level is (Kauss 1973). Regulation of the carbon flow in the IF

---

1 FB Biologie der Universität Kaiserslautern, Postfach 3049, D-6750 Kaiserslautern, FRG

Compartments in Algal Cells and Their Interaction
ed. by W. Wiessner, D. Robinson, and R.C. Starr

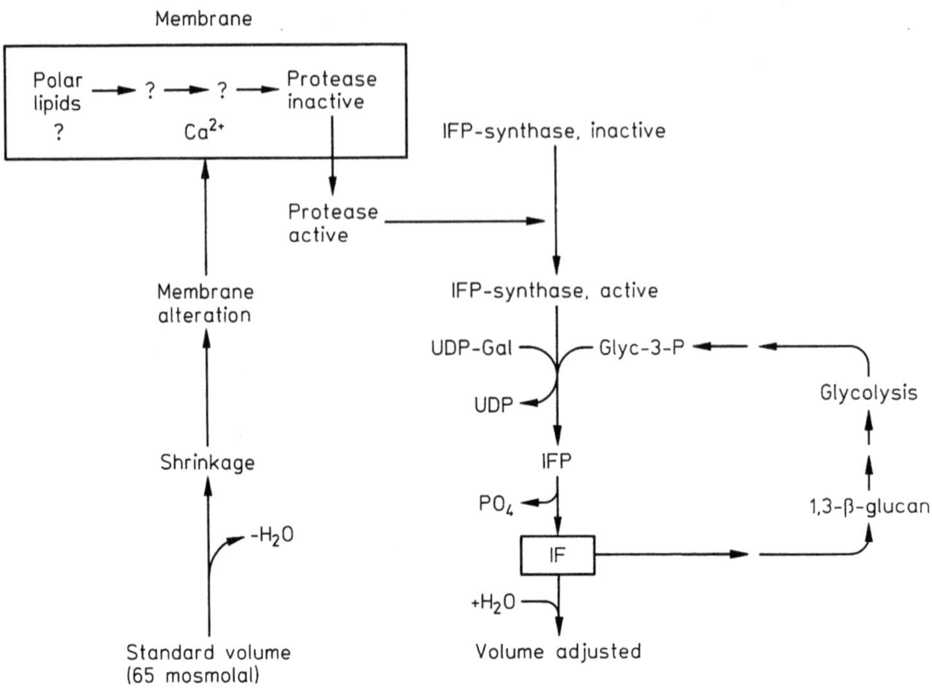

Fig. 1. Pathway of IF biosynthesis and a reaction sequence proposed for generating a membrane-borne protease which is capable of activating the IFP-synthase

pathway appears to occur at several enzymes and may also involve transport processes over the membrane of the storage vacuole which contains the 1,3-ß-D-glucan chrysolaminarin.

Our research has focused in recent years on the key enzyme IFP-synthase. We have suggested that one way to render this enzyme active is by a cascade of proteolytic steps which in vivo is triggered by cell shrinkage (Fig. 1). In crude cell homogenates, prepared under certain precautions, this reaction sequence can be initiated by addition of Ca^{2+} (Kauss 1983). Although many details of the membrane-bound part of the system remain obscure, it appeared that osmotically effected cell shrinkage could possibly lead to a transient increase in cytoplasmic Ca^{2+} and that this in turn may be part of the signal chain between volume or surface change and activation of the IFP-synthase. The data in this communication indicate on the one hand that Ca^{2+}-ions may be important for initiation of IF-synthesis after osmotically induced cell shrinkage. On the other hand they provide an example for the importance of Ca^{2+} compartmentation in algae and the problems involved in research relevant to this field.

Materials and Methods

Growth and preparation of cell suspensions of *Poterioochromonas malhamensis* (syn. *Ochromonas malhamensis*) has been previously described (Kauss 1983). An important deviation in our present work was that instead of diluted nutrient solution for suspension of the algae we had to use 70 mM MES/NaOH (65 mosmolal) in order to control [Ca^{2+}]. The density of the algal suspension was adjusted to give an absorbance of 0.8 after a 1:50 dilution, measured at 510 nm in a Zeiss PMQ II spectralphotometer. The pH of the buffer was 6.7 for experiments with the Ca^{2+}-ionophore A 23187, or 6.2 for all other experiments. It contained either 1.5 mM $CaCl_2$ for a [Ca^{2+}] of 1.5 mM or, alternatively, 3.5 mM EGTA which is sufficient to have a [Ca^{2+}] <0.1 μM in the final algal supension. The latter value was calculated according to Jewell and Rüegg (1966) assuming a $K_{CaL'} = 10^5$ at pH 6.2 (Ringboom 1979) and a [Ca^{2+}] of 25 μM, the maximum value measured in the supernatant using a Ca^{2+}-selectrode (Kauss 1981).

Cell suspensions were diluted 4.6-fold for labeling with CTC to optimize fluorescence efficiency. After incubation in buffer +Ca^{2+} for the indicated time, cells were washed twice in buffer + EGTA. Fluorescence was determined in a Jobin Yvon JY 3D spectrofluorometer (gain 100, excitation 390 nm, emission 530 nm) in a standard 10 x 10 mm stoppered cuvette, which was inverted once shortly before taking the measurements.

For Ca^{2+}-efflux experiments cells of *Poterioochromonas* were incubated for 2-3 h with 4 μCi $^{45}Ca^{2+}$/ml suspension in buffer + Ca^{2+} and then washed twice with buffer + EGTA. If indicated A 23187 was added at 0 time. At 2-min intervals aliquots of 0.5 ml were placed over a silicon-oil layer (AR 200/AR 20 = 5/1, v/v in an Eppendorf tube, centrifuged for 15 s and the radioactivity of 200 μl supernatant (water phase) was determined in 4 ml Triton-X-100/xylene scintillation cocktail in a BF 5001 A Betaszint (Berthold).

For Ca^{2+}-influx experiments, the algae were suspended in buffer +Ca^{2+}. $^{45}Ca^{2+}$ (1 μCi/ml) was added together with A 23187 or DCF at 0 time. After centrifugation of aliquots of the suspension as described above, the oil and water phases were removed, the algal pellet resuspended in 0.5 ml Triton-X-100 (1% w/v) and the radioactivity determined as described. Protein in the supernatant water phase was determined with the Coomassie method (Bradford 1976) using bovine serum albumin as a standard.

DCF was kindly provided by Bayer (Leverkusen). A 23187 was purchased from Calbiochem (Gießen) and CTC from Sigma (München). DCF and A 23187 were added as a 1000-fold concentrated solution in DMSO (controls: 0.1% DMSO, final conc.).

All other materials and methods were as described before (see Kauss 1983 for references).

Results and Discussion

Chlorotetracycline Fluorescence

A very convenient way to demonstrate Ca^{2+} cytochemically is to preincubate cells with the water-soluble antibiotic CTC, followed by fluorescence microscopy. When cells of *Poterioochromonas malhamensis* were incubated with CTC at low concentration (15 to 30 μM) for 30 to 60 min yellow fluorescence was mainly seen in one or a few relatively large spots (not shown). With higher concentrations (Fig. 2) the spots became more numerous and smaller ones appeared throughout the cytoplasm; the total cytoplasmic space also became fluorescent to some extent whereas no fluorescence was seen in the storage vacuole. As the Ca^{2+}/CTC-complex shows about 200-fold elevated fluorescence in polar environment, CTC mainly (but not exclusively) stains Ca^{2+} in vicinity of or in contact with membranes (Thomas 1982; Blinks et al.

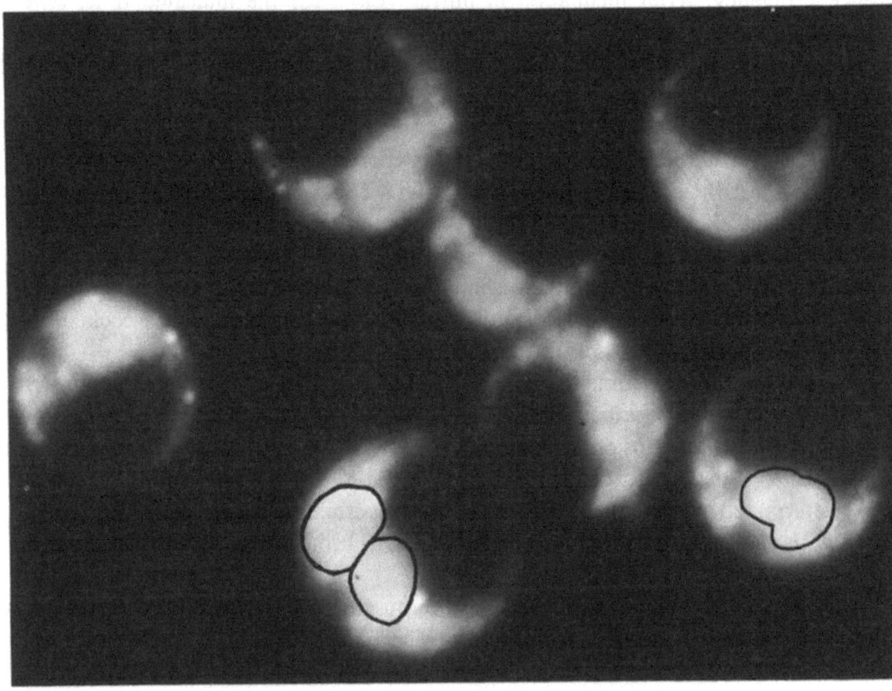

Fig. 2. Fluorescence of *Poterioochromonas* after incubation for 60 min with 500 μM CTC in diluted nutrient solution containing 1.5 mM Ca^{2+}. Photograph was taken in a Leitz Ortholox II/ fluorescence microscope with a Ploemopak D filter (excitation 355–425 nm) at a 1000 x enlargement. *Encircled areas* in the lower two cells indicate the position of the chloroplasts, which show deep red fluorescence. The vesicles and cytoplasm show bright yellow fluorescence. See text for comments on rounding up of the cells as a side-effect of the high CTC concentration and on staining with CTC at lower concentration

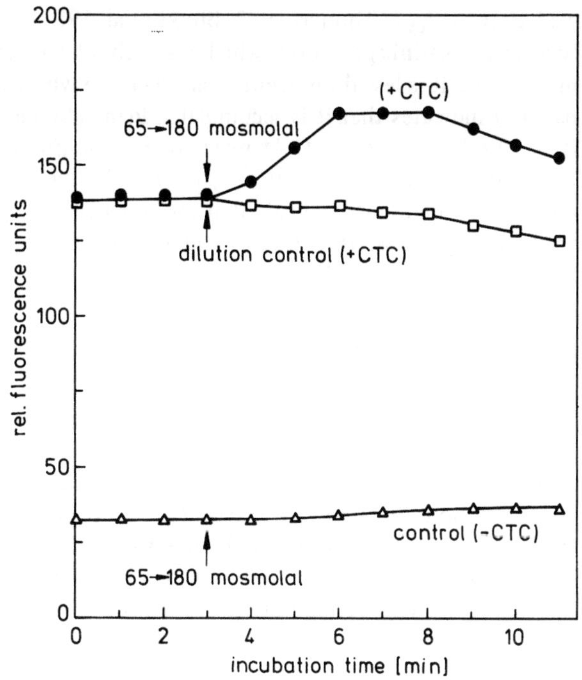

Fig. 3. Fluorescence signals of CTC-treated *Poteriochromonas* cells. The arrows indicate the time at which a cell suspension was either shrunken by addition of 1/10 volume of a 1.5 M sucrose (●, △) or diluted only to the same extent (dilution control, □). The cells were either preincubated in buffer + Ca^{2+} with CTC (15 μM, 60 min, ●, □) or remained untreated (△) for determination of autofluorescence as a control for possible changes in fluorescence quenching. Before taking the measurements the algae were washed with buffer + EGTA

1982). The fluorescent spots in CTC-stained cells thus indicate vesicles. Their identity is unknown in the above case of *Poterioochromonas malhamensis.*

The cells of *Poterioochromonas* exhibit a large storage vacuole and an elliptic form under the culture conditions used here. This is in some contrast to the "teardrop" form with a pronounced "tail" of cells grown under limiting nutrient conditions (e.g., Robinson and Quader 1980). With high CTC concentrations (500 μM) such as we have used (see Fig. 2), the cells showed a tendency to round up. They also moved more slowly in a trembling manner and appeared to become larger. The rounding up of *Poterioochromonas malhamensis* can also be observed on treatment with colchicine and is indicative of microtubule destruction (Robinson and Quader 1980).

In order to avoid the risk of cell damage or artificially elevated total Ca^{2+}-contents on incubation with CTC at higher concentrations or for a longer time (Blinks et al. 1982, Reiss et al. 1983) we continued the further investigations on cells incubated at lower CTC concentrations (Fig. 3). As the fluorescence measured remained low if buffer + EGTA was used (an indication that CTC can enter the cell mainly as Ca^{2+}/CTC-complex) the incubation was performed in buffer + Ca^{2+}. With CTC-labeled cells we observed on osmotic shrinkage an increase in fluorescence intensity, followed by a slower decline. The dilution control shows that this is not due to change in the suspension density (which is critical). The control minus CTC further shows that it is not due to an altered autofluorescence resulting from changes in cell volume or form.

Due to the influence of the polarity of the environment on the fluorescence of the Ca^{2+}/CTC-complex, the observed increase cannot directly be interpreted in regard to $[Ca^{2+}]_{cyt}$ (Thomas 1982, Blinks et al. 1982). It appears, however, that shortly after the shrinkage period (which takes about 1 min) Ca^{2+} moves from a compartment where it is less fluorescent to another one where it is more fluorescent. This apparently indicates that it is accumulated in membrane vesicles. As the external $[Ca^{2+}/$ is $<10^{-7}$ M, the moving Ca^{2+} must come from internal pools and it appears possible that it transiently increased $[Ca^{2+}]_{cyt}$. The decline of the fluorescence signal (Fig. 3) after 8 min might also indicate that the vesicular Ca^{2+} is undergoing turnover and, as an over-all process, is extruded and complexed by the outside EGTA, regardless whether the cells were shrunken or not.

Ca^{2+}-Ionophore A 23187

Another way to gain information on the importance of Ca^{2+} is to apply various conditions which alter the usual gradients, and to observe whether or not metabolic parameters are changed as a consequence. In previous experiments we used a diluted nutrient solution to suspend the algae (Schobert et al. 1972). Under these conditions $[Ca^{2+}]_{ext}$ cannot be controlled: Figure 4 shows that the cells also give the osmotically-induced increase in IF-content when suspended in MES-buffer. The extent of the increase is the same, regardless of whether Ca^{2+} is present or not. If Ca^{2+} is regarded

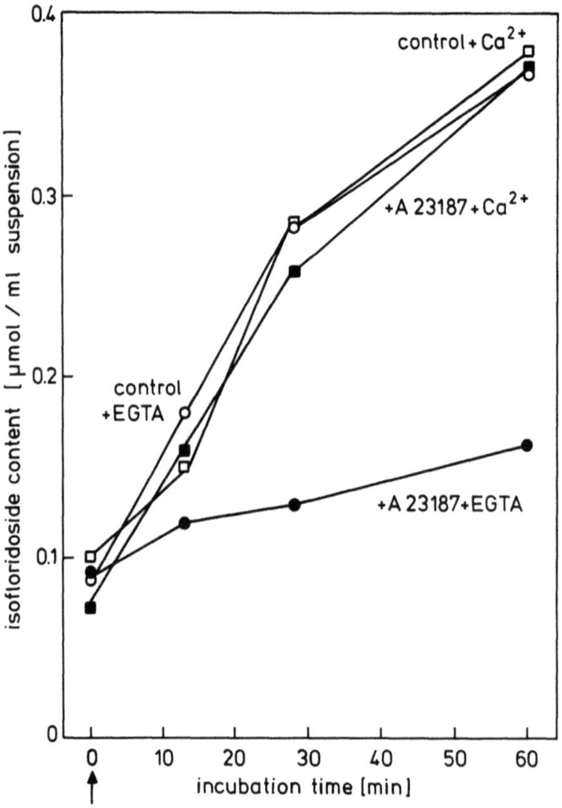

Fig. 4. Influence of external Ca^{2+} and ionophore A 23187 on the production of IF after osmotic shrinkage of *Poterio-ochromonas*. The cells were suspended in buffer + Ca^{2+} (■, □) or + EGTA (●, ○). The ionophore (100 μM) was added to the respective samples (●, ■) 15 min before the osmotic value was increased from 65 to 180 mosmolal by addition of sucrose solution (zero time, *arrow*). 0.1% DMSO had no influence on IF-production (data not shown)

Table 1. Influence of the Ca^{2+}-ionophore A 23187 (100 μM) on Ca^{2+}-fluxes. The gain or loss of $^{45}Ca^{2+}$ per time was constant up to about 18 min; the slope of curves similar to that of the control of Fig. 7 were used to calculate the values for efflux and influx

Additions	Ca^{2+}-efflux [%]	Ca^{2+}-influx [%]
None	100	100
A 23187 (in DMSO)	187	178
0.1% DMSO	102	124

to be important for the induction of IF-biosynthesis, then again it must come from internal pools from which it might be liberated as a consequence of cell shrinkage. Presence of the Ca^{2+}-ionophore A 23187 under conditions of high external Ca^{2+} concentrations had no significant influence upon either the IF-level at zero time and upon the speed of IF-formation (data not shown). It appears that the ionophore-induced increase in Ca^{2+}-influx (Table 1) is readily buffered by increased uptake into organelles. With an external EGTA plus ionophore, however, the rate of IF-formation after shrinkage is low (Fig. 4), indicating that the presumed increase of $[Ca^{2+}]_{cyt}$ caused by shrinkage cannot occur under these conditions as the Ca^{2+} is now flowing out of the cells with increased rates (Table 1) and is trapped by the EGTA.

Influence of the Fungicide DCF

Certain herbicides and fungicides affect Ca^{2+} transport of plant mitochondria (Hertel et al. 1980) and act upon the integrity of flagella in *Chlamydomonas,* a process assumed to be under control of $[Ca^{2+}]_{cyt}$ (Quader and Filner 1980). We investigated, therefore, whether such substances might have an influence upon osmotically induced IF-metabolism. The parameter observed was the activity of IFP-synthase because it had previously been demonstrated that this enzyme exhibits increased activity shortly after cell shrinkage (Kauss 1982). Best results were found with the fungicide DCF (Fig. 5). At concentrations around 80 μM this substance can partly mimic the osmotically induced shrinkage, and this is counteracted when cells are preincubated with nupercaine, a local anesthetic known to inhibit Ca^{2+}-fluxes by stabilizing membranes (e.g., Beatrice et al. 1980). The action of the fungicide can only be seen if Ca^{2+} is present outside at higher concentrations whereas it is absent in buffer with EGTA (Fig. 6). Immediately after addition of DCF the rate of $^{45}Ca^{2+}$ uptake increases dramatically (Fig. 7). These results can be taken as evidence that the fungicide might effect the IFP-synthase activation system by increasing the Ca^{2+}-influx from outside and thus leading to an increased $[Ca^{2+}]_{cyt}$.

The results of Figs. 5 and 6 should, however, be interpreted with caution. At concentrations where clear effects are seen on the activation system in vivo the substance already appears to be toxic to some extent. This is indicated by cell damage, a process which was quantitatively followed by measuring protein in the suspension buffer and was found to start about 4 to 5 min after the addition of the fungicide

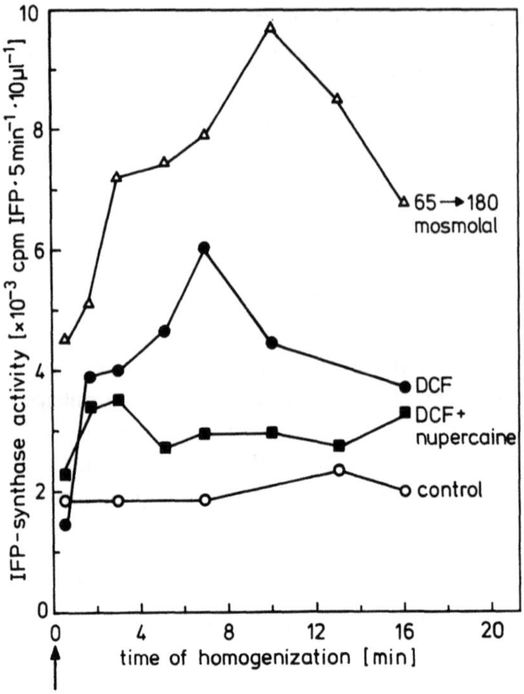

Fig. 5. Influence of DCF, nupercaine and osmotic shrinkage on the activity of the IFP-synthase. Cells in buffer + Ca^{2+} were used as such (○, ●, △) or preincubated for 15 min with 100 µM nupercaine (■). At zero time (*arrow*) 80 µM DCF (●, ■) in DMSO, or sucrose solution was added to increase the osmotic value from 65 to 180 mosmolal (△). Not shown are control samples, where DMSO (0.1%, v/v) was added alone or where cells were preincubated with nupercaine alone; these controls were similar to the control without any treatment (○)

(Fig. 7). The protein leakage is also decreased by nupercaine to an extent similar to IFP-synthase activation in Fig. 5, but the leakage remained the same in the presence and in the absence of Ca^{2+} (data not shown). It thus appears that DCF could also have non-specifically affected and stimulated the membrane-bound activation system. It might be that within the first 4 min, where protein leakage cannot be seen, the increase in Ca^{2+}-influx is responsible for the observed onset of activation and that this might be followed by non-specific membrane destruction and cell lysis which also appears to be responsible for the often sharp breakdown of the system (e.g., after 6 min in Fig. 6). We have tried to overcome these problems by using lower concentrations of DCF (8 to 10 µM), which were found by the $^{45}Ca^{2+}$-technique also to increase the overall Ca^{2+}-uptake (data not shown). However, under such conditions no effects on IFP-synthase activation were seen. Such negative results must not devaluate the above discussed findings, as they merely may mean that at low concentrations of DCF the increased influx of Ca^{2+} is rapidly nullified by Ca^{2+}-compartmentation which does not allow a transient increase of $[Ca^{2+}]_{cyt}$. This appears understandable as various Ca^{2+}-fluces cooperate to create a low $[Ca^{2+}]_{cyt}$. If one of them is manipulated experimentally, the others are likely to be regulated to counteract. Only at the limits of non-physiological conditions (e.g., the high DCF concentrations used in Figs. 5, 6 and 7) may the equilibrium be shifted to result in a change in $[Ca^{2+}]_{cyt}$ followed by effects on metabolism.

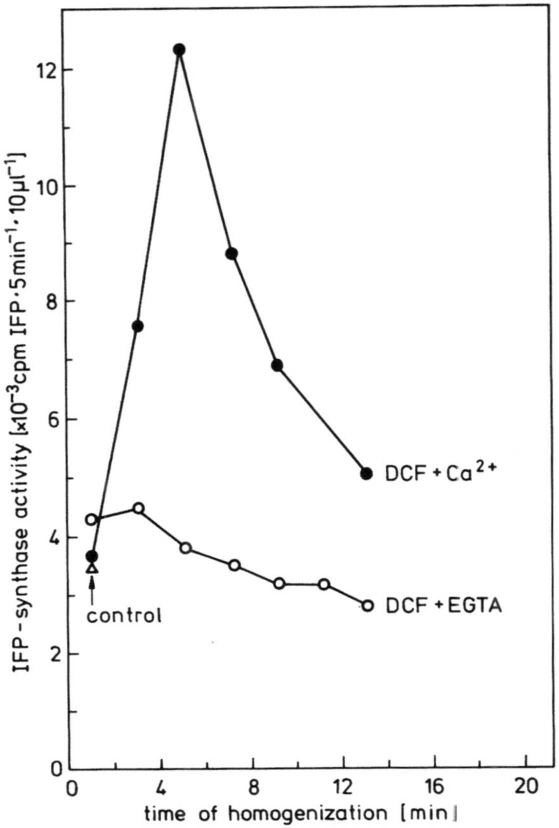

Fig. 6

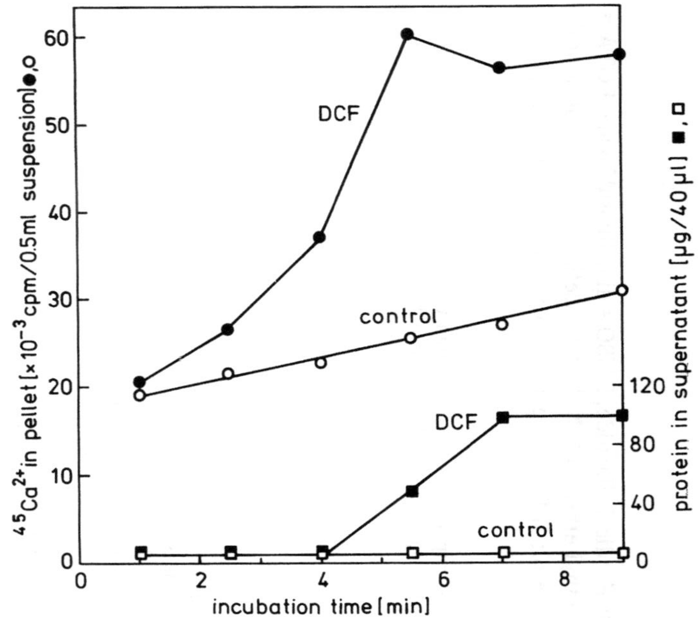

Fig. 7

Fig. 6. Influence of external Ca^{2+} on the activation of IFP-synthase by DCF. Same conditions as for Fig. 5. In one sample (○) the external Ca^{2+} was lowered to < 0.1 µM by addition of 3.5 mM EGTA, whereas the other (●) contained 1.5 mM Ca^{2+}. DCF at 100 µM was added at zero time; △ denotes a control sample without any treatment

Fig. 7. Influence of DCF on Ca^{2+}-influx and cell integrity. DCF (80 µM) was added at zero time together with $^{45}Ca^{2+}$ and the total radioactivity in the algal pellet determined (●, ○). The relatively high value for $^{45}Ca^{2+}$ which results when the control curve is extrapolated in zero time represents that part of the water phase which is carried by the cells through the silicon layer. Cell integrity was determined by measuring the protein content in the supernatant (□, ■)

Abbreviations

CTC = chorotetracycline, DCF = dichlofluanide, DMSO = dimethylsulfoxide, EGTA = ethylene glycol-bis-(2-aminoethyl ether)-N,N,N',N'-tetraacetic acid, IF = isofloridoside = α-D-galactosyl-1→1-glycerol; IFP = α-D-galactosyl-1 → 1-glycerol-3-phosphoric acid.

Acknowledgments. This work was supported by a grant from the Deutsche Forschungsgemeinschaft. The helpful discussions of H. Quader and the excellent technical assistance of W. Jeblick are gratefully acknowledged.

References

Beatrice MC, Palmer JW, Pfeiffer DR (1980) J Biol Chem 255:8663–8671
Blinks JR, Wier WG, Hess P, Prendergast FG (1982) Prog Biophys Mol Biol 40:1–114
Bradford MM (1976) Anal Biochem 72:248–254
Hertel H, Quader H, Robinson DG, Marmé D (1980) Planta (Berl) 149:336–340
Jewell BR, Rüegg JC (1966) Proc R Soc B 164:428–459
Kauss H (1973) Plant Physiol (Bethesda) 52:613–615
Kauss H (1981) Plant Physiol (Bethesda) 68:420–424
Kauss H (1983) Plant Physiol (Bethesda) 71:169–172
Kauss H, Thomson KS (1982) In: Marmé D, Marré E, Hertel R (eds) Plasmalemma and tonoplast: their function in the plant cell. Elsevier Biomedical Press, NY, pp 255–262
Quader H, Filner P (1980) Eur J Cell Biol 21:301–304
Reiss HD, Herth W, Schnepf E, Nobiling R (1983) Protoplasma 115:153–159
Ringboom AJ (1979) Complexation in analytical chemistry. Krieger Publishing Co, New York, p 42
Robinson DG, Quader H (1980) Planta (Berl) 148:84–88
Schobert B, Untner E, Kauss H (1972) Z Pflanzenphysiol 67:385–398
Thomas MV (1982) Techniques in calcium research. Academic Press, London, pp 125–128
Williamson RE (1981) What's New in Plant Physiology 12:45–48
Williamson RE, Ashley CC (1982) Nature (Lond) 296:647–651

Interaction Between Chloroplast-Cytoplasm Vacuoles with Respect to the Regulation of Nitrogen Metabolism in *Chlorella*

R. TISCHNER[1]

Introduction

Twenty percent of the electrons produced by the photosynthetic apparatus of plants are used for the reduction of nitrate. This pathway is thus the second most important electron-consuming process after carbon dioxide fixation. Therefore it is of general interest to clarify the problems involved in nitrogen metabolism.

Materials and Methods

All experiments presented were carried out with *Chlorella sorokiniana* (strain 211-8k from Göttingen algae collection) and *Chlorella vacuolata* (strain 211-11n). The algae were synchronized in LD (light-dark change) of 0705 h and 1212 h respectively. The mothercells produce 8 or 16 autospores per life cycle. Other methods have been published elsewhere: nitrate uptake measurements (Tischner and Lorenzen 1979); enzyme assays (Tischner 1976); antibody treatments (Tischner 1983) and the thioredoxin-activating procedure (Tischner and Schmidt 1982).

Results and Discussion

The uptake of nitrate into the cells has been demonstrated to be a metabolically regulated process (Tischner and Lorenzen 1981). There is some evidence that the uptake of nitrate is energy-dependent. Of course the uptake of this anion is also dependent on the nitrate concentration outside the cells. The K_m was estimated to be 4.3 μM (Fig. 1). A multiphasic uptake kinetic as reported by Nissen and Breteler (1978) was never obtained. This was also the case in *Chlorella vacuolata,* even after the vacuoles of these cells have been increased by limited CO_2 supply. The temperature dependent uptake rate (Fig. 2) is presented as an Arrhenius plot and the values for Q_{10} and the activation energy point to an energy dependent uptake process.

1 Pflanzenphysiologisches Institut, Universität Göttingen, Untere Karspüle 2, 3400 Göttingen, FRG

Compartments in Algal Cells and Their Interaction
ed. by W. Wiessner, D. Robinson, and R.C. Starr

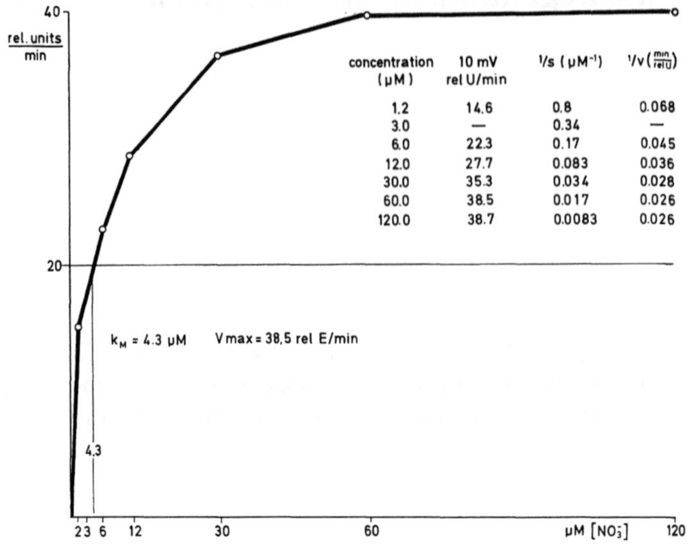

Fig. 1. Nitrate uptake in relation to nitrate concentration K_m and V_{max} are calculated

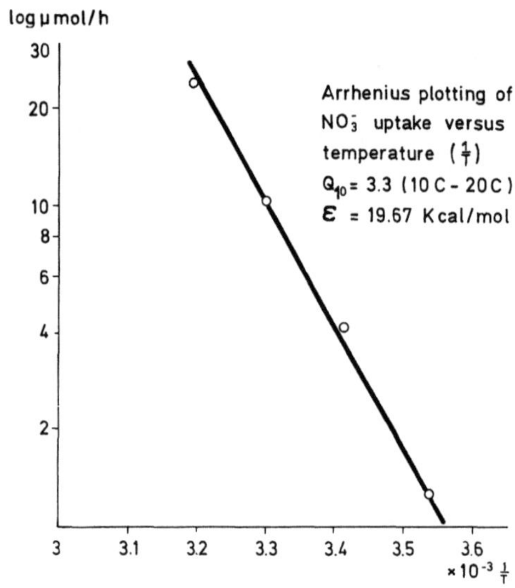

Fig. 2. Arrhenius plot of nitrate uptake, Q_{10} and ϵ are calculated

In *Chlorella vacuolata,* cells which were cultivated in the presence of nitrate, this anion was stored in the vacuoles similar to cells of higher plants. But this nitrate is not available at once for reduction to nitrite. If cells are transferred from nitrate medium into nitrogen free medium the activation of the nitrate reductase by light is inhibited. Only after the addition of nitrate, which is freshly taken up, is an activation of nitrate reductase to be detected (Table 1).

Table 1. Nitrate concentration dependency of light-dependent nitrate reductase activity

NAR[a]-activity in the dark	NAR-activity in the light	NO_3^- conc. in the medium	NO_3^- conc. in the cells
3%	100%	10 mM	20 nM
3%	4%	none	20 nM

[a]NAR: nitrate reductase

Such a separation of the nitrate content in a cell into a metabolic and a storage pool was also reported by Heimer and Filner (1971) for the cells of higher plants. The metabolic pool of nitrate is reduced to nitrite within the cytoplasm. The enzyme which catalyses the first reaction from nitrate to nitrite is nitrate reductase. This is a flavomolybdo-protein, which is capable of catalysing four different partial reactions. These are: NADH-NAR, FMN/dithionite-NAR, MV/dithionite-NAR and a diaphorase activity, commonly measured as cytochrome c reductase. After onset of the illumination the specific activity increased 30-fold in *Chlorella sorokiniana* (Tischner 1976). This increase is light-mediated without including de novo synthesis as we demonstrated earlier in density labeling experiments. Unfortunately this activation was never obtained in vitro. All the experiments reported by Pistorius et al. (1978), suggesting an oxidation/reduction of the enzyme as regulating steps or by Losada et al. (1979), who mentioned that NH_4^+ was the regulating intermediate of NAR, deal with NAR obtained from cells which were incubated with ammonium. They do not take into account the activation of NAR by light in the presence of nitrate. Therefore the activating procedures used by these workers were without effect on the activity of NAR from darkened cells in our experiments.

Thus we have followed another strategy with the intension of comparing both forms of NAR, that from cells after 1 h illumination and from darkened cells respectively.

The four partial reactions are present in both forms of the enzyme. But the rate of these reactions differ markedly and points to a molecular change in NAR after switching on the light (Tischner 1981).

This result was confirmed by the use of antibodies against the active form of NAR. In Ouchterlony double-diffusion tests the antibody did not precipitate with the inactive form of NAR. Even a wide range of concentrations was used (Fig. 3). Further we were able to demonstrate a precipitation with NAR from cells treated with a LD of 1:1 h. After extensive darkening no precipitation was obtained. From these results we assume that no structural differences exist in the forms of NAR which prevent their precipitation.

On the other hand, a precipitation of the active form was obtained independent of its redox state. The oxidation was obtained with ferricyanide, the reduction with dithionite (Fig. 3) Thus the regulation involving both processes is not identical to that occurring after LD.

We conclude that the light-mediated activation of NAR involves structural changes of the enzyme. This was supported by the effect of the antibody on the par-

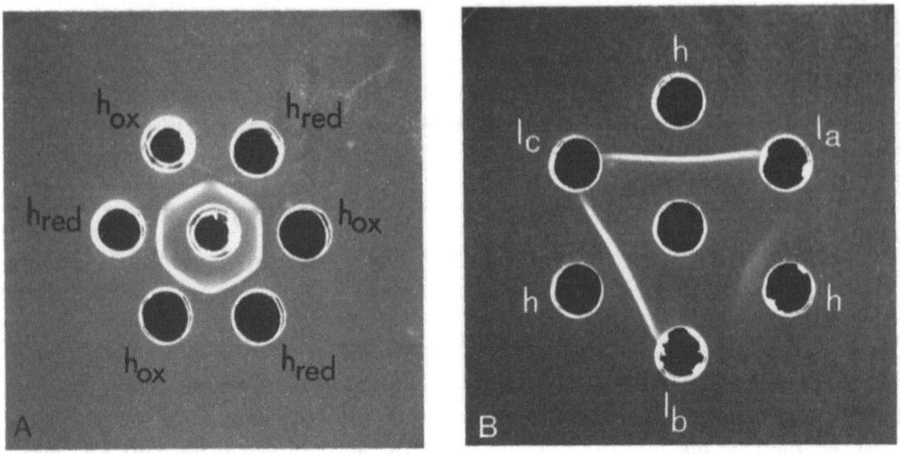

Fig. 3A,B. Diffusion of different concentrations of oxidized (h_{ox}) and reduced (h_{red}) nitrate reductase (high active form) against the antiserum (*central well*); **B** Diffusion of different concentrations of the low active form (*l*) and the high active form (*h*) of nitrate reductase against the antiserum (*central well*). l_a 10-fold concentrated; l_b the same protein content as (h); l_c the same activity as (h)

tial reactions. In both forms of the enzyme all reactions involving NO_3^- reduction were inhibited but that of diaphorase activity was not (Figs. 4, 5). In the inactive form the partial reactions were more sensitive to the antibody inhibition, especially those with NADH and FAD.

These results are contrary to those reported by Funkhouser (1980), who obtained no different sensitivity in the partial reactions against an antibody treatment. However, his experiments all dealt with a transition from NO_3^- to NH_4^+ medium without LD.

We suggest that after onset of the illumination the nitrate-binding site of the nitrate reductase is activated. This may be induced by an increase in the nitrate flux into the cells. An argument in favor of this suggestion was obtained from a preincubation of the enzyme with NO_3^- prior to the antibody treatment. Such a preincubation results in a protection of the enzyme and was more effective with NO_3^- (Fig. 6) than with NADH (Tischner 1983).

After the reduction of NO_3^- to NO_2^-, the latter is transported into the chloroplast. Within the chloroplast a further reduction to ammonium occurs, connected with an electron consumption of 6 e^- per NO_2^- reduced. The resulting NH_4^+ is immediately incorporated into aminoacids. The most important pathway is that via the GS/GOGAT system. The presence of this pathway in algae was reported by Lea and Miflin (1975). With onset of the illumination (1 h) the specific activity of GS increases nearly ten fold. From density labeling experiments we were able to demonstrate that the light-mediated increase is not due to a de novo synthesis. This increase is insensitive to lincomycin, rifampicin and cycloheximide and therefore not connected to 70S, 80S translation or chloroplastic transcription (Tischner and Hüttermann

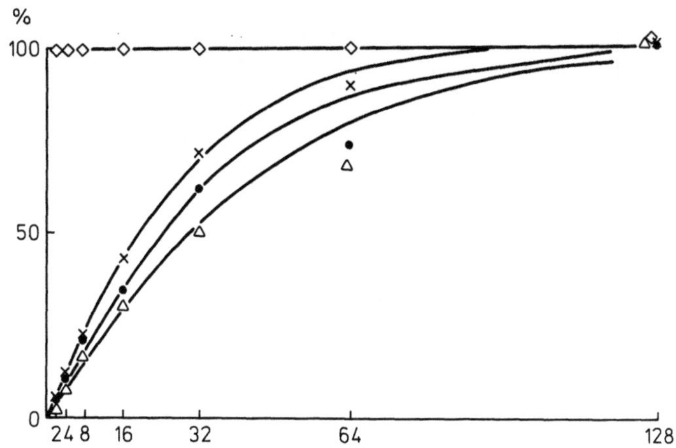

Fig.4. The effect of different dilutions of antiserum (*abscissa*) on the activity of the partial reactions of nitrate reductase (high active form) —●— NADH-NAR, — x — FMN-NAR MV-NAR —△— diaphorase —◇—

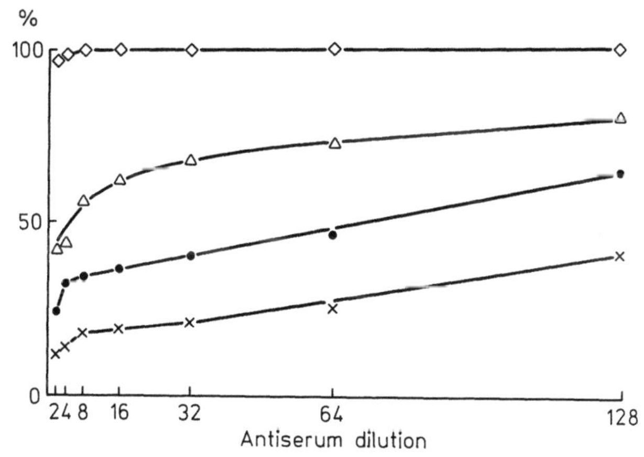

Fig. 5. The effect of different concentrations of antiserum on the activity of the partial reactions of nitrate reductase (low active form)
—●— NADH-NAR,
— x — FMN-NAR
—△— MV-NAR
—◇— diaphorase

1980). We have also succeeded in activating both GS and GOGAT in vitro. Crude extracts, as well as purified enzyme preparations from darkened cells, were incubated with both thioredoxin and dithioerythritiol (Fig. 7). Therefore we obtained an activation in the same range as observed in vivo. An activation of both NAR and nitrite reductase never occurred, even with different concentrations of thioredoxin or dithioerythritiol.

Our results presented here are combined to give a model of the regulation of nitrate metabolism in *Chlorella* (Fig. 8). We suggest that the direct reduction of thioredoxin from ferredoxin or by a NADP-thioredoxin-reductase, as shown by Tsang (1981), represents the direct effect of light on nitrogen metabolism. GS/GOGAT are immediately activated by thioredoxin and, due to the low K_m of GS for NH_4^+, the

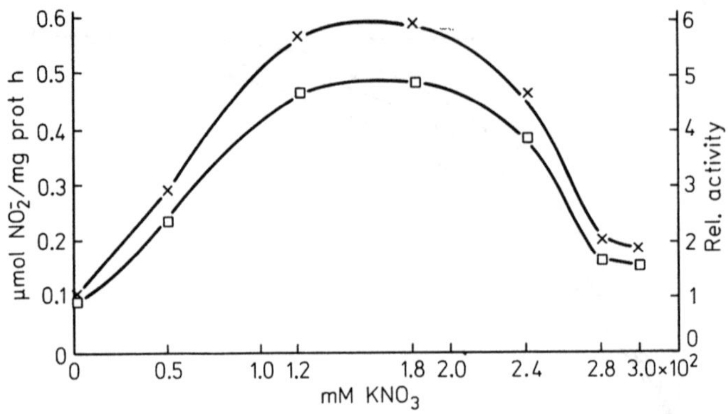

Fig. 6. The effect of a preincubation of NAR (high active form) with different concentrations of KNO3 on the inhibition by the antiserum. —x— specific activity, —□— relative activity

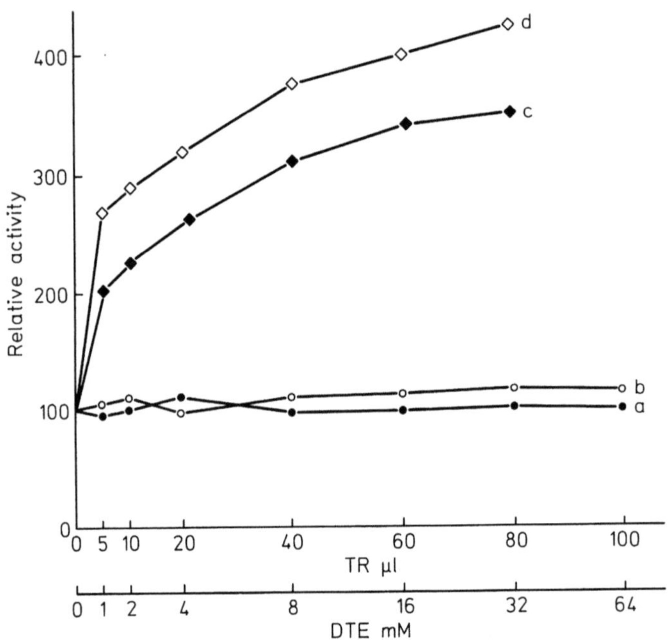

Fig. 7. The action of different amounts of thioredoxin, dithioerythreitol and both together on glutamine synthetase inactivated in vivo. *a* preincubation with thioredoxin; *b* preincubation with DTE; *d* preincubation with both (5 mM DTE). thioredoxin 1.3 mg/ml, GS 100% = 0.18 U/mg pool

NH_4^+ probably present in the cells becomes fixed. Thus N-deficiency stimulates the nitrate uptake mechanism. Freshly taken up nitrate activates the enzymes of the nitrate reducing system. Thus, after thioredoxin (red), the NH_4^+ pool if of importance

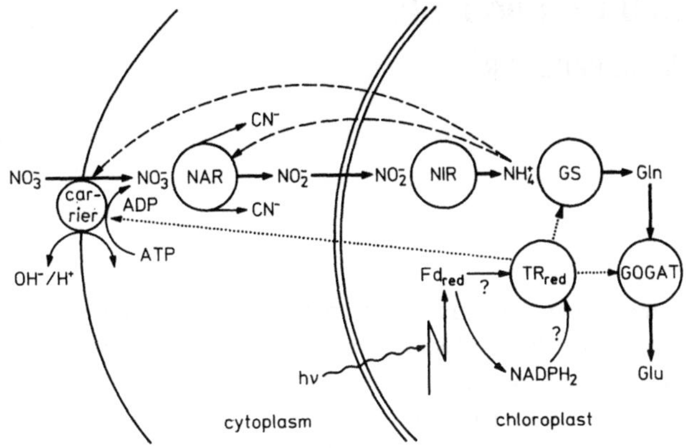

Fig. 8. A model for the regulation of NO_3^--metabolism via the action of thioredoxin on GS. For the mechanism please compare the text. *Dotted line* stimulation; *dashed line* inhibition

in our model, which shows that nitrate metabolism occurs by an interaction between three compartments of the plant cell, with the chloroplast as the regulating center.

Similar proposals have been made by Cullimore and Sims (1981) for the regulation of the nitrate metabolism of *Chlamydomonas*; although thioredoxin was not involved in their suggestion.

References

Cullimore JV, Sims AP (1981) Planta (Berl) 153:18–24
Funkhouser EA (1980) Plant Physiol (Bethesda) 65:944–948
Heimer YM, Filner P (1971) BBA 230:362–372
Lea PJ, Miflin BJ (1975) Biochem Biophys Res Commun 64:856–862
Losada M, Guerrero MG (1979) In: Barber J (ed) Photosynthesis in relation to model systems. Elsevier, Amsterdam, pp 365–408
Nissen P, Breteler H (1978) Poster 40A FESSP meeting Edinburgh
Pistorius E, Funkhouser EA, Voss H (1978) Planta (Berl) 114:279–282
Tischner R (1976) Planta (Berl) 132:285–290
Tischner R, Lorenzen H (1979) Planta (Berl) 146:287–292
Tischner R, Hüttermann A (1980) Plant Physiol (Bethesda) 66:805–808
Tischner R (1981) Ber Dtsch Bot Ges 94:635–645
Tischner R, Lorenzen H (1981) In: Bothe H, Trebst A (ed) Biology of inorganic nitrogen and sulfur. Springer, Berlin Heidelberg New York, pp 252–259
Tischner R, Schmidt A (1982) Plant Physiol (Bethesda) 70:113–116
Tischner R (1983) Planta, in press

Mitochondria and Their Interaction with Other Cell Components

W. DE PRIESTER[1]

Intracellular membrane systems can be considered to belong to either the endoplasmic or the explasmic space (de Duve 1969). Within each of these, compartments may interact by membrane fusion, enabling transport of their contents. Connections between compartments may have a permanent character, as appears to be the case with the nuclear envelope and the endoplasmic reticulum, or they may be discontinuous via transitory vesicles, budding of from one compartment and fusing with another, e.g., the shuttle vesicles in the ER-Golgi area. Compartments may also be subdivided into many elements, each of which may fuse with elements from a compartment belonging to the same space. One may think of individual lysosomes fusing with endocytic vacuoles. The postulated borderline between the endoplasmic and the exoplasmic space is located within the Golgi apparatus, which would serve as a one-way lock, allowing transport from the endoplasmic to the exoplasmic space only. One of the differences between these is the average thickness of their limiting membranes in material fixed with OsO_4-containing fixatives. The average thickness of the exoplasmic membrane is significantly greater than that of the endoplasmic membranes.

Mitochondria have been set apart in this system as a special case. Although on the basis of their membrane thickness they would fit in the endoplasmic space, the lack of consensus about their interactions with other compartments and their possible evolutionary origin does not allow a decision yet.

Some of the early literature reports on their ultrastructural interactions will be dealt with and compared with recent data. Since this is not to be an exhaustive review of the literature on this subject, a limited number of cases have been selected, and the references given for each topic reflect the author's personal interest only.

Interaction with the Endoplasmic Reticulum

One of the most intriguing problems in research on mitochondria is, no doubt, the biogenesis of these organelles. It is known from biochemical studies that mitochondrial proteins are degenerated rather rapidly (Doyle and Tweto 1975), and synthesis of such proteins is therefore necessary to maintain the normal mitochondrial volume and activity. It is generally assumed that this is taking place both within the mitochondrial compartment itself and in the cytoplasm, either on "free" or ER-bound

1 Laboratory of Zoology, University of Leiden, Kaiserstraat 63, 2311 GP Leiden, The Netherlands

Compartments in Algal Cells and Their Interaction
ed. by W. Wiessner, D. Robinson, and R.C. Starr

ribosomes. Although it has seldom been formulated explicitly, these two possibilities are often suggestively connected with mitochondriogenesis and growth, the "ER model" implicating transport of membrane-bound proteins from ER towards mitochondria, the "free ribosome model" favoring the concept of more or less autonomous mitochondrial behavior, without involvement of other cellular membrane systems. When reading the early and the more recent literature on these alternatives, it bebecomes clear that evidence for both can be derived from ultrastructural and from biochemical studies. It is also striking that most data have been collected from only two systems, namely the rat hepatocyte and *Saccharomyces.* This makes it hazardous to draw general conclusions, even if the observations on these systems fully agreed, which is not always the case.

Connections between mitochondria and endoplasmic reticulum membranes have been described in early studies on the ultrastructure of several cell types, notably in undifferentiated or fast-growing cells like oocytes, and have led to much speculation about the involvement of ER membranes in the de novo formation of mitochondria (Schjeide et al. 1963, 1964, Ruby et al. 1969). Experimental studies in systems with stimulated proliferation and cell growth were undertaken to further investigate this.

Early observations on regenerating rat liver (Bade 1964, Claude 1967) seemed to confirm the existence of continuities between the outer membrane of mitochondria and endoplasmic reticulum. In another experimental study Scarpelli et al. (1971) showed that the synthesis of outer mitochondrial membrane material exceeded that of the inner membrane in the presence of chloramphenicol, an inhibitor of the mitochondrial synthesizing apparatus. The resulting surplus membrane could clearly be seen as extensions of the outer mitochondrial membrane, suggesting transport from the endoplasmic reticulum. From a morphological point of view, it seemed an attractive idea that the outer membrane would be synthetized as a structural and functional entity and inserted in existing mitochondrial structures either for growth or replacement. We now know from many studies on the synthesis, transport, processing, and recycling of membrane components that this has been too static a view. Nevertheless, the notion that membrane material could be synthesized in the endoplasmic reticulum and subsequently transported via temporary connections to the mitochondria, was quite in keeping with many biochemical data from that period suggesting that the reticulum is involved in the synthesis of at least some "insoluble" mitochondrial membrane proteins (cf. Roodyn and Wilkie 1968). In later years, however, surprisingly little evidence was put forward in favor of an endoplasmic space extension towards the mitochondrial compartment. In itself, this does not say very much. If the transport took place via transitional elements, this would be extremely difficult to prove on the basis of "stationary" EM pictures. Autoradiography can, in principle, be used to demonstrate a spatial distribution of proteins from their site of synthesis, as has been done in the case of secretory proteins, by providing the cell with labeled precursor amino acids. In the case of mitochondrial synthesis, this technique was used in rat liver (Bergeron and Droz 1969) and *Neurospora* (Luck 1963, 1965) to obtain information about the temporal distribution of the incorporation in mitochondrial structure. The results, a relatively constant rate of incorporation over a long period implicating the entire population of mitochondria, were considered to indicate mitochondrial synthetic activity, compatible with the concept of mitochondria as autonomous organelles.

The situation in yeasts appeared to be just as confused. Under anerobic conditions *Saccharomyces* shows a reduced number of mitochondrial structures, in which the cristae are only poorly developed. Concomitantly, the number of cytoplasmic membranes is increased, suggesting some relation between these two sytems (Avers 1967, Kellerman et al. 1969). Moreover, the enzyme cytochrome peroxidase, which is present in mitochondria in aerobically grown wild-type cells, could be demonstrated cytochemically in the cytoplasmic membrane profiles both in anaerobic wild-type cells and in aerobic "petite" cells (a mutant with a much reduced mitochondrial population) (Avers 1967). Transfer to aerobic culture conditions restored the normal mitochondrial profiles in wild-type cells with a concomitant reduction of the cytoplasmic membrane strands. These results were in accordance with the hypothesis that the formation of mitochondrial structures would involve a rearrangement of cytoplasmic (ER type) membranes, in which mitochondrial components had been assembled already. The whole problem was reconsidered when freeze-substitution electron microscopy and freeze-etching technique showed that under low oxygen tension, small vesicles were present, closely resembling mitochondria, but lacking the clear definition of their crista structure (Plattner and Schatz 1969, Plattner et al. 1970). These "premitochondria" were shown to convert into normal mitochondria when the cells adapted to aerobic conditions. This lent support to the view that mitochondria do not arise from other membrane systems, but from preexisting (pre)mitochondrial structures.

It should be realized, when trying to evaluate these data, that both in the rat liver and in the yeast system one is dealing with a transitory phase, in which adaptive changes may be superimposed on normal processes. For a more extensive treatment of these problems the reader is referred to Dillon (1981), who also reviews the earlier literature.

From recent literature it becomes clear that the controversy still exists. Although both the biochemical and the morphological approach can now rely on more refined techniques, the crucial problem: are non-mitochondrial membranes in some way involved in the synthesis, transport and insertion of mitochondrial proteins in the living cell? has not been approached in a direct way, most investigations being carried out to collect information on the synthesis and insertion of the protein component only.

In in vitro studies on rat liver, for instance, it was shown by Shore (1979) that the synthesis of a number of outer mitochondrial membrane proteins is greatly enhanced by the presence of ER membrane, from which they can also be recovered. This would suggest that some ER-bound proteins are in some way transported to the mitochondrial compartment.

In similar studies on *Saccharomyces* by Riezmann et al. (1983), no proteins were found to be shared by the outer mitochondrial membrane and other membrane systems, such as the ER. Binding of newly synthesized precursors takes places rapidly and specifically on the mitochondrial outer membrane. Whereas these kinds of experiments will eventually lead to an understanding of the molecular composition of the protein component and its turnover, the membrane entity as a whole may now be studied by electron microscopy using low denaturing embedding techniques. Recent developments in the applicability of immunocytochemistry on the EM level, allowing a reliable in situ localization of specific proteins within the mitochondrial compartment (Bendayan and Shore 1982), may give an opening towards a closer combination of morphological and biochemical techniques.

Motility

Since the motility of organelles plays an important part in enabling contacts between compartments, it seems appropriate to mention some data on the motility of the mitochondria and the influence of nonmembranous cell components.

The changes in shape and motility of mitochondria in cultured amphibian cells have been the subject of extensive studies by Bereiter-Hahn et al. (Bereiter-Hahn and Morawa 1972, Bereiter-Hahn 1978, Bereiter-Hahn and Vöth 1983). They distinguish four types of mitochondrial locomotion. (1) Alternate extension and contraction of the two ends of rod-shaped mitochondria. (2) Lateral branching. (3) A kind of peristaltic movement in arbitrary parts of the organelle. (4) Transverse wave propagation along the organelle. The shape changes appear to be related to changes in the arrangement of the cristae; both are influenced by inhibitors of respiration and it is suggested that diminished motility is connected with a more rigid conformation of the cristae and the inner mitochondrial membrane. It is interesting that the structure of the outer mitochondrial membrane, which in normal conditions is of the "thin" type, may convert to a clearly defined "unit-membrane", with a greater width. A similar transition has been described in insect tissue under conditions of low oxygen tension (de Priester 1972, 1980). An arrangement of mitochondria along microfilament bundles and microtubules is common in normal cells and the typical mitochondrial movement is restricted to areas in between the ER lamellae. This could mean that the motility is brought about by processes taking place within the mitochondrion, and that the direction in which movement takes place, is imposed by interactions with the cytoskeleton.

Contacts with Microtubules and Filaments

Light microscopic demonstration of associations of mitochondria and cytoskeletal elements has been made possible by the use of specific antibodies in immunofluorescence. In the case of microtubules, the interaction has been demonstrated in several cultured cell types (Heggeness et al. 1978). Using the same technique, Couchman and Rees (1982) suggested a possible role of microtubules in directing mitochondrial movement in growing fibroblast. Ultrastructural connections between mitochondria and microtubules have been demonstrated in *Tetrahymena* (Aufderheide 1979, 1980), where the distribution of peripheral mitochondria is clearly related to that of the microtubules. They have also been demonstrated in lamprey and cockroach axons by thin sectioning and freeze-fracturing techniques (Smith et al. 1977). The possible functional implications of the movement of organelles in general may be regulated by a cytoskeletal lattice; on the other hand, it may be involved in the regulation of mitochondrial function.

The molecular basis for microtubule-organelle interactions has recently been demonstrated by Bernier-Valentin et al. (1983), who found tubulin receptors in a number of organelles, including mitochondria.

Associations with filaments are often found near plasma membrane specializations, especially desmosomes. A good example is given by Forbes and Sperelakis (1982) in myocardial cells. In these cells the nexus area, with its prominent filamentous component, shows a high density of mitochondria and both in thin sections and in freeze-fracture replicas connecting structures could be seen between the mitochondria and gap junctions. In some cases, as in rat hepatocytes, the connections could be demonstrated under experimental conditions only (Guillouzo et al. 1978), in others they occur regularly. Even in highly specialized glandular tissue, like mammary gland epithelium, their is no consistency in this respect, as stated by Lee et al. (1979), who also discuss the possible significance in cell function.

Associations between mitochondria and filaments within the cytoplasmic matrix have been encountered less frequently (David-Ferreira and David-Ferreira 1980).

If connections (either continuous or transient) between mitochondria and cytoskeletal elements will turn out to be a general phenomenon, the implications with regard to interaction with other organelles are evident. The cytoskeleton would then provide an ordered system of transport lines along which mitochondria would move and make contact with other membrane systems. Such contacts would then not be random, and might be modulated by factors affecting adhesion and motility.

Fusion with Nonlysosomal Membranes

Fragmentation and fusion of mitochondrial structures are quite common in some cells and may be observed under the light microscope. It would not seem to be an obvious characteristic of tissue cells like striated muscle, in which the mitochondria maintain a fixed position. Nevertheless, it has been shown that in insect flight muscle the dimensions of individual mitochondria increase with age, whereas their number decreases (Sohal 1972). This appears to be related to changes in the lipid composition of the mitochondrial outer membrane.

A very interesting case is the fusion of mitochondria and plastids in barley leaf cells, described by Wellburn and Wellburn (1980), and with chloroplasts (Crotty and Ledbetter 1973). Although the evolutionary origin of the chloroplast outer membrane is by no means certain, there is a tendency to interpret it as derived from an exoplasmic "host" membrane. This would place it in the category of lysosomal and plasmamembranes.

Degeneration and Lysis

Whereas the mechanisms of mitochondrial biogenesis are still incompletely understood, the phenomena relative to degeneration and degradation have been described quite clearly. The changes in cellular ultrastructure ensuing from unphysiological or pathologic conditions, for instance under oxygen deprivation, cyanide or sublimate poisoning, virus infection, antibody treatment, have been treated by Trump et al. (1971). The consecutive stages of mitochondrial damage can be distinguished quite clearly from the normal "orthodox" state, on the basis of condensation or swelling of the mitochondrial subcompartments and formation of flocculent densities in the matrix. Some of these morphological characteristics are regularly found in thin sections of "healthy" cells, even when the fixation conditions have been optimal. This suggests that within the mitochondrial population "aging" may take place; the resulting variation in physiological conditions would then be reflected in the morphology.

The concept of aging in mitochondria may seem incompatible with the notion that mitochondrial material is continuously fed into the organelle,the growth of mitochondrial volume apparently exceeding the loss of functional capacity by protein degradation. It must be realized, however, that we have no evidence that mitochondrial components can be recycled completely within the compartment. This means that degradation products may accumulate, and adversely affect mitochondrial function.

However this may be, there is little doubt that entire mitochondria are eliminated even in normal conditions in many cells. The main problem to be solved concerns the question whether this is done selectively or not.

Autophagic Vacuoles

There is evidence from electron microscopic morphometric studies on autophagic activity in rat liver (Pfeifer 1978) that cytoplasmic organelles are degraded in autophagic vacuoles rather selectively. For instance, the rate at which mitochondria are degraded is 3.9% per day, whereas for microbodies a value of 8.9% was estimated. This means that autophagic vacuoles are not formed at random in the cytoplasm. The isolating membranes operating in the early stages of autophagy (Locke and Collins 1965) evidently distinguish between the two organelles. The different rates at which mitochondria and microbodies are degraded is in good agreement with the differences found with biochemical techniques by Doyle and Tweto (1975) for the protein breakdown. The absolute values given by these authors, however, are higher (10-15% per day for mitochondria and 20-30% for microbodies). This implies that autophagic vacuoles can account for roughly 50% of organelle turnover in terms of protein degradation. Other mechanisms must then be responsible for the other half. One may think, for instance, of degradation within the mitochondrial compartment itself. This would imply the presence of enzymes that are generally assigned to the lysosomal compartment. Although catalytic enzymes do occur in the mitochondria (e.g., phospholipase) mitochondria show only very little morphological signs of degradation when kept in vitro, even for prolonged periods. In contrast, when injected intravenously in rats, they are taken up by macrophages in the liver and digested in the lysosomal system very fast (Glaumann et al. 1975). This suggests that intramitochondrial degradation is either highly selective or takes place at a low rate.

Direct Fusion?

Although the possibility cannot be ruled out that katalytic activity connected with protein turnover is a normal phenomenon in mitochondria, an alternative may be found in the transport of lytic enzymes to the mitochondria by direct fusion with lysosomes. In that case only a part of the population would be affected and worn-out mitochondria could be selectively eliminated. This mode of elimination has been described among others in insect midgut cells (de Priester 1972, 1980) in a specific developmental stage of the blowfly *Calliphora.* A decrease in the number of mitochondria is accompanied by an increase in lysosomal structures (characterized cytochemically by their acid phosphatase activity). No isolation bodies or structures resembling autophagic vacuoles are found at this stage and lytic activity can be demonstrated in structures closely resembling mitochondria (Fig. 1). Concomitantly, some of these structures show an increase in the thickness of their outer limiting membrane, which then falls within the same dimensions as the lysosomal membrane. Areas of mitochondrial membrane of the thick variety may represent lysosomal membrane material inserted during a fusion or, alternatively, the acquisition of a thick membrane may be a prerequisite for fusion with lysosomal elements.

This mechanism of elimination of mitochondria will, of course, easily escape detection, especially when occurring in conjunction with autophagic vacuoles. Fusion with lysosomes is seldom seen in EM pictures and the resulting secondary lysosomes will closely resemble the lamellated bodies, known to be one of the varieties of residual bodies in which undigested membrane material may be stored.

Fig. 1. Aryl sulphatase reaction performed on *Calliphora* (Diptera) midgut epithelium to demonstrate lysosomal enzyme activity in mitochondrial structures. The organization of the mitochondrial membranes suggests incipient degradation. Since these structures did not contain other recognizable cell components, and since isolation bodies or autophagic vacuoles have not been found, it is assumed that a direct fusion of mitochondria and small, primary lysosomes takes place in these cells. (x ca. 34,000)

Fig. 2. Homogenate of *Phormia* (Diptera) salivary gland tissue. The electron-dense material represents secretory granules. Whereas in fixed-tissue sections these are round and do not exceed ca. 1 μ in diameter, here they exhibit irregular outlines and, as judged by their volume, they tend to fuse with one another. The coalescent granules are often in close contact with mitochondria, one of which appears to be engulfed during the process. (x ca. 26,000)

Fig. 3. Engulfment of organelles is not restricted to mitochondria, as illustrated by this picture of an endoplasmic reticulum profile, surrounded by granule material. (x ca. 52,000)

Fig. 4. Fusion of a mitochondrion (characterized here by a dense precipitate of cytochrome oxidase reaction product) with a protein granule in *Calliphora* fat body under anaerobic condition. The fusion process involves the outer mitochondrial membrane only. (x 14,500)

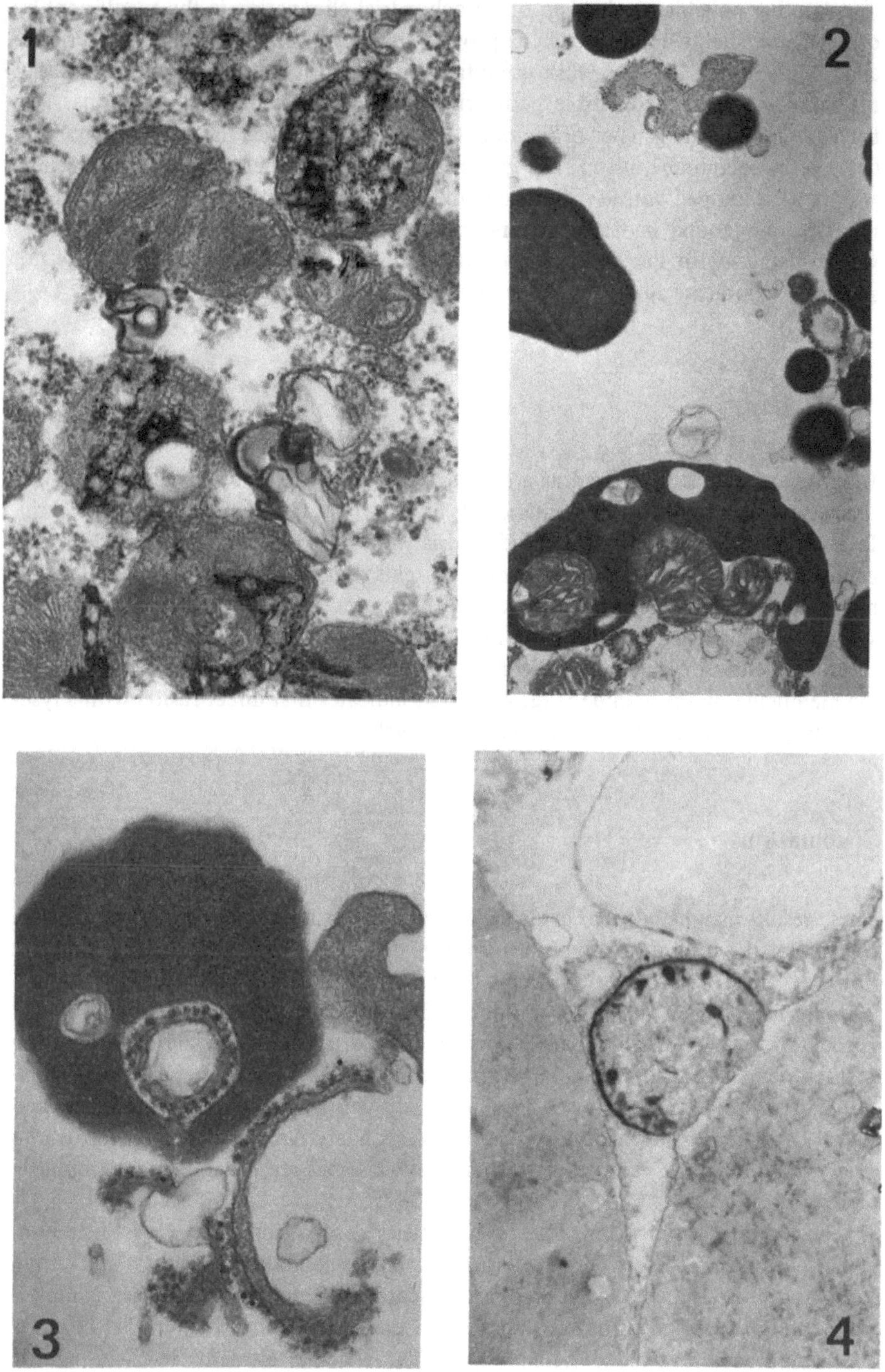

Figs. 1-4. (captions see opposite page)

Another possible mechanism of mitochondrial elimination is the engulfment by other membrane structures, such as protein granules in insect fat body (de Priester and van der Molen 1979). In this case intact mitochondria are enclosed within another membrane-bounded exoplasmic compartment. In tissue homogenates this has also been found. In homogenates of *Phormia* (Dipt.) salivary glands the secretory granules coalesce by membrane fusion. The mitochondria closely adhere to these coalescences and may be trapped within (Fig. 2, unpubl., Persoon and Rothert). In vivo, however, this was only found in tissue undergoing histolytic changes. Since similar pictures were obtained with the endoplasmic reticulum (Fig. 3), these phenomena are not supposed to indicate an organelle-specific interaction.

Anoxia

Whereas anoxia has been used as a means to induce autophagic activity in vertebrate tissue, attempts to obtain the same results in *Calliphora* fat body were not successful. Instead, in a number of cases mitochondria were found closely apposed to protein granules. These granules contain storage proteins, but, in addition, they exhibit acid phosphatase activity. They may therefore be considered as representing secondary lysosomes, having fused already with preexisting primary or secondary lysosomal structures. The outer mitochondrial membrane was sometimes continuous with the limiting membrane of the granule (Fig. 4), which indicates that direct fusion with a lysosome may occur in experimental situations as well.

Speculations

These results may be taken as evidence that a direct fusion process can be responsible for part of the total mitochondrial turnover. If it amounts to 50% remains doubtful however, and it will be very difficult to investigate with morphological techniques. Autophagic vacuoles are clearly defined structures with an average lifetime in rat hepatocytes of ca. 15 h (Pfeifer 1978), which makes their detection and the estimation of their abundance quite possible, albeit time-consuming. Fusion between mitochondria and lysosomes, in contrast, gives no clearly defined morphologic structure, and it will be extremely difficult to prove that a secondary lysosome contains mitochondrial structures only, especially so when the lysosomes are engaged in heterophagic activity as well.

So far, it has been assumed that lytic activity derives from lysosomes. Since lysosomal enzyme precursors are synthesized in the endoplasmic reticulum and subsequently processed in the Golgi area, a transient connection between these compartments and the mitochondria (as described in the context of mitochondrial biogenesis), would acquire a broader significance. The vesicular elements in the Golgi-ER area, the protein contents of which vary according to their destination, must show a highly selective recognition mechanism in order to avoid mistakes when delivering their contents in other compartments. Little is known about the nature of these recognition mechanisms, but, hypothesizing, one might envisage a situation in which defects in the mitochondrial outer membrane structure make it recognizable to lysosomes, eliciting a membrane fusion. This would result in what seems to be the most efficient way of dealing with defect organelles, namely selective elimination by degradation in the lysosomal compartment.

References

Avers CJ (1967) J Bacteriol 94:1225–1235
Aufderheide K (1979) J Cell Sci 39:299–312
Aufderheide K (1980) J Cell Sci 42:247–260
Bade EG (1964) Z Zellforsch 61:754–768
Bendayan M, Shore GC (1982) J. Histochem Cytochem 30:139–147
Bereiter-Hahn J (1978) J Cell Sci 30:99–115
Bereiter-Hahn J, Morawa G (1972) Cytobiologie 6:447–467
Bereiter-Hahn J, Vöth M (1983) Biol Cell 47:309–322
Bergeron M, Droz B (1969) J Ultrastruct Res 26:17–30
Bernier-Valentin F, Aunis D, Rousset B (1983) J Cell Biol 97:209–216
Claude A (1967) Protoplasma 63:275–282
Couchman JR, Rees DA (1982) Eur J Cell Biol 27:47–54
Crotty WJ, Ledbetter MC (1973) Science (Wash DC) 182:839–841
David-Ferreira KL, David-Ferreira JF (1980) Cell Biol Int Rep 4:655–662
De Duve C (1969) In: Dingle JT, Fell HB (eds) Lysosomes. North-Holland Amsterdam, pp 3–40
De Priester W (1972) Z Zellforsch 129:430–448
De Priester W (1980) In: Schwemmler W, Schenk HEA (eds) Endocytobiology. Endosymbiosis and Cell Biology. De Gruyter, Berlin New York, pp 763–770
De Priester W, van der Molen LG (1979) Cell Tissue Res 198:79–93
Dillon LS (1981) Ultrastructure, Marcomolecules, and Evolution. Plenum Press, New York, Chapter 9
Doyle D, Tweto J (1975) Methods Cell Biol 10:235–260
Forbes MS, Sperelakis N (1982) Tissue & Cell 14:25–37
Glaumann H, Berezesky JK, Ericsson JLE, Trump BF (1975) Lab Invest 33:239–251
Glaumann H, Trump BF (1975) Lab Invest 33:262–272
Guillouzo A, Guillouzo C, Boisnard M (1978) Biol Cell 31:315–318
Heggeness MH, Simon M, Singer SJ (1978) Proc Nat Acad Sci USA 75:3863–3866
Kellerman GM, Biggs DR, Linnane AW (1969) J Cell Biol 42:378–391
Lee CS, Morgan G, Wooding FBP (1979) J Cell Sci 38:125–135
Locke M, Collins JV (1965) J Cell Biol 26:857–884
Luck DJL (1963) J Cell Biol 16:483–499
Luck DJL (1965) J Cell Biol 24:445–460
Pfeifer U (1978) J Cell Biol 78:152–167
Plattner H, Schatz G (1969) Biochem 8:339–343
Plattner H, Salpeter MM, Saltzgaber J, Schatz G (1970) Proc Nat Acad Sci USA 66:1252–1259
Riezman H, Hay R, Gasser S, Daum G, Schneider G, Witte C, Schatz G (1983) EMBO J 2:1105–1111
Riezman H, Hay R, Witte C, Nelson N, Schatz G (1983) EMBO J 2:1113–1118
Roodyn DB, Wilkie D (1968) The Biogenesis of Mitochondria. Methuen, London
Ruby JR, Dyer RF, Skalko RG (1969) Z Zellforsch 97:30–37
Scarpelli DG, Chiga M, Haynes E (1971) In: Cell Membranes. Biological and Pathological Aspects. Williams and Wilkins, Baltimore

Schjeide OA, McCandless RG, Munn RJ (1963) Growth 27:111–123
Schjeide OA, McCandless RG, Munn RJ (1964) Nature (Lond) 203:158–160
Shore GC (1979) J Cell Sci 38:137–153
Shore GC, Power F, Bendayan M, Carignan P (1981) J Biol Chem 256:8761–8766
Smith DS, Jarlfors U, Cameron BF (1975) Ann N Y Acad Sci 253:472–506
Smith DS, Jarlfors U, Cayer ML (1977) J Cell Sci 27:255–272
Sohal RS, McCarthy JL, Allison VF (1972) J Ultrastruct Res 39:484–495
Trump BF, Croker BP, Mergner WJ (1971) In: Richter GW, Scarpelli DG (eds) Cell Membranes. Biological and Pathological Aspects. Williams and Wilkins, Baltimore, pp 84–128
Wellburn FAM, Wellburn AR (1980) In: Schwemmler W, Schenk HEA (eds) Endocytobiology. De Gruyter, Berlin New York, pp 839–848

Interaction of the Golgi Apparatus and the Plasmalemma in the Cytomorphogenesis of *Micrasterias*

O. KIERMAYER and U. MEINDL[1]

Introduction

Cytomorphogenesis of the green alga *Micrasterias* is a highly differentiated process, which takes place according to a characteristic and strongly determined temporal sequence. Because of the specific cell pattern, which develops continuously during cell growth, *Micrasterias* is especially suitable for the study in detail of the individual steps in cytomorphogenesis. A special fixation procedure for electron microscopy of *Micrasterias* was introduced by Kiermayer (1968) and this made it possible to investigate how ultrastructural components relate to the process of cytomorphogenesis. Thus interactions have been found between the Golgi apparatus and the plasma membrane during particular steps of cell pattern formation (Kiermayer 1970a,b, 1973, 1981), which are mediated by various kinds of vesicle.

Cell growth and hence cytomorphogenesis in *Micrasterias* starts with the formation of a septum, which grows inward centripedally from the isthmus invaginations during mitosis. This process, which has the effect of separating the two semicells from one another, is complete within 15 min. Subsequently the primary wall begins to grow, forming a bulge which develops in strict sequence the lobe invaginations of the new half cell, producing the characteristic cell shape. Primary wall growth lasts approximately 4 h and is followed by the formation of a thick rigid secondary wall and the simultaneous development of cell wall pores (Kiermayer and Staehelin 1972, Dobberstein 1973). These last processes take about 8 h. Finally the primary wall is shed (Drawert and Mix 1962b, Dobberstein 1973).

The following article is intended to describe the ultrastructural aspects of the distinct steps of cytomorphogenesis in *Micrasterias* in relation to interactions between the golgi apparatus and the plasmalemma. Most emphasis is placed on the morphology and function of the different products of the Golgi apparatus which arise during cell development.

Septum Formation

Light microscopic studies of osmotically treated *Micrasterias* cells at the stage of septum formation (Kiermayer 1964, 1967) have demonstrated that instead of cell elong-

1 Botanisches Institut, Universität Salzburg, Lasserstr. 39, A-5020 Salzburg, Austria

ation, accumulations of primary wall material occur on the septem wall. The accumulations form a pattern, which corresponds to the basic three-lobed symmetry of the *Micrasterias* cell and has been termed the "septum initial pattern". From these observations it can be concluded that even at the early stage of septum formation a prepattern for cytomorphogenesis is present. The template for this pattern is probably located within the plasma membrane (Kiermayer 1967, 1981).

Ultrastructural investigations revealed that septem formation is indicated by the formation of a girdle of wall material lying around the isthmus lobe invaginations (Lacalli 1973), followed by the widening of the isthmus. At metaphase the three-layered septum (Drawert and Mix 1962a) starts to grow inward toward the middle of the cell surrounded by a great number of vesicles, which are characterized by their asymmetric, electron-dense contents and a diameter of 100-160 nm (Kiermayer 1981). These vesicle have been called "septum vesicles" (SEV) and are obviously pinched off from the numerous dictyosomes, each containing 10 or 11 cisternae. Since incorporated SEV are clearly visible in the material of the growing septum they are obviously functional in septum formation. It seems likely that the contents of the SEV produce the septum material while their membranes build up the adjacent plasma membrane. It may be suggested, therefore, that the "septum initial pattern" which has been assumed from light microscopic experiments may be established via the membranes of the septum vesicles. The information for the prepattern for morphogenesis is probably already present in the SEV-membranes, which arise from the Golgi apparatus (Kiermayer 1981). An interaction of the Golgi apparatus and the plasma membrane mediated by the SEV therefore seems to be the basis for pattern formation during this first step of development.

Primary Wall Growth

Primary wall growth and thus cell pattern formation in *Micrasterias* is characterized by the presence of another, quite distinct kind of vesicle, the "dark vesicles" (DV; Kiermayer 1970a). These have a diameter of about 200 nm and electron-dense contents which give positive reactions for acid polysaccarides (Menge 1973, 1976). DV are produced by the Golgi apparatus throughout the formation of the primary wall and obviously fuse with the plasma membrane, thus delivering their contents to the growing wall (Kiermayer 1970a,b).

In light microscopic experiments under conditions of reduced turgor, each developmental stage of the *Micrasterias* cell was found to be characterized by distinct growth areas, which could be visualized as accumulations of wall material (Kiermayer 1964, 1965, Tippit and Pickett-Heaps 1974, Ueda and Yoshioka 1976). It was therefore supposed that, depending on the developmental stage, incorporation of DV into the plasma membrane is possible only at distinct predetermined areas (Kiermayer 1970b). Kiermayer (1970a,b, 1973, 1981) assumed that only at these areas "membrane-membrane recognition" processes take place between the plasma membrane and the DV membranes.

Recent studies with chlorotetracycline, a fluorescent probe for membrane-associated Ca^{2+} (Caswell 1979), indicate patterned accumulations of Ca^{2+} according to

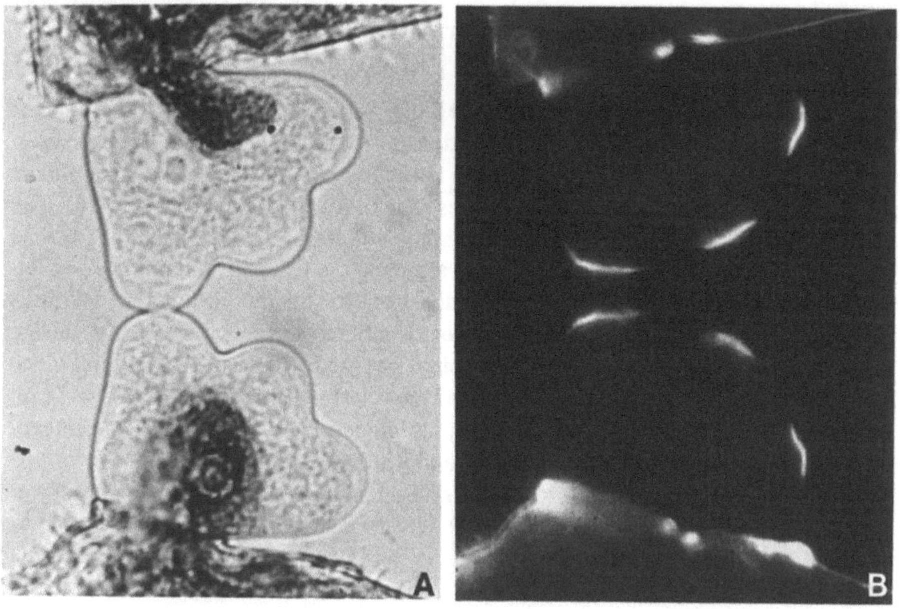

Fig. 1A,B. Developmental stages of *Micrasterias thomasiana* f. *uniradiata* (Kallio) treated with 4 x 10^{-3}% chlorotetracycline dissolved in 0.18 mol glucose. *A* Micrograph taken using normal light. *B* Same cells as in *A* micrograph taken using ultraviolet light. Only at the patterned face of the growing half cell are three fluorescent sites (corresponding to the growth areas of the cell) visible

the distribution of the growth areas of the *Micrasterias* cell (Meindl 1982a). Thus, since Ca^{2+} has been demonstrated to favor fusions in various systems (Gratzl 1980) it may be concluded that membrane-associated Ca^{2+} in *Micrasterias* mediates the fusion between the DV membranes and the plasma membrane, hence producing the distinctive cell shape (Meindl 1982a). This hypothesis has been supported recently by McNally et al. (1983). Further evidence has been obtained from studies on *Micrasterias thomasiana f. uniradiata,* a mutant of *M. thomasiana* (Waris 1950, Kallio 1951) lacking any pattern at one side of the cell. After treatment of this alga with chlorotetracycline during primary wall formation no Ca^{2+} accumulations are visible on the unpatterned face of the cell; the other patterned side, however, exhibits the characteristic distribution of fluorescent areas (Fig. 1A,B).

In summary primary wall formation and thus cytomorphogenesis in *Micrasterias* is mediated by fusion of distinct Golgi products, the D-vesicles with special areas of the plasma membrane (membrane-membrane recognition) which are characterized by an accumulation of Ca^{2+}.

Secondary Wall Growth and Pore Formation

After primary wall growth is complete, the Golgi apparatus produces another kind of vesicle, which is functional in the formation of the secondary wall. These vesicles, termed "flat vesicles" (FV) have a disc-like shape, have an unusually thick membrane and carry sack-like structures at their edges (Kiermayer and Dobberstein 1973). The inner surface of the thick FV membrane is highly differentiated and exhibits special "globular particles" which develop during a membrane differentiation process (Kiermayer 1981). When pinched off from the distal pole of the dictyosome, the FV lacks any clear visible differentiation of the inner membrane surface; nor are any sac-like appendages present (Fig. 2A,B). While being transported toward the plasma membrane they obviously develop the "globular particles" which are about 20 nm in diameter and are oriented in rows. The fully developed FV also exhibits a helical striation visible in tangential sections (Meindl and Kiermayer 1982).

The FV are incorporated into the plasma membrane during secondary wall formation, so that their inner surface with their "globular particles", are transferred to the upper surface of the plasma membrane (Kiermayer and Dobberstein 1973). The "globular particles" are arranged at a distance of approximately 25 nm (center to center spacing) from each other. This corresponds to the distance between the parallel oriented microfibrils of the secondary wall and it has therefore been suggested that the "globular particles" are active in the formation and orientation of the microfibrils ("Bildungsapparate" Kiermayer and Dobberstein 1973). In freeze-etch studies of cells forming their secondary walls the "globular particles" could be demonstrated to be rosettes composed of six subunits (Kiermayer and Sleytr 1979, Giddings et al. 1980). The rosettes themselves are arranged in a hexagonal lattice.

FV are produced not only by dictyosomes of the developing half cell, where they are incorporated into the plasma membrane, but also in the old half cell which does not perform secondary wall growth (Fig. 3A,B). Here obviously no fusion with the plasma membrane takes place as remnants of FV are visible at the periphery of the old half cell even days after cell division (Fig. 3C-F).

At the same time as the secondary wall is growing cell wall pores are formed by special "pore vesicles" (PV). The origin of these vesicles, which have a diameter of 0.3 μm, is not yet known but possibly they too are produced by the Golgi apparatus. The PV fuse with the plasma membrane forming plugs in the growing secondary wall. The plugs subsequently develop into pores (Dobberstein 1973, Neuhaus and Kiermayer 1981). Treatment of *Micrasterias* cells at the stage of pore formation with chlorotetracycline leads to a pattern of fluorescent dots at the periphery of the cell which corresponds to the distribution of the fully developed pores (Meindl 1982b). It can therefore be assumed that pore formation as well as primary wall formation (see above) is mediated by fusion processes between the vesicle membranes and areas of the plasma membrane which are characterized by an accumulation of Ca^{2+} (Meindl 1982a,b).

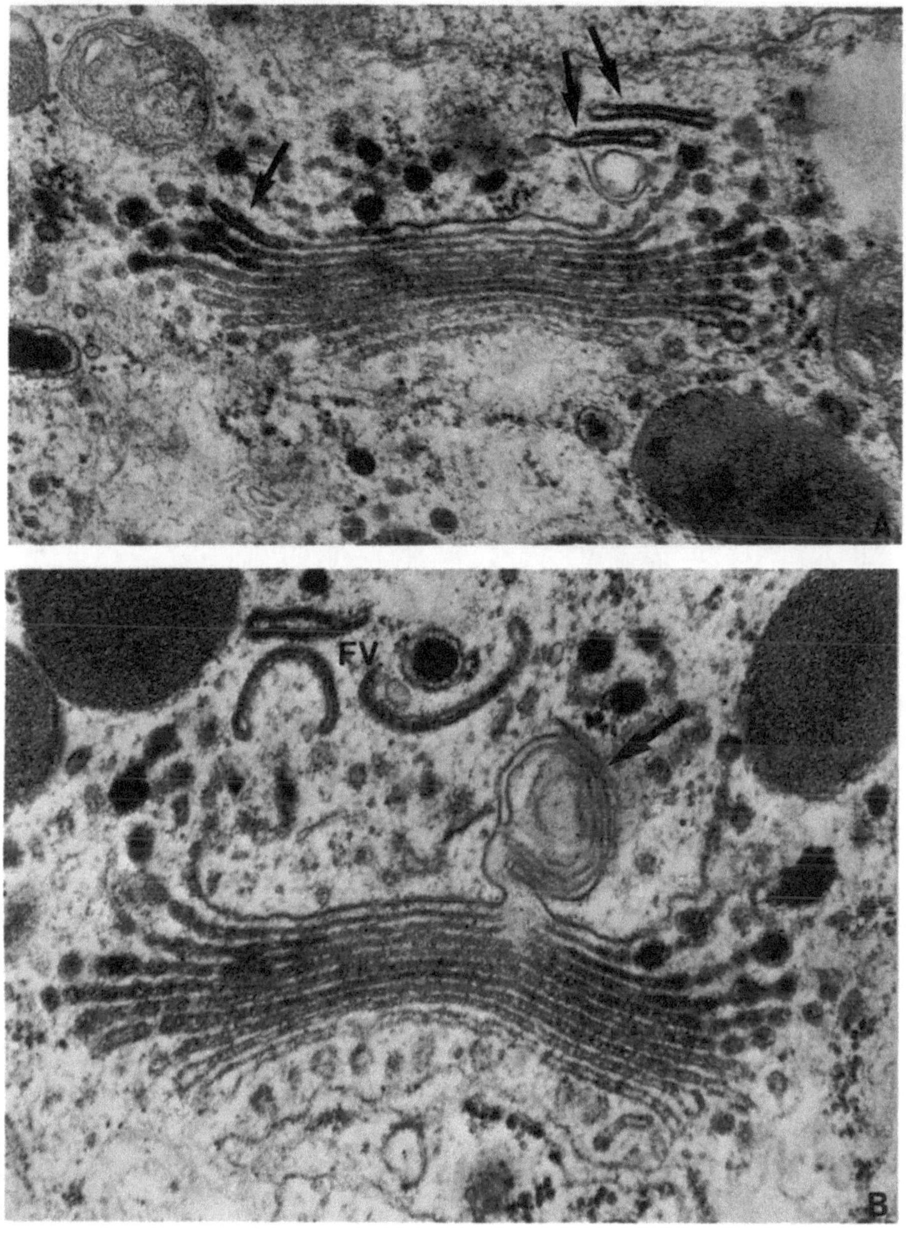

Fig. 2A,B. Dictyosomes in the young half cell of *Micrasterias denticulata* at the stage of secondary wall formation. *A* "Flat vesicles" (FV) are visible in close proximity to the distal pole of the dictyosome from which they are pinched off (*arrows*) x 48,500; *B* The dictyosome exhibits an bladder-like protrusion which seems to arise from the most distal cisterna (Burgstaller-Getzinger and Kiermayer 1981). Most of the "young" FV visible near the dictyosome lack any clear inner differentiation or appendages, x 55,000

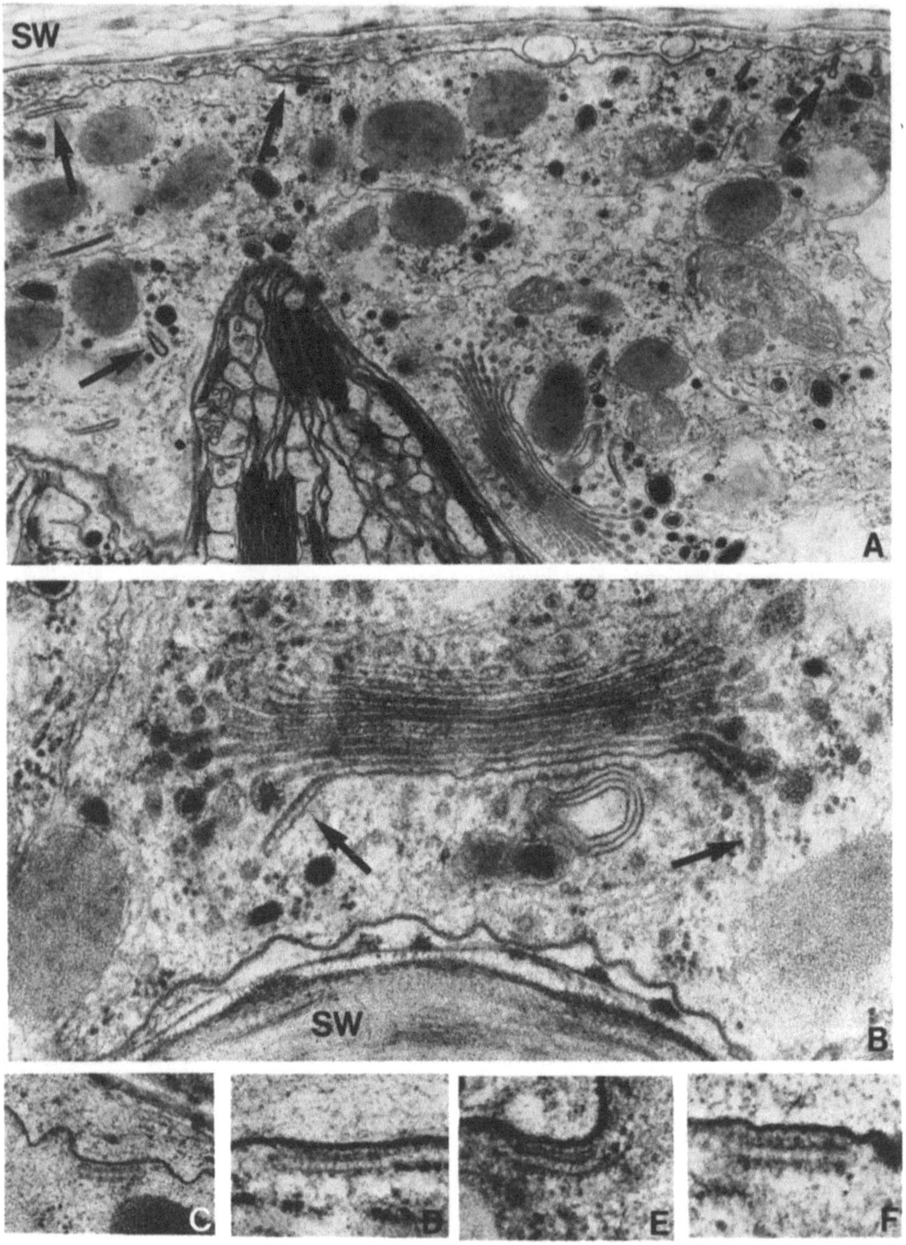

Fig. 3A-F. Parts of the old half-cell of *Micrasterias denticulata* during secondary wall formation in the young half-cell. *A* At the periphery of the cell numerous "flat vesicles" are visible (*arrows*) x 18,500; *B* Dictyosome of the old half cell producing "flat vesicles" (*arrows*) **SW** secondary wall. x 49,000; **C-F** "Flat vesicle" – remnant at the cell periphery of the old half cell days after secondary wall formation has finished. Although the "flat vesicles" are located beneath the plasma membrane they are not incorporated. *C* x 31,500; *D* x 47,000; *E* x 63,000; *F* x 86,000

Conclusion

Interactions between different products of the Golgi apparatus and the plasma membrane occur throughout morphogenesis in *Micrasterias.* Depending on the developmental stage of the cell, the Golgi apparatus switches to produce SE-vesicles, D-vesicles, F-vesicles and possibly P-vesicles. The membranes of these vesicles arise from the membranes of the Golgi cisternae and then pass through "membrane differentiation" and "membrane transformation" processes before fusing with the plasma membrane at distinct "membrane-membrane recognition" sites. An accumulation of Ca^{2+}, possibly mediated by calcium-binding proteins, has been demonstrated to be characteristic of these receptor sites during both primary wall formation and pore formation (Fig. 4).

It is interesting that the whole Golgi apparatus of the cell, both in developing and in non-developing halves, produces the vesicles required at a particular develop mental stages. In developing half cells the vesicles fuse with the plasma membrane, but in the old half cell they are not incorporated and remain at the cell periphery as has been demonstrated for secondary wall formation. This again supports the hypothesis that the plasma membrane controls local vesicle fusion thus functioning as a template for cytomorphogenesis in *Micrasterias.*

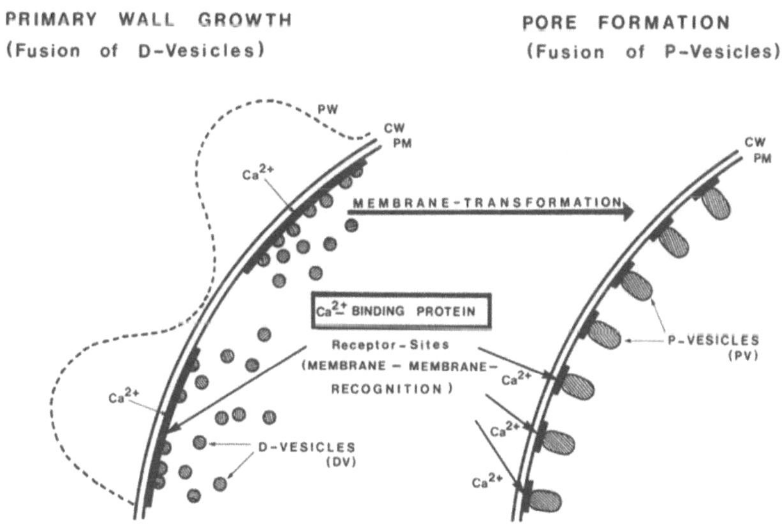

Fig. 4. Hypothetical representation of the possible interaction processes between Golgi products and the plasma membrane during primary wall growth and pore formation. **CW** cell wall; **PM** plasma membrane. (After Meindl 1982a,b)

References

Burgstaller-Getzinger Ch, Kiermayer O (1981) Sitzungsber Österr Akad Wiss Math -nat Kl Abt 1, 189:233–247
Caswell AH (1979) Int Rev Cytol 56:145–181
Dobberstein B (1973) Nova Hedwigia 42:83–90
Drawert H, Mix M (1962a) Planta (Berl) 58:448–452
Drawert H, Mix M (1962b) Arch Mikrobiol 42 (1):96–109
Giddings TH, Brower DL, Staehelin LA (1980) J Cell Biology 84:327–339
Gratzl M (1980) In: Bücher Th, Sebald W, Weiss H (eds) Biological chemistry of organelle formation. Springer, Berlin Heidelberg New York
Kallio P (1951) Ann Bot Soc Zool Bot Fenn Vanamo 24:1–122
Kiermayer O (1964) Protoplasma 59:382–420
Kiermayer O (1965) Planta (Berl) 66:216–220
Kiermayer O (1967) Protoplasma 64:481–484
Kiermayer O (1968) Planta (Berl) 83:223–236
Kiermayer O (1970a) Protoplasma 69:97–132
Kiermayer O (1070b) Ann NY Acad Sci 175:686–701
Kiermayer O (1973) Ber Dtsch Bot Ges 86:287–291
Kiermayer O (1981) In: Kiermayer O (ed) Cytomorphogenesis in Plants. Springer, Wien New York
Kiermayer O, Dobberstein B (1973) Protoplasma 77:437–451
Kiermayer O, Sleytr U (1979) Protoplasma 101:133–138
Kiermayer O, Staehelin LA (1972) Protoplasma 74:227–237
Lacalli TC (1973) Protoplasma 78:433–442
McNally JG, Cowan JD, Swift H (1983) Dev Biol 97:137–145
Meindl U (1982a) Protoplasma 110:143–146
Meindl U (1982b) Protoplasma 112:133–141
Meindl U, Kiermayer O (1982) Phyton (Horn) 22:115–135
Menge U (1973) Thesis, Köln
Menge U (1976) Protoplasma 91:115–123
Neuhaus G, Kiermayer O (1981) In: Kiermayer O (ed) Cytomorphogenesis in Plants. Springer, Wien New York
Tipitt DH, Pickett-Heaps JD (1974) Protoplasma 81:271–296
Ueda K, Yoshioka S (1976) J Cell Sci 21:617–631
Waris H (1950) Physiol Plant 3:236–245

Microbodies in Different Algae

H. STABENAU[1]

Microbodies with distinctive morphological features in electron micrographs were first described in 1954. Since then they have been seen, not only in many animal and plant tissues but also in protozoa, fungi and algae, indicating that their occurrence is ubiquitous (Hruban and Rechcigl 1969, Gerhardt 1978).

With regard to the latter group, microbodies have been observed by electron microscope studies in species of almost every taxonomic division with the exception of the Cyanophyta. Commonly, they are identified by a roughly spherical shape with a diameter in the range from 0.2 to 0.5 μm which corresponds well to the size of the higher plant microbodies (Frederick et al. 1975). The organelle is limited by a single membrane which encloses a granular matrix of variable but usually moderate electron density, in most cases without a distinctive internal structure. Associations of the algal microbodies with profiles of the endoplasmic reticulum frequently were observed, suggesting that the microbodies originate from the endoplasmic reticulum by the process of budding. Close spatial associations also have been seen between algal microbodies and either lipid bodies, chloroplasts, mitochondria or pyrenoids (for details see Silverberg 1975a). In *Pyramimonas,* a single microbody containing catalase has been identified which appears to be closely attached to the basal bodies in this alga (Norris and Pearson 1975).

The number of microbodies in the different algae is reported to vary greatly, even in closely related species. Generally, it appears that they are much less numerous than mitochondria. For several species the number of microbodies was estimated to be one tenth of those of mitochondria. In *Netrium digitus* there might be even 40 times more mitochondria than microbodies, and in the filamentous algal species *Stichococcus* and *Klebsormidium* only one microbody has been seen in each cell which invariably lies between the chloroplast and the nucleus (Floyd et al. 1972, Silverberg 1975a). In some species tentative examinations have not revealed the presence of microbodies at all (Silverberg 1975a).

Microbodies may be characterized also biochemically by their content of enzymes. They are defined as peroxisomes if they contain at least one H_2O_2 producing oxidase in addition to catalase, an enzyme which reduces the H_2O_2 to water (de Duve 1969). Under certain conditions catalase shows also a peroxidative activity and in the presence of 3,3'-diaminobenzidine (DAB) an osmiophilic reaction product can be formed by which finally a staining of the microbodies may be achieved. Using this

1 Universität Oldenburg, Fachbereich Biologie, Ammerländer Heerstr. 67–99, D–2900 Oldenburg, FRG

Compartments in Algal Cells and Their Interaction
ed. by W. Wiessner, D. Robinson, and R.C. Starr

procedure catalase could be demonstrated to be present in the microbodies of more than 20 algal species. Negative results for DAB cytochemistry also were obtained. However, in some cases, staining was observed merely by slightly modifying the DAB-procedure and therefore, negative results have to be interpreted with caution (Gerhardt and Berger 1971, Silverberg 1975b). Nevertheless, in some algal species catalase may be absent from microbodies (Bibby and Dodge 1973, Silverberg 1975a) which cannot then be regarded as peroxisomes. With the electron microscopy only the microbodies of *Klebsormidium* have been demonstrated to be peroxisomes because they contain both catalase and glycolate oxidase as shown by cytochemical tests (Stewart et al. 1972, Gruber and Frederick 1977).

Since there are limited possibilities for demonstration of enzymes cytochemically, the isolation and biochemical characterization of algal microbodies is necessary in order to obtain more comprehensive data. However, the isolation of organelles from algae presents problems. For example, breaking the algal cells often requires rough methods which also may break the organelles to a large extent. This is particularly serious because there are apparently only few microbodies per cell. Moreover, the activities of microbody enzymes are lower than in higher plants, e.g., on the average algal cells contain only 10% of the catalase compared with higher plants (Frederick et al. 1973).

Up to the present, microbodies have been isolated from only a few algal species. Considering the different results, it seems that there are two groups of algae, one with so-called unspecialized microbodies containing mainly catalase and uricase and another one with organelles highly similar to glyoxysomes and leaf peroxisomes of higher plants (Gerhardt 1978, Beevers 1982, Tolbert 1982). Organisms of both groups have been investigated in our laboratory. The most comprehensive data were obtained from experiments for which *Eremosphaera* representing the first, or *Mougeotia*, representing the second group were used. In the following, results are presented in order to compare the different types of algae.

Microbodies from *Mougeotia*

Microbodies of the leaf peroxisomal type were isolated first from *Spirogyra* (Stabenau 1976). But culturing this alga is difficult, and *Mougeotia*, which can be grown in glass tubes like unicellular algae (Stabenau 1978), is a more convenient organism. The growth rate of *Mougeotia* is relatively high, even when the CO_2 concentration in the aeration mixture is decreased from normally 2% to 0.03% which in combination with a high light intensity increases the activities of the microbody enzymes by a factor of 2 to 5 (Stabenau and Säftel 1981, Winkler et al. 1982).

Cells of *Mougeotia* can be disrupted under gentle conditions. By this procedure most cells are merely punctured and chloroplast particles as well as cytoplasm, with constituent organelles, are completely released. Cell walls and larger chloroplast particles then can be easily removed from the crude homogenate by low speed centrifugation and subsequent filtration steps. Finally, the homogenate is placed on a linear sucrose gradient and centrifuged for 4 h at 65,000 *g*. During centrifugation all par-

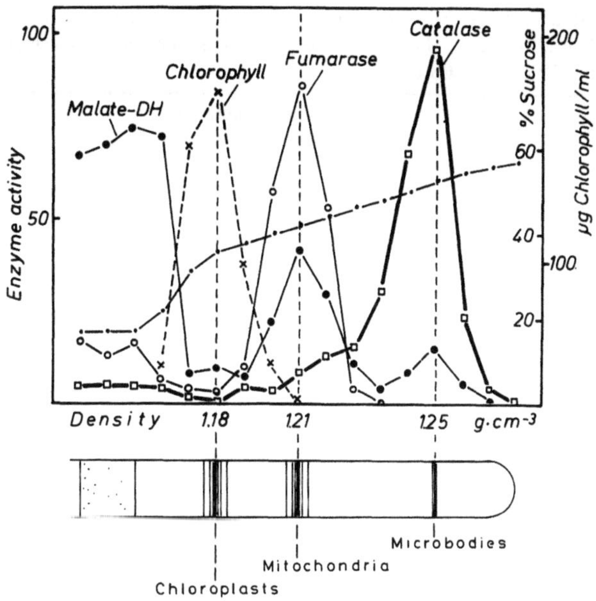

Fig. 1. Distribution of enzymes in a sucrose gradient after separation of organelles from *Mougeotia*

ticles move to their specific density, while soluble constituents of the cytoplasm but also from broken organelles remain in the top fractions of the gradient (Stabenau and Säftel 1981).

Although there are many different organelles, usually only two distinct bands are seen in the gradient after centrifugation representing chloroplast particles and mitochondria, respectively. Only in a very few cases a third band of microbodies can be seen around density 1.25 g/cm^3. However, even if this band is not visible, electron micrographs demonstrate that there are microbodies which apparently are less numerous than mitochondria.

The presence of microbodies in this region of the gradient is also demonstrated by the distribution of enzymes characteristic for these organelles (Fig. 1). Catalase, for example, was found mainly in the microbody band at density 1.25 g/cm^3, suggesting that the percentage of unbroken organelles was very high. Since most of the fumarase was detected also within the gradient, the bulk of the mitochondria (density 1.21 g/cm^3) were intact, too. As shown in Fig. 1, the microbodies contain malate dehydrogenase (peak at density 1.25 g/cm^3) clearly separated from the malate dehydrogenase in the mitochondria (peak at density 1.21 g/cm^3). But most of this enzyme was found at the top of the gradient and must therefore be cytosolic or chloroplastic. The distribution pattern of the different enzymes indicates that the microbodies and mitochondria are well separated and contamination in the microbody fraction at density 1.25 g/cm^3 is very low. Therefore, enzymes found in this fraction are expected to be constituents of the microbodies.

By our investigations all the peroxisomal enzymes which are known to be directly involved in the glycolate metabolism of higher plants could be demonstrated to be present also in *Mougeotia* microbodies, whereas serine hydroxymethyl-transferase, the enzyme for conversion of glycine to serine, as expected, was found exclusively in the mitochondria (Table 1).

Table 1. Enzymes of the glycolate metabolism which have been demonstrated in *Mougeotia* and their activities in the different organelles. Specific activities were determined from cells grown at 20,000 lx and 0.03% CO_2

	Enzyme activity nmoles substrate/min x mg protein	
	Mitochondria	Microbodies
1 Catalase	-	836 500
2 Glycolate oxidase	-	204
3 Alanine-glyoxylate aminotransferase	-	457
4 Glutamate-glyoxylate aminotransferase	-	375
5 Serine hydroxymethyl-transferase	26	
6 Serine-pyruvate aminotransferase	-	74
7 Hydroxypyruvate reductase	-	1290

The different enzymes have been biochemically characterized. Except for alanine-glyoxylate aminotransferase and serine-pyruvate aminotransferase the enzymes of the glycolate pathway in *Mougeotia* are very similar to those of higher plants (Winkler et al. 1982). It is noteworthy in this regard that the glycolate oxidizing enzyme in *Mougeotia* is an oxidase rather than a dehydrogenase as it is in several unicellular algae (Stabenau and Säftel 1982).

All steps of peroxisomal reactions as far as have been proved for *Mougeotia* are demonstrated in Fig. 2. Differences from the corresponding scheme for higher plants (Tolbert 1982) are apparent only in the transamination steps. In *Mougeotia* peroxisomes glycine is formed from glyoxylate with alanine as amino donor and the pyruvate resulting from this reaction can be used for the conversion of serine to hydroxypyruvate. Therefore, it seems that in this alga there is an alanine-pyruvate shuttle which is not known from leaf peroxisomes (Winkler et al. 1982).

As has been demonstrated by Gerhardt (1981) enzymes of β-oxidation are constituents not only of glyoxysomes but also of leaf peroxisomes from higher plants. From our recent work results were obtained indicating that in *Mougeotia,* enzymes of β-oxidation are also located in the microbodies. After gradient centrifugation about 60% of the hydroxyacyl-CoA dehydrogenase was found in the microbody band, whereas 40% was soluble (Stabenau et al. in preparation). There are indications that acyl-CoA oxidase is also located in the peroxisomes.

Besides *Mougeotia* and *Spirogyra, Klebsormidium, Netrium, Coleochaete* and *Nitella* all contain a glycolate oxidase and therefore in these algae there may be the same kind of microbodies (Chevalier et al. 1974, Gruber et al. 1974). On the other hand, most unicellular algae possess a glycolate dehydrogenase which has no capability to transfer electrons directly to oxygen (Nelson and Tolbert 1970, Frederick et al. 1973). Therefore, H_2O_2 is not formed during glycolate oxidation in these algae. After isolation of organelles from *Chlorogonium* and *Chlamydomonas* it could be

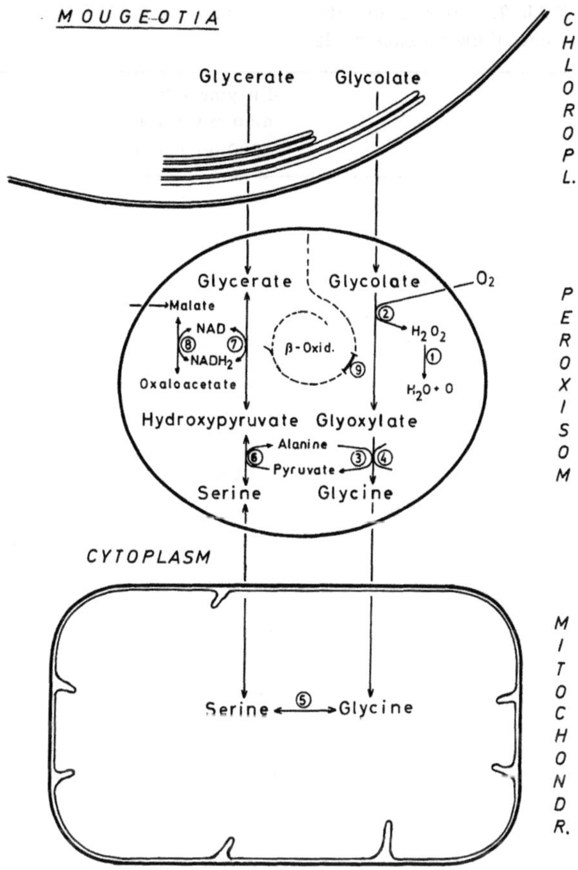

Fig. 2. Glycolate pathway and peroxisomal reactions demonstrated in *Mougeotia*. Enzymes: *1-7* see Table 1; *8* malate dehydrogenase; *9* hydroxyacyl-CoA dehydrogenase

demonstrated that the glycolate dehydrogenase is located in the mitochondria rather than in microbodies (Stabenau 1974a,b). Furthermore, the microbodies isolated were different from leaf peroxisomes or glyoxysomes by the absence also of other characteristic enzymes (Stabenau 1974, Stabenau and Beevers 1974). Meanwhile more data were obtained from *Eremosphaera viridis* which apparently possesses the same kind of microbodies.

Microbodies and Metabolism of Glycolate in *Eremosphaera*

Eremosphaera is an unusual large unicellular alga (diameter: 0.1–0.2 mm) which is of advantage for gentle breaking of the cells in order to isolate organelles. After centrifugation in a sucrose gradient microbodies were found by electron microscopy at the lower density 1.23 g/cm^3. Again, no visible band of organelles was present at density 1.23 g/cm^3 suggesting that there are also few microbodies in *Eremosphaera.*

Table 2. Influence of culture conditions on the activities of enzymes in the different compartments of *Eremosphaera* cells

	Enzyme activity nmoles substrate/min x mg protein	
	6000 lx; 2% CO_2	20 000 lx; 0.03% CO_2
Mitochondria		
2 Glycolate dehydrogenase	37	79
4 Glutamate-glyoxylate aminotransferase	159	161
5 Serine hydroxymethyl-transferase	17	27
6 Serine-glyoxylate aminotransferase	156	294
7 Hydroxypyruvate reductase	576	1022
Microbodies		
Catalase	62 830	56 540
Uricase	5-20	5-20

As shown in Table 2, the microbodies contain catalase and uricase. No other enzymes have been found in the microbodies of *Eremosphaera* (Stabenau et al. in preparation). Instead, all enzymes necessary for reactions of the glycolate pathway known to be peroxisomal in *Mougeotia* are located in the mitochondrial band at density 1.19 g/cm^3 as is the serine hydroxymethyl-transferase (Table 2).

With the exception of glutamate-glyoxylate aminotransferase the activities of all enzymes related to the glycolate pathway are increased almost two-fold when the light intensity was increased and the CO_2-concentration was decreased during growth of the cells (Table 2). Since the synthesis of glycolate is enhanced by this alteration of culture conditions the enzymes identified in the mitochondria apparently participate in the metabolism of glycolate as demonstrated in Fig. 3.

There are indications that the glycolate oxidation is coupled to the function of the respiration chain as in some other algae (Paul et al. 1975, Paul and Volcani 1975, Collins and Merrett 1975). Oxygen then would be an electron acceptor during glycolate oxidation even in *Eremosphaera,* though not directly, and the oxygen would be reduced to water rather than to H_2O_2 with concomitant synthesis of ATP (Collins et al. 1975).

D-lactate was found to be an alternate substrate for the glycolate oxidizing enzyme from *Eremosphaera* as from other algae (Gruber et al. 1974). The activitiy is even two times higher than with glycolate suggesting that this enzyme is a D-lactate dehydrogenase rather than a glycolate dehydrogenase (Stabenau et al. 1983). This result seems not to be fortuitous since a pyruvate reductase was demonstrated in the

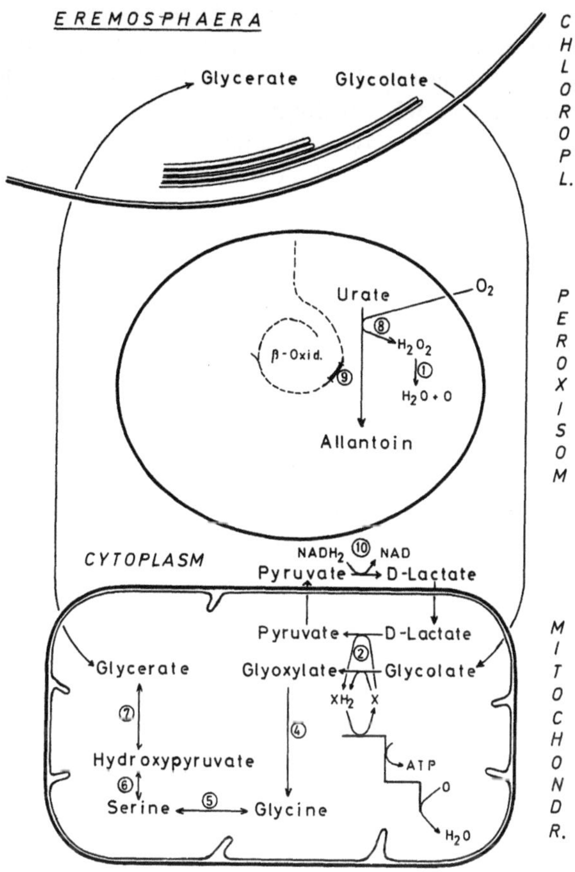

Fig. 3. Scheme of possible reactions in the different compartments of *Eremosphaera* cells. Enzymes: *1* catalase; *2-7* see Table 2; *8* uricase; *9* hydroxyacyl-CoA dehydrogenase; *10* pyruvate reductase

cytoplasm of *Eremosphaera* with high activity, forming D-lactate exclusively (Gruber et al. 1974, Stabenau et al. 1983). As suggested in Fig. 3, there might be a pyruvate/lactate shuttle in this alga by which cytoplasmic NADH is oxidized via the mitochondrial respiration chain. In this case different reactions would be catalyzed by the same enzyme.

As is evident from Fig. 3, microbodies from *Eremosphaera* are not at all involved in glycolate metabolism. By the possession of mainly catalase and uricase they are similar to the microbodies isolated from other unicellular algae (Gerhardt 1978). The microbodies from *Chlorogonium* and *Polytomella* also did not contain enzymes of the glyoxylic acid cycle which are characteristic for glyoxysomes (Gerhardt 1971, Cooper and Lloyd 1972, Stabenau and Beevers 1974). Generally it appears that the microbodies isolated from unicellular algae, except *Euglena* (Graves et al. 1972, Graves and Becker 1974, Collins and Merrett 1975), are neither of the glyoxysomal nor of the leaf peroxisomal type. Thus, they were regarded as unspecialized peroxisomes of unknown function. However, from our recent work data were obtained indicating that enzymes of β-oxidation are located also in the unspecialized microbodies from algae. For example, after gradient centrifugation about 60% of the hydroxyacyl-CoA dehydrogenase from *Eremosphaera* was detected in the microbody band whereas 40% of the enzyme was found to be soluble in the top fraction of the gradient (Stabenau et al. in preparation). Since the distribution pattern of this enzyme clearly shows only one distinct peak at density 1.23 g/cm^3, microbodies are the only organelles in *Eremosphaera* containing hydroxyacyl-CoA dehydrogenase and, therefore, may be specialized for degradation of fatty acids.

Acknowledgments. The author would like to express his thanks to Dr. Uwe Winkler and Werner Säftel for their support.

References

Beevers H (1982) Ann N Y Acad Sci 386:243–253
Bibby BT, Dodge JD (1973) Planta (Berl) 112:7–16
Chevalier E, Vasconcelos A, Floyd GL (1974) Bull N J Acad Sci 19:28–29
Collins N, Merrett MJ (1975) Biochem J 148:321–328
Collins N, Brown RH, Merrett MJ (1975) Biochem J 145:373–377
Cooper RA, Lloyd D (1972) J Gen Microbiol 72:59–70
de Duve C (1969) Proc R Soc Ser B 173:71–83
Floyd GL, Stewart KD, Mattox KR (1972) J Phycol 8:176–184
Frederick SE, Gruber PJ, Tolbert NE (1973) Plant Physiol (Besthesda) 52:318–323
Frederick SE, Gruber PJ, Newcomb EH (1975) Protoplasma 84:1–29
Gerhardt B (1971) Arch Mikrobiol 80:205–218
Gerhardt B (1978) Cell Biology Monographs, vol 5. Microbodies/peroxisomes of plant cells. Springer, Wien New York
Gerhardt B (1981) FEBS Lett 126:71–73
Gerhardt B, Berger C (1971) Planta (Berl) 100:155–166
Graves LB, Trelease RN, Grill, Becker WM (1972) J Protozool 19:527–532
Graves LB, Becker WM (1974) J Protozool 21:771–773
Gruber PJ, Frederick SE (1977) Planta (Berl) 135:45–49
Gruber PJ, Frederick SE, Tolbert NE (1974) Plant Physiol (Bethesda) 53:167–170
Hruban Z, Rechcigl M (1969) Microbodies and related particles. Morphology, biochemistry and physiology. Academic, New York
Nelson EB, Tolbert NE (1970) Arch Biochem Biophys 141:102–110
Norris RE, Pearson BR (1975) Arch Protistenkd 117:192–213
Paul JS, Volcani BE (1975) Plant Sci Lett 5:281–285
Paul JS, Sullivan CN, Volcani BE (1975) Arch Biochem Biophys 169:152–159
Silverberg BA (1975a) Protoplasma 83:269–295
Silverberg BA (1975b) Protoplasma 85:373–375
Stabenau H (1974a) Plant Physiol (Bethesda) 54:921–924
Stabenau H (1974b) Planta (Berl) 118:35–42
Stabenau H (1976) Plant Physiol (Bethesda) 58:693–695
Stabenau H (1978) Ber Dtsch Bot Ges 91:251–255
Stabenau H, Beevers H (1974) Plant Physiol (Bethesda) 53:866–869
Stabenau H, Säftel W (1981) Ber Dtsch Bot Ges 94:59–64
Stabenau H, Säftel W (1982) Planta (Berl) 154:165–167
Stabenau H, Winkler H, Säftel W (1983) Plant Physiol (Bethesda) 72:S–975
Stewart KD, Floyd GL, Mattox KR, Davis ME (1972) J Cell Biol 54:431–434
Tolbert NE (1982) Ann N Y Acad Sci 386:254–268
Winkler U, Säftel W, Stabenau H (1982) Plant Physiol (Bethesda) 70:340–343

Cytological Aspects of Blue-Green Algal Endosymbiosis

L. KIES[1]

Introduction

This contribution is confined to cytological aspects of symbiosis between eukaryotic unicellular organisms and blue-green prokaryotic photosynthesizing endosymbionts. It covers the following genera: *Cyanophora, Glaucosphaera, Gloeochaete, Glaucocystis, Rhopalodia, Geosiphon,* and *Paulinella.* Results that are discussed are taken both from our own investigation (all genera mentioned except *Rhopalodia* and *Geosiphon*) and from the literature.

Blue-green endosymbionts, termed cyanelles by Pascher (1929), are thought to be phylogenetically derived from free-living Cyanophyceae which in the past invaded or were phagocytosed by the host and which became established (Whatley and Whatley 1980) in its cytoplasm. The subsequent coevolution of both partners ensured the survival of the endosymbiont in the extreme habitat of the host cytoplasm and its transmission in the reproductive cycle of the host which also covers the regulation of the symbiont population by the host organism (Smith 1980). This has only been possible through an increasing physiological and biochemical integration of the symbiotic partners.

It is postulated that the degree of integration achieved in a particular endocyanome is reflected in the morphology and ultrastructure of its cyanelles as compared to that of free-living Cyanophyceae and chloroplasts. The author regards the view that stages of increasing integration of the endosymbiotic partner as revealed by ultrastructural and biochemical investigations mark the true way of the evolution of plastids with caution.

Materials and Methods

For our investigations we have used the following cultures: 1555 *Cyanophora paradoxa,* 2323 *Gloeochaete wittrockiana,* 2344 *Glaucocystis nostochinearum,* all isolated by the author and maintained in the Institut für Allgemeine Botanik, Universität Hamburg; UTEX LB 1662 *Glaucosphaera vacuolata. Paulinella chromatophora* was

1 Institut für Allgemeine Botanik und Botanischer Garten, Universität Hamburg, D–2000 Hamburg 52, FRG

Compartments in Algal Cells and Their Interaction
ed. by W. Wiessner, D. Robinson, and R.C. Starr

taken from a ditch near Hamburg (Kies 1974). Culture and processing for electron microscopy was as described earlier (Kies 1976).

Cytology of Cyanelles

With the exception of the endosymbiont of *Geosiphon pyriforme* which is a species of *Nostoc* (Schnepf 1964), and which can be cultivated outside the host, a phycomycete, all other endocyanomes under consideration cannot be assigned with certainty to any existing genera of free-living Cyanophyceae and cannot be cultivated outside their host (Trench 1982). The number of cyanelles to be regulated by the host cell varies from one to four (*Cyanophora*), one to two (*Paulinella*) to many (*Glaucocystis*, in *Gloeochaete* approximately 100). They may be spherical, rod-shaped, wedge-shaped or sausage-shaped. The smallest cyanelles are encountered in *Gloeochaete* (3 μm), the largest in *Paulinella* (20 μm long and curved) (Fig. 10).

The arrangement of thylakoids in cyanelles of *Cyanophora, Glaucosphaera, Gloeochaete,* and *Paulinella* (Figs. 2–6) is peripheral and concentric. There is a clear distinction between chromatoplasm and centroplasm as seen in coccoid free-living Cyanophyceae. In the nearly colourless cyanelles of *Rhopalodia gibba* the few thylakoids are arranged radially mostly in a peripheral zone (Drum and Pankratz 1965, Floener 1982). The endosymbiotic *Nostoc* in *Geosiphon* shows no clear distinction between chromatoplasm and centroplasm, the thylakoids which show a remarkable degree of branching are distributed across the entire cell (Schnepf 1964). Phycobiliproteins are organized in interthylakoidal phycobilisomes (Fig. 6). They have been shown to occur in all cyanelles besides *Rhopalodia* (Floener 1982) which needs reinvestigation.

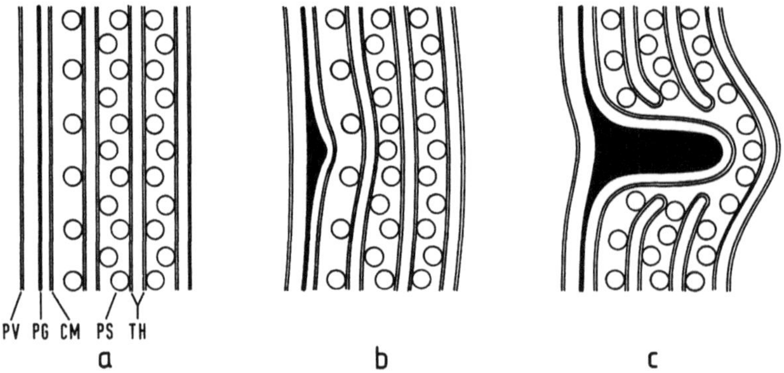

Fig. 1a-c. Diagram of cyanelle division in *Cyanophora, Gloeochaete, Glaucocystis* and *Paulinella* which follows the constrictive type of coccoid Cyanophyceae. *a* Prior to cyanelle division; *b* wall material (peptidoglycan) is accumulated in the plane of division forcing the cell membrane to invaginate; *c* the ingrowing cross wall cuts through the thylakoids. **PV** membrane of phagsome vesicle; **PG** peptidoglycan cyanelle wall; **CM** cyanelle membrane; **TH** thylakoid; **PS** phycobilisome

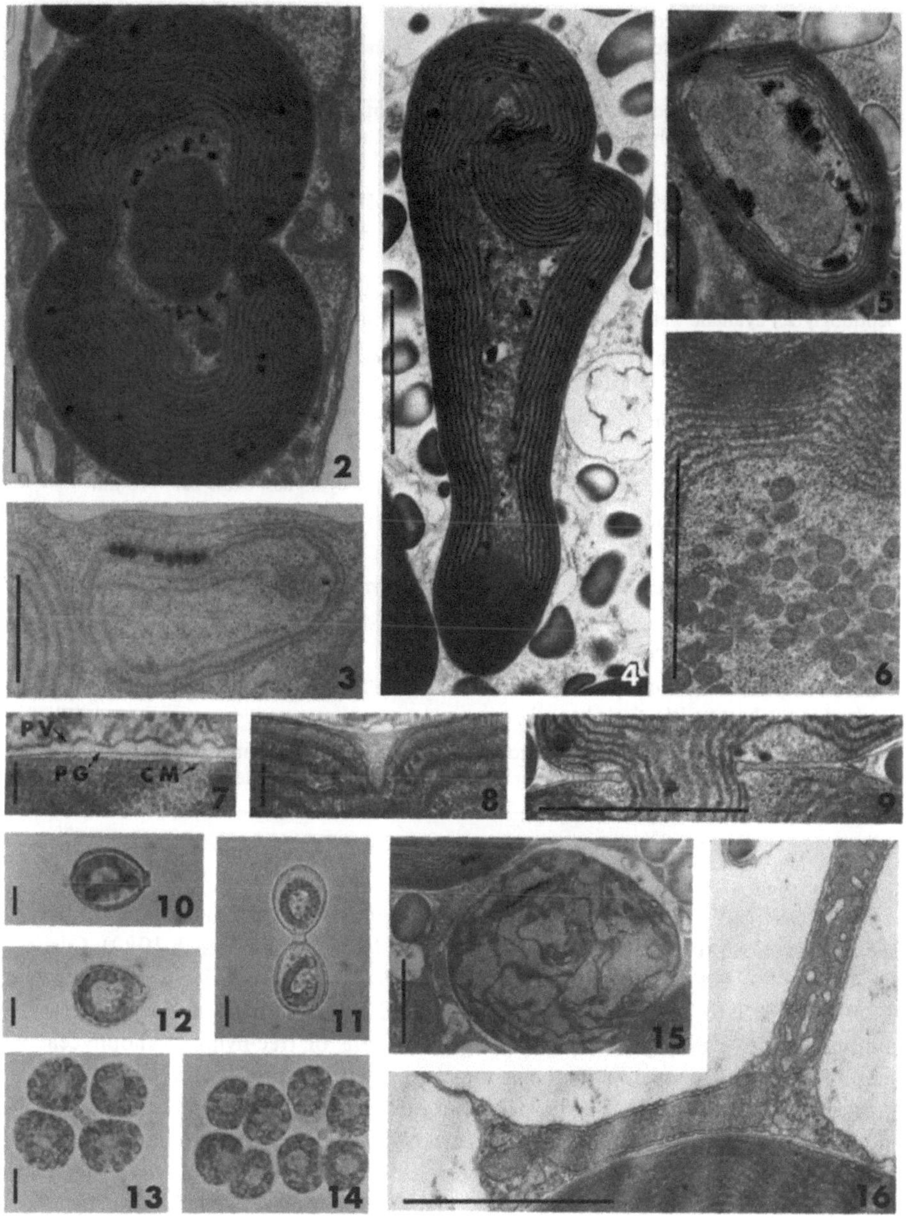

Figs. 2–16 (captions for figs. 4–16 see page 194)

Fig. 2. Cyanelle of *Cyanophora paradoxa* during division. Between concentrically arranged thylakoids dark staining lipid droplets. Large central body (carboxysome ?) surrounded by polyphosphate granules. Scale bar = 1 μm

Fig. 3. Cyanelle of *Glaucosphaera vacuolata* showing eye-spot-like arrangement of lipid globules. Scale bar = 1 μm

Fig. 4. Cyanelle of *Glaucocystis nostochinearum.* Chromatoplasm with lipid globules, centroplasm with polyphosphate granules, narrow end of cyanelle with pyrenoid-like body (carboxysome ?). Cyanelle surrounded by starch grains lying freely in the host cytoplasm. Scale bar = 1 μm

Fig. 5. Cyanelle of *Gloeochaete wittrockiana.* In the centroplasm large central body (carboxysome ?) surrounded by polyphosphate granules. Adjacent to the cyanelle starch grains. Scale bar = 1 μm

Fig. 6. Cyanelle of *Paulinella chromatophora.* Chromatoplasm with concentrically arranged thylakoids and interthylakoidal phycobilisomes. Centroplasm with hexagonal particles (carboxysome ?). Scale bar = 1 μm

Fig. 7. Cyanelle of *Paulinella chromatophora* enclosed in a vesicle of the host (PV) showing the peptidoglycan cyanelle wall (PG) outside the cyanelle membrane (CM). Scale bar = 0.1 μm

Fig. 8. Division furrow in cyanelle of *Cyanophora paradoxa* showing accumulation of wall material (peptidoglycan) and invagination of cyanelle membrane. Scale bar = 0.1 μm

Fig. 9. Cyanelle division in *Glaucocystis nostochinearum.* Division furrow cuts through the thylakoids. Scale bar = 1 μm

Fig. 10. *Paulinella chromatophora* with two sausage-shaped cyanelles. Scale bar = 10 μm

Fig. 11. During division of the thecamoeba the cyanelles are evenly distributed to the daughter individuals. Scale bar = 10 μm

Fig. 12. Cross division of cyanelle in young *Paulinella chromatophora.* Scale bar = 10 μm

Fig. 13 and 14. Loss or extrusion of cyanelles during colony formation in *Gloeochaete wittrockiana.* Scale bar = 10 μm

Fig. 15. Degeneration or digestion by the host of cyanelle of *Glaucocystis nostochinearum* still retained in the host vesicle. Scale bar = 1 μm

Fig. 16. *Paulinella chromatophora.* Microbody (peroxisome ?) associated with a cyanelle and a mitochondrium suggesting metabolic interaction of compartments. Scale bar = 1 μm

The centroplasm contains the cyanelle DNA which can be visualized by the DNA-specific fluorescent stain 4'6-diamido-2-phenylindole (DAPI) (Trench 1982). Cyanelles operate the reductive pentose phosphate cycle (Trench et al. 1978, Kremer et al. 1979, Kremer and Kies 1979). Assimilation of CO_2 in the light involves ribulose-1, 5-bis-phosphate carboxylase which in vegetative cells of free-living Cyanophyceae is present in the centroplasm in the form of hexagonal particles, the carboxysomes (Stewart and Codd 1975). The centroplasm of cyanelles usually contains dark staining bodies which are also thought to be carboxysomes, though these bodies have not been isolated yet. Presence of RuBP carboxylase has been confirmed only for cyanelles of *Cyanophora* (Codd and Stewart 1977), both subunits are coded by the cyanelle genome (Heinhorst and Shively 1983). Hexagonal particles 130–160 nm in diameter and bound by a nonunit membrane in the centroplasm of cyanelles from *Paulinella* (Fig. 6) resemble the typical cyanophycean carboxysomes. In cyanelles of *Gloechoaete* (Fig. 5) there is only one large hexagonal central body which is mostly bound by a nonunit membrane, probably a large carboxysome. In the cyanelle of *Cyanophora* (Fig. 2) the dark staining central body is not surrounded by a nonunit membrane, the same is true for a similar structure in cyanelles of *Glaucocystis* (Fig. 4) located at the narrow end and sometimes (Schnepf and Brown 1971) called a pyrenoid. Dark staining central bodies are also found in some cyanelles of *Glaucosphaera* whereas none are reported for cyanelles of *Rhopalodia*.

Prokaryotic and eukaryotic cells that change from a free to an endosymbiotic mode of life show an enhanced release of assimilation products into their new environment, the host organism. They become leaky (Whatley and Whatley 1980) and no longer accumulate reserve material.

The ultrastructural equivalent to this is the lack of polyglucoside (or glycogen) granules in cyanelles (Schnepf and Brown 1971, Kies 1980). These granules are a carbon reserve and are typically present in free-living Cyanophyceae. Most of the assimilated carbon is transported from the cyanelles to the host in the form of glucose, synthesized to maltose in the host and finally laid down as starch granules in the host cytoplasm (Trench et al. 1978, Kremer et al. 1979). Cyanelles seem to have lost the ability to build the highly branched polyglucans of free-living Cyanophyceae (Frederic 1977a,b).

The pathway of assimilatory nitrate reduction has been thoroughly investigated in *Cyanophora* (Floener and Bothe 1982, Floener et al. 1982, Böttcher et al. 1982) and shown to be similar to that of eukaryotic cells with chloroplasts. The host reduces nitrate to nitrite which enters the cyanelle and is there reduced to NH_4^+ which in turn is exported to the host. This is probably the reason why cyanelles do not store nitrogenous compounds. They lack the cyanophycin granules of free-living Cyanophyceae (Schnepf and Brown 1971, Kies 1980), which are a copolymer of arginine and aspartic acid (Simon 1971).

Another type of inclusion found in Cyanophyceae are the polyphosphate granules lying in the centroplasm. They are usually identified under the electron microscope by their mottled appearance because they evaporate partially under the electron beam. Polyphosphate granules which may function as a source of phosphates and as an energy source (Shively 1974) have been seen in cyanelles of *Cyanophora* (Fig. 2), *Gloeochaete* (Fig. 5) and *Glaucocystis* (Fig. 4) (Hall and Claus 1967).

Cyanelles usually contain lipid droplets (Figs. 2–4) located between thylakoids in the chromatoplasm. All inclusions found in cyanelles have been identified by the various authors only by their electron microscopic appearance; no biochemical or histochemical investigations on isolated inclusion bodies have been performed.

The cell envelope of free-living Cyanophyceae which serves to retain the shape of the cell, to protect it against damage, to control to a certain extent the influx and efflux of substances and which takes part in cell division and cell growth, resembles that of Gram-negative bacteria. It consists of a 1-10 nm thick peptidoglycan layer (LII layer) adjacent to the cell membrane, followed by a 7–10 nm thick outer membrane (L IV layer) which both constitute the cell wall. Outside this is a mucilaginous sheath up to 1 μm thick (Drews and Weckesser 1982). It may be expected that due to their endosymbiotic mode of life the cell envelope of cyanelles became reduced or was entirely lost.

In the endosymbiotic *Nostoc* of *Geosiphon* all layers of the cyanophycean cell envelope are present (Schnepf 1964). In cyanelles of *Rhopalodia*, (Drum and Pankratz 1965) report the presence of a 28-nm-thick cell envelope outside the cell membrane made up of four to five layers. From this it may be concluded that cyanelles of *Rhopalodia* also possess the typical cell wall but without the mucilaginous sheath of free-living Cyanophyceae. In contrast to this, the cyanelles of *Cyanophora, Gloeochaete, Glaucocystis,* and *Paulinella* are surrounded by only one 8–10 nm thick electron dense layer (Figs. 7–9), which corresponds to the peptidoglycan layer. In cyanelles of *Cyanophora* this layer is sensitive to lysozyme (Schenk 1970), its peptidoglycan has been analyzed by Aitken and Stanier (1979). The cyanelles of *Glaucosphaera* are devoid of a remnant cell envelope (Fig. 3) thus resembling true chloroplasts. McCracken et al. (1980) think that all cyanelles in *Glaucosphaera* are part of one deeply lobed chloroplast and that *Glaucosphaera* is not an endocyanome but a red alga, but *Glaucosphaera* gives a similar pattern of primary assimilation products as *Cyanophora, Gloeochoaete,* and *Glaucocystis* and the heterosides such as glycerol galactosides and mannosides which are typical for Rhodophyta are lacking in *Glaucosphaera* (Kremer et al. 1979).

It is postulated that during coevolution of the symbiotic partners first the external mucilaginous sheath and the outer membrane, then the peptidoglycan layer of cyanelles were lost because there was no more need for a thick and costly protective cell envelope. The transport of substances between the symbiotic partners is facilitated if cyanelles are surrounded by only a thin wall layer or have no wall at all. Cyanelles are enclosed in a vesicle. Its membrane is usually thought to be homologous with a phagosome membrane (Schnepf 1964). Cavalier-Smith (1982) discusses the possibility that the vesicle membrane is derived from the outer layer of the procaryotic endosymbiont whereas the original phagosome membrane was lost during coevolution.

Division of Cyanelles

Endocyanomes such as *Cyanophora* and the other glaucophytes are of high phyletic age. This can be concluded from the high degree of biochemical (Kremer et al. 1979, Floener et al. 1982, Trench 1982) and genetical (Herdman and Stanier 1977, Löffelhard et al. 1983) integration of both symbiotic partners. It seems surprising that cyanelles of *Cyanophora, Gloeochaete, Glaucocystis,* and *Paulinella* have still retained a peptidoglycan wall. It is therefore postulated that this wall plays a significant role during cyanelle division which in the cyanelles mentioned above follows the constrictive type of coccoid Cyanophyceae (Fig. 1a-c; Drews and Weckesser 1982). In cyanelles the plane of division is always perpendicular to their long axis. At the beginning of cross wall formation synthesis and deposition of peptidoglycan material takes place in the division plane. A ring-shaped wall structure is formed which forces the cyanelle membrane to invaginate (Fig. 8). As the cross wall formation proceeds centripetally (Fig. 9) the thylakoids of the chromatoplasm are cut through.

After the cyanelle division is completed the cross wall splits, the daughter cyanelles separate and the host vesicle divides to that each cyanelle is enclosed in a separate vesicle. It is reasonable to suggest that in free-living Cyanophyceae and in cyanelles that are surrounded by a complete or reduced wall a different division mechanisms is in operation than in cyanelles that lack a wall and in true chloroplasts. The aquisition of a new type of division mechanism rendering the peptidoglycan wall obsolete may have been a prerequisite for the evolution of chloroplasts.

Regulation of the Number of Cyanelles per Host Cell

The regulation of cyanelle number per host cell necessitates the interaction of the host with its symbiotic compartment. The number of cyanelles is species-specific and more or less constant. There is only one report by Pascher (1929) on *Peliaina,* a flagellate endocyanome of very rare occurrence, where, due to defective regulation during cell division, flagellates devoid of cyanelles were formed which survived and divided again. In *Geosiphon* the host organism can be freed experimentally from its endosymbiotic Nostoc (Knapp 1933) and cultivated separately. In obligate endocyanomes with a low number of cyanelles per host cell, regulation is achieved by coordinate division of both partners. In *Paulinella* two cyanelles are present. They are evenly distributed during cell division to the daughter thecamoebae, thereafter each cyanelle divides in two (Figs. 10–12). In *Cyanophora* the division of host and cyanelles is not in synchrony during all phases of growth of a culture. During the logarithmic phase flagellates contain 1 or 2 cyanelles, during the stationary phase 2 to 4, seldom 8 (Trench et al. 1978).

In endocyanomes with a large number of cyanelles per host cell, accessory mechanisms for the regulation of cell number are encountered. During cell division and colony formation in *Gloeochaete,* cyanelles single or in small groups are lost or extruded from their host cell (Figs. 13, 14). They are still enclosed in the common mucilaginous sheath surrounding the colony where they finally disintegrate.

Degeneration or digestion by the host of single cyanelles is seen in *Gloeochaete* (Kies 1976) and *Glaucocystis* (Fig. 15). In thin sections of healthy vegetative cells of *Gloeochaete* up to one third of the cyanelles present are found in all stages of degeneration. No histochemical investigations have been performed to demonstrate the action of lysosomal enzymes or acid hydrolases on degenerating cyanelles. It is not known whether cyanelles that are weakened in some way are digested, or whether this is an active way of regulating the number of cyanelles. Nevertheless the number of cyanelles per host cell becomes reduced. In *Peliaina,* Pascher (1929) reported the digestion of single cyanelles which were finally extruded from their host.

Spatial Relationships Between Cyanelles and Host Organelles

To deduce a metabolic interaction on the basis of a close spatial relationship between compartments is highly speculative. The examples given here are therefore not conclusive. In *Gloeochaete* single cyanelles are encircled by a cisterna of the endoplasmic reticulum (Fig. 29 in Kies 1976) which originates from the nuclear membrane and which is a transport system in eukaryotic cells. In *Cyanophora* a microbody (probably a peroxisome) is frequently found adjacent to the cyanelle. It is present in the cleavage furrow of dividing cyanelles and divides synchronously with it (Pickett-Heaps 1972). A similar association of a cyanelle with a microbody (peroxisome ?) plus a mitochondrium is frequently seen in *Paulinella* (Fig. 16). It resembles the association of chloroplast, peroxisome and mitochondrium found in some algae and higher plants where these compartments interact in photorespiration (Frederick and Newcomb 1969).

Remarks on the Taxonomic Treatment of Endocyanomes

From the high degree of integration of both partners (which means at the same time a low degree of autonomy for the cyanelle) in *Cyanophora* it follows that its cyanelles cannot be treated as a separate taxonomic unit named *Cyanocyta korschikoffiana* (Hall and Claus 1963). The same is true for cyanelles of *Gloeochaete, Glaucocystis* and *Paulinella*. Cyanelles take an intermediate position between Cyanophyceae and chloroplasts.

The establishment of cyanelles probably has occurred several times, in a host of different systematic positions. However, electron microscope investigations (Kies 1979, 1980) have revealed that the endocyanomes *Cyanophora, Gloeochaete, Glaucocystis* and to a lesser extent, *Glaucosphaera* share most systematically important characters such as the presence of a lacunae system just below the cytoplasma membrane, dorsoventral construction of flagellated forms and two heterodynamic flagellae with nontubular hairs, arising in an apical depression, presence of a cruciate flagellar root system with two (in *Cyanophora,* Melkonian 1983) or four (in *Gloeochaete* and *Glaucocystis,* Kies 1980) multilayered structures, formation of an open spindle during mitosis, lack of centrioles and of a phycoplast, starch granules lying freely in the host cytoplasm, mitochondria with flattened cristae as well as biochemical characters such as the pattern of primary assimilation products (Kremer et al. 1979). This particular combination of characters, which is not encountered in any other common algal class, is an argument in favor of the revival of the Glaucophyceae Bohlin 1901 (= Glaucocystophyceae Schaffner 1922) and the phylum *Glaucophyta Skuja* 1954 (Kies and Kremer, in press).

References

Aitken A, Stanier RY (1979) J Gen Microbiol 112:219–223
Böttcher U, Brandt P, Müller B, Tischner R (1982) Z Pflanzenphysiol 106:167–172
Cavalier-Smith T (1982) Biol J Linn Soc 17:289–306
Codd GA, Stewart WDP (1977) FEMS (Microbiol) Lett 1:35–38
Drews G, Weckesser J (1982) In: Carr NG, Whitton BA (eds) The biology of Cyanobacteria. Blackwell, Oxford London Edinburgh Boston Melbourne, pp 333–357
Drum RW, Pankratz S (1965) Protoplasma 60:141–149
Floener L (1982) Disseration Univ. Köln
Floener L, Dannenberg G, Bothe H (1982) Planta (Berl) 156:70–77
Floener L, Bothe H (1982) Planta (Berl) 156:78–83
Frederic JF (1977a) Phytochemistry (Ox) 16:1489–1490
Frederic JF (1977b) Phytochemistry (Ox) 16:1571–1573
Frederic SS, Newcomb EH (1969) J Cell Biol 43:343–353
Hall WT, Claus G (1963) J Cell Biol 19:551–563

Hall WT, Claus G (1967) J Phycol 3:37–51
Heinhorst S, Shively JM (1983) Nature (Lond) 304:373–374
Herdman M, Stanier RY (1977) FEMS Microbiol Lett 1:7–12
Kies L (1974) Protoplasma 80:68–89
Kies L (1976) Protoplasma 87:419–446
Kies L (1979) Ber Dtsch Bot Ges 92:445–454
Kies L (1980) In: Schwemmler W, Schenk HEA (eds) Endocytobiology: Endosymbiosis and cell research. De Gruyter, Berlin, pp 7–19
Kies L, Kremer BP (in press) In: Margulis L, Chapman D, Corliss J (eds) Handbook of Protoctista
Knapp E (1933) Ber Dtsch Bot Ges 51:210–216
Kremer BP, Kies L, Rostami-Rabet A (1979) Z Pflanzenphysiol 92:303–317
Löffelhardt W, Mucke H, Crouse EJ, Bohnert HJ (1983) Current Genetics 7:139–144
McCracken DA, Nadakavukaren MJ, Cain JR (1980) New Phytol 86:39–44
Melkonian M (1983) In: Schenk HEA, Schwemmler W (eds) Endocytobiology II. De Gruyter, Berlin, pp 1003–1007
Pascher A (1929) Jahr wiss Bot 71:386–462
Pickett-Heaps J (1972) New Phytol 71:561–567
Schenk HEA (1970) Z Naturforsch 25b:640–656
Schnepf E (1964) Arch Mikrobiol 49:112–131
Schnepf E, Brown RM Jr (1971) In: Reinert J, Ursprung II (eds) Origin and continuity of cell organelles. Springer, Berlin Heidelberg New York, pp 299–322
Shively JM (1974) Annu Rev Microbiol 28:167–187
Simon RD (1971) Proc Nat Acad Sci USA 68:265–267
Smith DC (1980) In: Schwemmler W, Schenk HEA (eds) Endocytobiology: Endosymbiosis and cell research. De Gruyter, Berlin, pp 317–332
Stewart WDP, Codd GA (1975) Br Phycol J 10:273–278
Trench RK, Pool RJ Jr, Logan M, Engelland A (1978) Proc R Soc Lond Biol Sci 202:423–443
Trench RK (1982) In: Round FE, Chapman DJ Jr (eds) Progress in Phycological Research 1: 257–288
Whatley FR, Whatley JM (1980) In: Schwemmler W, Schenk HEA (eds) Endocytobiology: Endosymbiosis and cell research. De Gruyter, Berlin, pp 848–865

Endosymbioses: Evolutionary Trends in Cellular Compartmentation

W. REISSER[1]

Introduction

Ultrastructure of eukaryotic cells is characterized by compartmentation, e.g., by the existence of special areas surrounded by membranes. These areas have special functions in cell metabolism and are partially endowed with their own genetic apparatus which is different from the nuclear one. This discontinuity in cellular organization has been intriguing to biologists since the beginning of microscope studies, and has initiated many speculations on development of eukaryotic features. Two main hypotheses trying to explain compartmentation are being discussed nowadays:

a) endogenous differentiation of membrane enclosed areas (Cavalier-Smith 1981) and
b) uptake of foreign cells and subsequent symbiosis formation (endosymbiosis theory) (Margulis 1981).

Although the case is still not settled, there are sufficiently valid arguments for the symbiotic origin of plastids. They have probably evolved from some prokaryotic predecessors with an oxygen evolving type of photosynthesis which have been taken up by host cells.

Study in the field of organelle evolution is mainly based on the comparison of ultrastructure, biochemistry, and genetics in organelles and presumably related prokaryotes, but is hampered by the fact that a status quo situation is described and fossile records are lacking. Yet symbiosis formation is a dynamic process and hence characterized not only by morphological but also by physiological and cybernetic features. Therefore a somewhat new approach to the study of eukaryote's evolution is proposed: Different types of contemporary endosymbiotic systems should be compared with each other and screened for common features from which basic principles of symbiosis formation and evolution per se can be deduced. Thus it is the aim of this paper to discuss this way of looking at endosymbiotic systems. The theme is deliberately restricted to endosymbiotic photosynthesizing ones, e.g., to units in which the smaller photosynthesizing partner lives permanently within a unicellular or multicellular host. Smaller symbiotic partners can show different kinds of reductions but can always be distinguished from cell organelles because they possess a completely autonomous genetic apparatus.

---

1 FB Biologie der Universität Marburg, Lahnberge, D-3550 Marburg, FRG

Compartments in Algal Cells and Their Interaction
ed. by W. Wiessner, D. Robinson, and R.C. Starr

Formation of Endosymbiotic Systems

Although knowledge of mechanisms involved in symbiosis formation is fundamental for the understanding of symbiotic relationships, it has up to now been only rudimentary. The reason for this information gap is the fact that contemporary endosymbiotic systems are well established units consisting of highly adapted partners with an unknown evolutionary history. Fortunately it is possible to circumvent this problem to some extent by a "trick". In some endosymbiotic units partners can be separated from each other and cultured independently. Reinfection experiments in these systems can mimic the "historic" process of symbiosis formation. Nevertheless, it should always be kept in mind that experiments are run with highly specialized organisms and this means a sort of artificial impact on the system as compared to natural contemporary conditions. The same reservations hold for experiments with systems which need a regular infection of host's eggs by specially adapted algal partners.

In general, infection events procede in the same way in different kinds of hosts: The smaller symbiotic partner is phagocytosed by the presumptive host cell and sequestered in an individual special so-called perialgal vacuole. Cells not suitable for symbiosis formation get into food vacuoles and are digested. These events have been studied best in green *Paramecium* (Reisser 1981a) and in green *Hydra* (Muscatine et al. 1975). In perialgal vacuoles symbiotic partners are protected, at least for some time, against host digestive enzymes by still unknown mechanisms, presumably either by inactivation of attacking enzymes or by a special structure of the perialgal vacuole membrane preventing fusion of primary lysosomes, as has been proposed for the *Paramecium bursaria-Chlorella* sp. system (Meier et al. 1980). Thus endosymbiotic partners get access to their presumptive host by a natural process. They are isolated from each other, being stored within the host in special individual membrane surrounded areas, e.g., compartments, where they remain topologically outside and have maximum surface contact to their partner. These two mechanisms, isolation and compartmentation, are probably basic in endosymbiosis formation.

Application of the above rules to the evolution of organelles leads to the assumption that it originated with host cells which must have been able to phagocytose, e.g., to form plasmalemma-derived membranes. As a matter of fact, there exist no contemporary endosymbiotic systems between prokaryotes. So most probably presumptive host cells were already eukaryotic and able to perform some digestive process with engulfed material. In addition, it is reasonable to postulate that concomitant to digestion via phagocytosis some recognition system must have evolved in order to avoid autophagy. Probably specific recognition systems in eukaryotes developed only slowly, so that uptake of partners could have been a common event during an early "experimental" phase of evolution. There is good evidence that contemporary plastids have evolved several times independently via symbiosis formation, e.g., in *Rhodophyta, Chlorophyta, Chromophyta,* and *Euglenophyta* (Gibbs 1981; Whatley et al. 1979).

It is also conceivable that during the "experimental" phase one host cell could have harbored more than one symbiotic partner species, at least for some time, but that finally only the best-adapted one persisted. A somewhat similar situation has been observed during infection experiments with aposymbiotic *Paramecium bursaria:*

Ciliates can be successfully infected with different species of *Chlorella*, but the most stable symbiotic system is formed only with the original symbiotic alga (Reisser 1981a).

The time schedule of organelle evolution is not known. The only experimentally successful endosymbiosis formation between formerly nonsymbiotic organisms which has been reported (Jeon 1972) was achieved between an amoeba and a bacterium. It took about 200 generations of host cells until a symbiotic relationship was established.

Evolutionary Trends in Development of Endosymbiotic Systems

The problem is how to define evolution or development in the context of endosymbiotic systems. Contemporary algae-invertebrate associations hold a rather isolated position and no line of development of one type of partner within the same host range seems to exist. Certainly some basic trends in evolution do exist, which can probably be deduced best by comparing different types of contemporary systems although their age is not known.

The evolutionary advantage in symbiosis formation between a heterotrophic and a photosynthesizing organism is the fact that the resulting system is able to settle new ecological niches which can be depleted of organic nutrients. Suggesting an evolutionary trend toward increasing effectivity, some basic postulates can be deduced: The symbiotic system should be

a) permanent, i.e., hereditary, and should consist of
b) specific partners which thus can be highly adapted to each other by
c) an increasing refinement of the special properties which they contribute to the association.

It is quite obvious that permanency and specificity are prerequisites for specialization. As to permanency, in stable systems with unicellular hosts, e.g., in ciliate-algae associations, smaller partners are distributed to daughter cells during host cell division (Reisser 1981a). In multicelluar hosts the situation is far more complicated. Sexuality and hence the formation of generative cells work as a sort of filter. In some systems, e.g., in green *Hydra,* algae can obtain access to eggs (Muscatine and McAuley 1982). This same tendency of maternal transmission can be observed also with chloroplasts, e.g., in angiosperms (Whatley 1982). In other systems, e.g., in *Convoluta roscoffensis,* symbiotic algae have to pass a short free-living phase before infection of host eggs (Oschman 1966). Probably the *Hydra* type is the more advanced one in the sense of increasing effectivity, although in the established system the algae in *Convoluta* seem to be more adapted by losing their theca than partners in *Hydra* which retain a prominent wall.

Permanency also requires the establishment of mechanisms protecting partners from host digestion. In general, this seems to be achieved by the formation of the above-mentioned perialgal vacuoles, e.g., by compartmentation.

A fairly high specificity of both partners is characteristic for all contemporary stable endosymbiotic systems. Among ciliates usually only one species of a genus forms a symbiotic association with an algal partner which is specific for one host, i.e., algae isolated from green *Paramecium* cannot easily replace algae in green *Clima-*

costomum and vice versa. This specificity results from the action of elaborate cell recognition systems which stabilize the association and protect it against the invasion by less-adapted organisms (Pool 1979; Reisser et al. 1982).

A third important force in evolution of endosymbiotic systems is probably the tendency to increase effectivity by concentration on partner-specific morphological and physiological features. In general, symbiotic algae differ from free-living counterparts by excretion of significant amounts of organic substances. In green ciliates and in green *Hydra,* excretion of sugars by the symbiotic chlorellae can be observed also after isolation from the system (Muscatine et al. 1967), e.g., this means an evolutionary drawback for competition in an aposymbiotic environment. For these algae the symbiotic way of life offers an ecological niche and thus is a good example for the evolutionary advantage of "metabolic loss" (Zamenhof and Eichhorn1967).

It is admissable to speculate that refinement of special features allows a reduction of other properties which are of minor value for the persistence of the symbiotic association. Accordingly, in the *Convoluta roscoffensis-Tetraselmis convolutae* system the host stops feeding after infection with suitable algae (Muscatine et al. 1974), and in green *Paramecium* uptake of bacteria decreases with increasing light intensity (Weis 1974).

As to the algal partner, there seems to be a tendency in morphological reduction although there are only few contemporary examples for this process. Whereas in green ciliates and in green *Hydra* the cell wall of the algal partner remains intact and probably is absolutely necessary for the formation of perialgal vacuoles, in the *Convoluta roscoffensis* system the alga loses both theca and flagellar apparatus, living within the host as a sort of protoplast (Oschman 1966). Yet is is difficult to decide whether the *Convoluta* system is a more advanced one because of its more primitive transmission mechanism. In foraminifers and in cyanellae associations also a tendency toward reduction of cell walls can be observed (Kies 1980; Lee 1983).

A very exciting example for the reduction of not symbiosis-essential features and concentration on the photosynthetic apparatus can probably be observed in cryptomonads, where a eukaryotic photosynthesizing symbiont has been reduced to a nearly organellar status during evolution. Besides its plastid, the so-called nucleomorph, a special type of ribosomes, and its plasmalemma remind of its symbiotic origin, the contemporary cryptomonad plastid being surrounded by four membranes. In dinoflagellates and in euglenoids, the third membrane around plastids can be interpreted as a remnant of either a symbiotic eukaryotic alga's plasmalemma or a phagocytotic membrane around an engulfed plastid. In *Mesodinium rubrum,* a cryptomonad-like alga and in *Peridinium balticum* a chrysophycean one live endosymbiotically at a rather reduced stage (Gibbs 1981, Hibberd 1977, Lefort-Tran 1982, Margulis 1976, Whatley et al. 1979).

It is very interesting to note that during reduction processes membranes are obviously an extremely conservative feature, e.g., the compartmentation principle proves to be very successful in evolution. The reason for this conservatism is probably the multifunctional task of membranes. They serve as borderlines of compartments, protect them against digestion and regulate exchange of metabolites between host and symbiont (see also Schnepf, this Vol.).

Concomitant to morphological, also physiological reductions can be observed in algal symbionts. Whereas symbiotic chlorellae isolated from ciliates, green *Hydra* or sponges can be cultured in a standard medium (Muscatine et al. 1967, Reisser 1981b) isolation and independent growth of partners from cryptomonads or dinoflagellates is no longer possible. Reduction can be observed also on the genetic level leading to an ultimate change of partners from symbionts to organelles sensu stricto. In *Euglena* sp. chlorophyll synthesis could be shown to be coded in the nucleus, in cyanellae systems there are some conclusive hints to a host-symbiont cooperation in pigment synthesis (Ortiz and Stutz 1980, Trench 1982).

Application of developmental strategies in endosymbiotic systems to evolution of contemporary plastids such as chloroplasts allows speculation on the following scenario:

A photosynthesizing, possibly *Prochloron*-like, and photosynthate-excreting prokaryote is phagocytosed by a heterotrophic nucleate cell. It is enclosed individually by a protective membrane which regulates metabolic exchange between the partners. After the association has become permanent, e.g., hereditary, in further evolutionary steps features of the minor partner which are not essential for photosynthesis are eliminated. Concomitant to these processes, the whole system develops an immune barrier against infection by further potential partners. Taking over coding of some symbiont properties by the nuclear apparatus then means the final shift from symbiotic partner to organelle.

Cybernetics of Endosymbiotic Systems

A basic problem of endosymbiotic systems is how to coordinate growth rates of partners in order to maintain stability. Although only few data are available on regulatory processes in contemporary systems preliminary results on ciliate-algae associations show truly cybernetic features in the sense of feedback control systems. Indeed, it is of some heuristic value to look at those associations as microecological units consisting of genetically independent partners sharing a common habitat which usually cannot be invaded by other organisms.

Green *Paramecium* is a good example for the study of control mechanisms: symbiotic algae in perialgal vacuoles are usually not digested but grow and divide within the unicellular host according to light, water, and available nutrients, e.g., algal population size depends solely on environmental conditions which are typical for each ecological situation (Reisser 1980). A similar regulation type may be conceivable for chloroplasts in higher plants, as has been shown by studies on *Phaseolus vulgaris* (Whatley 1980).

In stable endosymbiotic systems partner functions can be coordinated and integrated to such a degree that the whole unit reacts to an external stimulus as one organism. This qualitatively new way of behavior means an exciting example of the evolutionary value of symbiosis formation, the "mere living together of dissimilar organisms" triggers the formation of new entities. The most exciting example for such a shift is photobehavior of symbiotic units which is reported for such systems as cyanellae associations (Pascher 1929), *Convoluta roscoffensis, Anthopleura elegantissima,* and for foraminifera-zooxanthellae systems. The best-studied example is photoaccumulation of ciliate-algae associations (Niess et al. 1982, Reisser and Häder 1984). Neither alga-free *Paramecium bursaria* nor the isolated *Chlorella* show any photoaccumulation, but integration of both organisms by forming a symbiotic unit leads to a qualitatively new way of behavior. It is most important to stress that this process is a genuine cybernetic one. There is no establishment of any new morphological or physiological structures, but the formation of a new type of informational network. Analogous events are conceivable for the evolution of photobehavior in flagellated algae. It is interesting to note that in those organisms the locomotory apparatus is very conservative, whereas photoreceptors are of different type (Kivic and Walne 1983). Hence it is admissable to speculate that evolution started with heterotrophic flagellated cells which aquired photosynthesizing partners of different kind. After symbiosis formation then interactions between the host locomotory apparatus and the algal light perceiving system could be established.

Conclusions

It is quite evident that the study of contemporary endosymbiotic systems can only give indirect access to the understanding of organelle evolution. One of the most serious reservations results from the fact that photosynthesizing partners in recent systems are usually eukaryotic, whereas plastid evolution ultimately started with prokaryotes. Probably integration, at least on the genome level, can work with prokaryotes in another way than with eukaryotes. Yet in some organisms, such as cryptomonads, reduction of an eukaryotic partner can be observed.

Another reservation which should be always borne in mind is the fact that nowadays the ecological situation is probably quite different from that during the start of plastid evolution, e.g., evolutionary pressure toward development of photosynthesizing eukaryotes was much greater in former times. For contemporary associations with photosynthesizing partners only few ecological niches have been left.

There is some evidence that the evolutionary impact of symbiosis formation has shifted to combinations with nonphotosynthesizing partners, e.g., to partners meeting special requirements of "nowadays" host cells and their ecological situation. What is more, evolution of ectosymbiotic associations seems to be favored, perhaps because of the well-established immune barriers in contemporary eukaryotes.

Probably for the evolution of higher land plants mycorrhiza formation was of decisive value (Pirozynski and Malloch 1975) and in many contemporary plant-prokaryote associations the latter's capability to fix molecular nitrogen is exploited, as is the case in rhizobia- and cyanobacteria-systems. Accordingly, ectosymbiotic plant-cyanobacteria associations do not rely primarily on the prokaryote's photosynthesis but on its N_2-fixing capability. Interestingly, in these ectosymbiotic associations evolutionary trends similar to those in endosymbiotic units can be observed: Smaller symbiotic partners are enclosed in special regions such as the culture chambers in *Blasia* sp. (Duckett et al. 1977), development of both host and symbiont can be synchronized as in *Azolla* sp. (Hill 1975), in some cases symbiotic cyanobacteria seem to be host-specific (Grilli-Caiola 1980).

Thus it can be concluded that symbiosis formation is still an important, powerful factor in evolution.

References

Cavalier-Smith T (1981) In: Carlile MJ, Collins JF, Moseley BEB (eds) Molecular and cellular aspects of microbial evolution. Cambridge University Press, pp 33–84

Duckett JG, Prasad AKSK, Davis DA, Walker S (1977) New Phytol 79:349–362

Gibbs SP (1981) NY Acad Sci 361:193–208

Grilli-Caiola M (1980) New Phytol 85:537–544

Hibberd DJ (1977) J Mar Biol Assoc UK 57:45–61

Hill DJ (1975) Planta (Berl) 122:179–184

Jeon KW (1972) Science (Wash DC) 176:1122–1123

Kies L (1980) In: Schwemmler W, Schenk HEA (eds) Endocytobiology vol 1. De Gruyter Berlin New York, pp 7–19

Kivic PA, Walne PL (1983) Bio Systems 16:31–38

Lee JJ (1983) Int Rev Cytol 14:49–77

Lefort-Tran M (1982) Cell Function and Differentiation 8:67–76

Margulis L (1976) Exp Parasit Rev 39:277–349

Margulis L (1981) Symbiosis in cell evolution. Freeman & Co San Francisco

Meier R, Reisser W, Wiessner W, Lefort-Tran M (1980) Z Naturforsch 35c:1107–1110

Muscatine L, McAuley PJ (1982) Cytobios 33:111–124

Muscatine L, Boyle JE, Smith DC (1974) Proc R Soc Lond B Biol Sci 187:221–234

Muscatine L, Karakashian SJ, Karakashian MW (1967) Comp Biochem Physiol 20:1–12

Muscatine L, Cook CB, Pardy RL, Pool RR (1975) Symp Soc Exp Biol 29:175–203

Niess D, Reisser W, Wiessner W (1982) Planta (Berl) 156:475–480

Ortiz W, Stutz E (1980) FEBS Lett 116:298–302

Oschman JL (1966) J Phycol 2:105–111

Pascher A (1929) Jahrb wiss Bot 71:386–462

Pirozynski KA, Malloch DW (1975) Bio Systems 6:153–164

Pool RR (1979) J Cell Sci 35:367–379

Reisser W (1980) In: Schwemmler W, Schenk HEA (eds) Endocytobiology vol 1. De Gruyter Berlin New York, pp 97–104

Reisser W (1981a) Ber Dtsch Bot Ges 94:557–563

Reisser W (1981b) Protoplasma 105:273–284

Reisser W, Häder D-P (1984) Photochem Photobiol 39:673–678

Reisser W, Radunz A, Wiessner W (1982) Cytobios 33:39–50

Trench RK (1982) In: Round FE, Chapman DJ (eds) Progress in phycological research vol 1. Elsevier Biomedical Press B V Amsterdam, pp 257–288

Weiss DS (1974) Trans Am Micros Soc 93:135–140

Whatley JM (1980) New Phytol 86:1–16

Whatley JM (1982) Biol Rev 57:527–569

Whatley JM, John P, Whatley FR (1979) Proc R Soc Lond B Biol Sci 204:165–187

Zamenhof S, Eichhorn HH (1967) Nature (Lond) 216:456–458

General Aspects of Location and Interaction of Respiratory and Photosynthetic Electron Transport in Blue-Green Algae

G. SANDMANN, S. SCHERER and P. BÖGER[1]

Introduction

Blue-green algae (cyanobacteria) are photoautotrophic organisms with oxygenic photosynthesis (for review see Krogmann 1977). Their main difference to other algae is their prokaryotic organization – they lack particular organelles such as chloroplasts or mitochondria, but have different types of membranes (Lang 1968, Golecki and Drews 1982). Nevertheless, some filamentous species can differentiate special cells known as heterocysts which serve for reductive fixation of atmospheric nitrogen (Haselkorn 1978).

Blue-green algae respire carbohydrates in the dark (van Liere et al. 1979) as well as in the light (Scherer and Böger 1982). There are some early observations about partial inhibition of respiration by light in blue-green algae (Brown and Webster 1953, Hoch et al. 1963, Jones and Myers 1963) which were interpreted by interference of respiration with photosynthesis. There has been increasing interest in this topic during the last years; Binder (1982) has reviewed the development up to 1981. Thus, this survey intends to present the progress of the last three to four years in this field and to give an outline on the current status of interaction between respiration and photosynthesis in vegetative cells and heterocysts of blue-green algae. Some older references, although important, should be obtained from the more recent publications quoted here.

Light and Respiration

In blue-green algae, respiration starts with carbohydrate degradation primarily via the oxidative pentose-phosphate pathway (Stewart 1980, Smith 1982). NADPH is provided by NADP-specific glucose-6-phosphate and 6-phosphogluconate dehydrogenases. Both NADPH and NADH can be oxidized by membrane preparations, but the rates for the former are higher than for the latter substrate (Biggins 1969, Leach and Carr 1970, Peschek 1980a, Sandmann and Malkin 1983a, Stürzl et al. 1982, 1984).

During measurement of respiration in the light by monitoring either CO_2 release or oxygen consumption, care must be taken to avoid interference by refixation of

1 Lehrstuhl für Physiologie und Biochemie der Pflanzen, Universität Konstanz, D-7750 Konstanz, FRG

Compartments in Algal Cells and Their Interaction
ed. by W. Wiessner, D. Robinson, and R.C. Starr

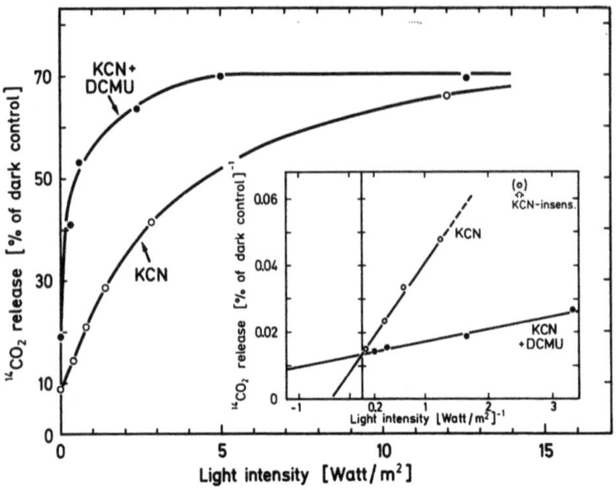

Fig. 1. Interaction of photosynthetic and respiratory electron transport. Light-dependent $^{14}CO_2$ release of prelabeled *Anabaena variabilis* is increased by DCMU inhibiting the electron flow from PS II. CO_2 release is insensitive to KCN, which inhibits both cytochrome oxidase and ribulose-1, 5-bisphosphate carboxylase. Concurrent light-dependent oxygen uptake with KCN plus DCMU present exhibits about the same light-intensity curve. Strict competition is apparent between electrons delivered from photosystem II vs electrons from carbohydrate breakdown (see *inset*). Both pathways feed into the photosynthetic electron-transport system at the same site between photosystems II and I. DCMU, 10 μM; KCN, 2.5 mM; 100% is equivalent to 4200 dpm. (See Scherer et al. 1982, for details)

CO_2 or oxygen evolution, respectively. The problem of refixation can be overcome by complete prelabeling of cellular carbohydrates by $^{14}CO_2$ assimilated in the light and measuring $^{14}CO_2$ release after addition of excess unlabeled bicarbonate (Scherer and Böger 1982). Figure 1 demonstrates such a $^{14}CO_2$-release experiment with intact cells, measured at different light intensities when either cytochrome oxidase (and CO_2 fixation) were inhibited by KCN or when both cytochrome oxidase and photosystem II were blocked (with KCN plus DCMU present). Both curves exhibit a hyperbolic shape with increasing light intensities. The double-reciprocal plot indicates competition for a common site used by reducing equivalents originating from respiration or photosystem II (Scherer et al. 1982).

O_2 uptake by the cytochrome-oxidase reaction of intact illuminated cells can be measured in the presence of DCMU which inhibits photosynthetic oxygen evolution (Hoch et al. 1963, Imafuku and Katoh 1976, Sandmann and Malkin 1984). In the light, the (dark) rate of cytochrome-oxidase activity was inhibited to about 75%. With *Aphanocapsa* a closer examination of this light effect at increasing light intensities showed that inhibition of cytochrome-oxidase coincided with the onset of photosynthesis (Sandmann and Malkin 1984). Both inhibition of the cytochrome-oxidase reaction, as well as stimulation of carbohydrate degradation in the light with cytochrome-oxidase inhibited by KCN (Scherer et al. 1982), suggest an interaction of respiratory and photosynthetic electron transport. This is further supported by the observation that NADH or NADPH oxidation is unaffected by light or dark conditions, whereas the oxygen consumption by cytochrome oxidase is inhibited in the light (Sandmann and Malkin 1983a; "Kok" effect, see also Scherer et al. 1982, for *Anabaena variabilis*).

The proposed regulation of glucose-6-phosphate dehydrogenase through inhibition of NADPH and ATP (negative feedback) generated in the light seems to be of minor importance (for details of this enzyme regulation see Smith 1982).

Partial Reactions and Components of the Respiratory Chain

In eukaryotic organisms, the respiratory electron-transport chain consists of a subsequent arrangement of NADH dehydrogenase (complex I), the ubiquinone pool, the cytochrome b-c complex (complex III), soluble cytochrome c-550, and cytochrome oxidase (complex IV) (cf. Boyer et al. 1977).

Membrane preparations from various species of blue-green algae oxidize NADH or NADPH with concurrent electron transfer to oxygen. The electron-transport reaction was shown to be sensitive to rotenone (Horton 1968, Peschek 1980a), to KCN (Horton 1968, Biggins 1969, Peschek 1980a, Stürzl et al. 1982, 1984, Sandmann and Malkin 1984), and to azide (Horton 1968). These results coincide with the inhibitory action of the compounds mentioned on mitochondrial respiration. However, an inhibitory effect of antimycin A, an inhibitor of the mitochondrial b-c complex (Slater 1973) was not observed with blue-green algae (Horton 1968, Peschek 1980a). Measurement of the activities of assumed respiratory and photosynthetic cytochrome b-c/b-f complexes in membranes of *Aphanocapsa* with duroquinol as electron donor and reduced mammalian cytochrome c as acceptor gave no indication for contribution of an antimycin A-sensitive respiratory cytochrome b-c complex (Sandmann and Malkin 1984). The reaction was sensitive to o-phenanthroline, DBMIB, and DNP-INT.

As was demonstrated with several blue-green algae, soluble c-type cytochromes from various organisms can serve as electron donors to cytochrome oxidase with high efficiency (Lockau 1981, Kienzl and Peschek 1982). Furthermore, plastocyanin, a blue copper protein which is formed in exchange of cytochrome c-553 in some blue-green algae depending on the copper supply of the medium (Sandmann and Böger 1980, 1981), is also a good electron source for cytochrome oxidase (Lockau 1981). Reconstitution experiments of the complete photosynthetic and respiratory electron-transport chains of *Nostoc muscorum* (Stürzl et al. 1982, Alpes et al. 1984) provide evidence that (soluble) cytochrome c-553 is involved in both pathways (see next Section). Noteworthy, these electron transports were reconstituted specifically with *Nostoc* cytochrome c-553 only.

Attempts were made to isolate the components of the respiratory electron-transport chain. A dehydrogenase was obtained from *Aphanocapsa*. The solubilized and partially purified preparation had affinity for both NADPH and NAD with k_m values 77 and 220 μM, respectively (Sandmann and Malkin 1983b). This enzyme was less sensitive to rotenone than mitochondrial NADH dehydrogenase. Another feature of this algal dehydrogenase is the presence of only one redox-active iron-sulfur center (Sandmann 1984) as compared to eukaryotic NADH dehydrogenases which contain 5 to 7 iron-sulfur centers (Onishi 1973, Ragan 1976). Furthermore, kinetic analyses of oxygen uptake have been reported recently with reference to nucleotide concentrations exhibiting different k_m and v_{max} values for NADH (and NADPH) in the light and in the dark (Stürzl et al. 1984), which may indicate light modulation of their activity.

Unlike mitochondria, blue-green algae lack ubiquinones (Lester and Crane 1959, Carr et al. 1967, Peschek 1980b). The only quinones detectable were plastoquinone-9 and phylloquinone (vitamin K_1). There is no doubt about the involvement of the plastoquinone pool in photosynthetic electron transport (Lightbody and Krogmann 1966), whereas the nature of the respiratory quinone pool is still controversial. Hirano et al. (1980) and Aoki and Katoh (1982) interpreted their results from flash-induced absorbance changes and chlorophyll-fluorescence kinetics in favor of participation of plastoquinone in respiratory electron flow. In contrast, reconstitution experiments (Peschek 1980b) and the higher affinity of isolated NAD(P)H dehydrogenase for naphthoquinones vs. benzoquinones (Sandmann and Malkin 1983b) suggest an involvement of phylloquinone in respiratory electron transport.

A purified cytochrome b_6-f complex has been obtained from *Anabaena.* It contains two cytochromes b-564 per one cytochrome f-557 and one "Rieske" iron-sulfur center (Krinner et al. 1982). This complex showed characteristic photosynthetic-type inhibition by DBMIB (I_{50} = 0.3 μM) and DNP-INT (I_{50} = 12.6 μM). Furthermore, plastoquinol-9 was a better electron donor than ubiquinol-9. The authors emphasize that they found no indication of an additional cytochrome b-c complex.

Cytochrome oxidase in cyanobacteria has been identified by spectral properties in acetone-extracted membranes, as an aa_3-type cytochrome-containing enzyme complex (Peschek 1981a,b). The reduced minus oxidized difference spectrum showed the typical α-band at 605 nm and a β-band at 445 nm. These peaks shifted to 590 and 430 nm in CO-difference spectra. There is a recent report on proton-translocating activity of cytochrome oxidase from *Anacystis* (Peschek et al. 1983).

Interaction of Respiratory and Photosynthetic Electron Transport

The experiment described in Figure 1 was interpreted in terms of a common sequence for electrons from photosystem II and respiratory carbohydrate degradation (Scherer et al. 1982). By means of electron-transport measurements with isolated membrane material it was also shown that both NADH and NADPH act as reductants for photosystem I (Binder et al. 1981, Sandmann and Malkin 1983a, Stürzl et al. 1984).

Figure 2 illustrates how experimental evidence for an interconnection of respiration and photosynthetic electron transport has been obtained. Oxygen uptake mediated by cytochrome oxidase was recorded with shocked spheroplasts from *Aphanocapsa.* As the experiments were carried out in the light, oxygen evolution had to be inhibited by either DCMU or hydroxylamine (upper and middle traces). In the latter case, hydroxylamine instead of water was the electron donor to photosystem II. The

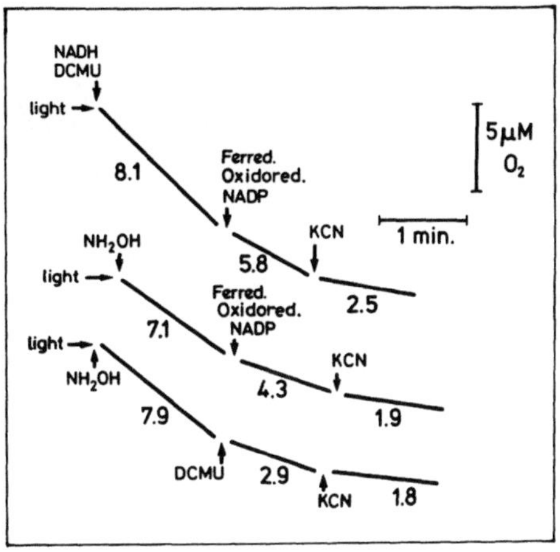

Fig. 2. Respiratory oxygen uptake of *Aphanocapsa* spheroplasts with either photosystem II or NAD(P)H dehydrogenase as electron source. *Upper trace* shows cytochrome-oxidase activity with NADh as electron source after inhibition of PS II with DCMU (10 μM). The rate is decreased when NADP (together with ferredoxin and ferredoxin-NADP oxidoreductase, E.C. 1.18.1.2) is added as electron acceptor for a concurrent PS-I reaction. KCN inhibits cytochrome oxidase. *Middle trace* demonstrates a similar electron flow with NH_2OH as electron donor, which feeds into the electron-transport chain via PS II in a DCMU-sensitive reaction (*lower trace*). Rates are expressed in μmol O_2/mg CHl x h (Sandmann and Malkin 1984). Addition of catalase had no effect. The rate of 2.9 μmol O_2/mg Chl x h is due to endogenous donors (see lower trace). Spheroplasts from this organism are especially useful since they retain light inhibition of NADH-dependent respiration, similar to intact cells, after osmotic shock. At the moment this finding is not evident using material from filamentous blue-green algae

activity of photosystem I was found comparably low and was enhanced by addition of NADP, ferredoxin, and ferredoxin-NADP oxidoreductase (cf. lower and middle traces). Both, NADH oxidation (upper trace) as well as electron flow from photosystem II (middle trace) drive terminal oxidation, which can be markedly inhibited by KCN. Regardless of the electron donor, stimulation of photosystem-I activity in both cases decreased the cytochrome-oxidase reaction. In the dark, NADH-dependent respiration was about 3 times higher than in the light (Sandmann and Malkin 1983a). Thus, competition of cytochrome oxidase and photosystem I for reducing equivalents must be assumed. Experiments are under way in our laboratory to strengthen this hypothesis.

The redox state of plastoquinone was examined by monitoring the early transient of chlorophyll fluorescence induction in *Synechococcus* cells (Aoki and Katoh 1982). In starved cells, plastoquinone was reduced after addition of fructose in the dark. This reduction was enhanced in the presence of KCN. Apparently, NADPH from carbohydrate degradation can reduce the plastoquinone pool especially when terminal oxidation is impaired.

Most likely, the "photosynthetic" cytochrome b-f complex is the component reduced by quinone. Its participation as a common link was concluded from redox changes of cytochromes b-563 and f-556, both oxidized by photosystem I (in the light) as well as by cytochrome oxidase during anaerobic to aerobic transition (Peschek and Schmetterer 1982). Photooxidation changes of cytochrome f-556 with shocked *Aphanocapsa* spheroplasts in the presence of either NADH or duroquinol as electron donors was increased by KCN-inactivation of cytochrome oxidase (Sandmann and Malkin 1984). The same was observed for the extent of cytochrome c-553 oxidation after illumination. Both observations indicate partial oxidation of cytochromes f and c by cytochrome oxidase in the dark steady state. A similar approach to demonstrate an involvement of cytochrome c-553 in respiration was per-

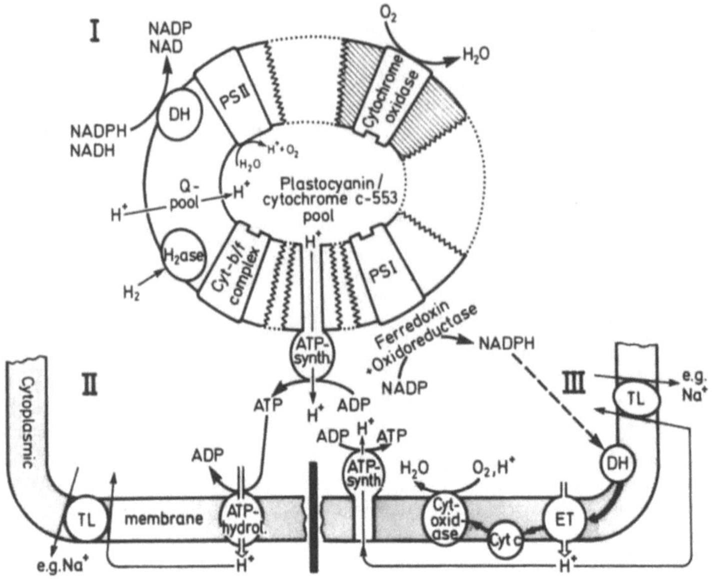

Fig. 3. Proposed arrangement of redox components and ATPases in blue-green algae. *Part I* shows respiratory and photosynthetic electron transport on thylakoid membranes. An NAD(P)H dehydrogenase (DH) donates electrons into a quinone pool. This Q-pool is oxidized by an antimycin-A insensitive cytochrome b-f complex. A reaction site is marked at the inner side of three complexes to indicate that soluble cytochrome c-553 or plastocyanin serve as electron acceptors or as donors (to PS I and cytochrome oxidase). The lower part indicates two different modes of proton efflux through the cytoplasmic membrane. In *II* a cytoplasmic ATP-hydrolase is indicated pumping protons out of the cell. No ATP-synthase is present. In *III* a tentative respiratory electron-transport chain with an ATP-synthase on the cytoplasmic membrane accounts for proton extrusion (see text for further explanation). No sizes of complexes or membrane thickness are intended to be indicated here

formed by Hirano et al. (1980). They looked at the re-reduction of cytochrome c-553 after photooxidation by a short flash in starved *Synechococcus* cells, which was stimulated either by addition of fructose or by inhibition of cytochrome oxidase with KCN.

Figure 3, part I, summarizes the current view on the arrangement of respiratory and photosynthetic pathways in thylakoids of blue-green algae. However, further work, especially reconstitution of the respiratory chain from isolated individual components, is necessary for unequivocal acceptance. From the current state of knowledge, a different localization of respiratory and photosynthetic electron-transport chains on the thylakoids cannot be excluded. Hydrogenase (H_2ase) is included as a third electron donor to the quinone pool. This link can be concluded from donor and inhibitor studies with membranes from *Anacystis* (Peschek 1980) and *Anabaena* heterocysts (Eisbrenner and Bothe 1979, Houchins and Hind 1982, Ernst et al. 1983), respectively. Details about uptake hydrogenase in heterocysts are given in the Section "Bioenergetics of Heterocysts".

Localization of Respiratory Electron Transport

There is disagreement as to whether respiratory chains are present both on thylakoids and cytoplasmic membranes of blue-green algae. Ultrachemical detection of cytochrome oxidase gave inconsistant results with different species (Bisalputra et al. 1969, Peschek et al. 1981). Different break-points in temperature profiles for respiratory and photosynthetic activities were interpreted in terms of separate membrane-lipid environments for both processes (Scherer et al. 1981, Peschek et al. 1982a). Oxidation of different exogenously added c-type cytochromes (from several bacteria and eukaryotes) by intact spheroplasts was taken as evidence for association of respiration with the cytoplasmic membrane (Peschek et al. 1982b).

Dark oxygen-dependent proton extrusion of cells or spheroplasts as demonstrated by Scholes et al. (1969) and Peschek et al. (1983) appears to be doubtful evidence for respiratory activities of cytoplasmic membranes. As illustrated in Fig. 3, parts II and III, proton efflux can also be due to a plasmalemma ATP-hydrolase. Experiments on Na^+ and glucose uptake coupled to proton translocation were not helpful in deciding between the two possibilities (Paschinger 1977, Raboy and Padan 1978, Reed et al. 1981). Closer examination of proton efflux from several blue-green algae, which included concurrent analysis of oxidative phosphorylation and determination of low H^+/e ratios (0.2 to 0.3) led to the conclusion that respiratory electron transport of *Anabaena variabilis* is located essentially on thylakoids, and not on the cytoplasmic membrane, as outlined in Fig. 3, part II (Scherer et al. 1984). The oxygen-dependent (dark) H^+-efflux in *Anabaena variabilis* is suggested to be due to a proton-pumping ATP-hydrolase. Lockau and Pfeffer (1982, 1983) isolated cytoplasmic membranes of *Anabaena* containing a Ca^{2+}-transporting ATPase, its distribution in sucrose-density gradients did not follow that of thylakoids. However, cytochrome oxidase as well as b- and c-type cytochromes were exclusively associated with the thylakoid fraction. Evidence for a proton-pumping ATP-hydrolase localized on the cytoplasmic membrane comes from the vanadate sensitivity of light-induced proton efflux of *Anabaena variabilis* and *Nostoc muscorum* (Scherer et al. 1983, Scherer and Böger 1984). An important step to solve the problem of whether respiratory electron transport is present on cytoplasmic membranes would be the isolation and purification of these membranes. First attempts in this direction, although inconclusive, have been made (Murata et al. 1981, Lockau and Pfeffer 1982).

Bioenergetics of Heterocysts

Most filamentous blue-greens form heterocysts to fix molecular nitrogen. In these cells, photosystem-II reactions are absent (Bradley and Carr 1971, Donze et al. 1972, Tel-Or and Stewart 1977). In a detailed study, Almon and Böhme (1980) have shown that the remaining components are the plastoquinone pool (decreased in size), the cytochrome b-f complex, and photosystem I, which includes P700 and the iron-sulfur centers A and B (Cammack et al. 1976). Some enzymes for photosynthetic carbohydrate assimilation, especially ribulose-1,5-bisphosphate carboxylase, are also missing in heterocysts (Wolk 1982). Therefore, these cells depend on supply of organic compounds from the connected vegetative cells.

Available reductants for ferredoxin, the immediate electron donor to nitrogenase, are the pyridine nucleotides. There is evidence of a functional replacement of photosystem II by NADH as electron donor to photosystem I by feeding electrons into the plastoquinone pool (Houchins and Hind 1982, Schrautemeier et al. 1984a). However, nitrogenase reduction by NADPH and ferredoxin seems to be more complicated. There are indications of regulation by light. As shown with heterocyst homogenates, reduction of ferredoxin (most probably by ferredoxin-NADP oxidoreductase) takes place in the dark provided a high ratio of NADPH/NADP is present. In the light, however, direct ferredoxin reduction by the oxidoreductase is inhibited, but NADPH is oxidized like NADH through a DBMIB-sensitive PS-I reaction (Schrautemeier et al. 1984b). Higher light-dependent respiratory CO_2 release from KCN-treated filamentous blue-green algae (with cytochrome oxidase and ribulose-1,5-bisphosphate carboxylase inhibited) was observed under nitrogen-fixing conditions as compared to cultures grown on combined nitrogen (Scherer et al. 1982). This is feasible, since nitrogenase has to be supplied via photosystem I with reductants originating by (glycolytic) degradation of endogenous carbohydrates.

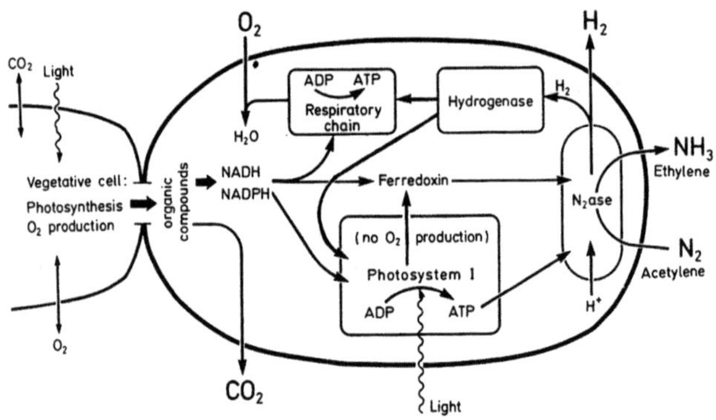

Fig. 4. Assumed interactions of bioenergetic pathways in heterocysts to support the nitrogenase (N_2ase) reaction. Organic compounds imported from vegetative cells are degraded and resulting pyridine nucleotides are used as electron donors for the respiratory chain and photosystem I. As in vegetative cells electron supply to these two pathways and by hydrogenase as well may occur via a quinone pool (see Fig. 3, part I). Also a cytochrome b-f complex and soluble cytochrome c-553 or plastocyanin may be used by all reductants in common. Alternatively to photoreduction through photosystem I, ferredoxin is directly reduced by NADPH in the dark, most probably by NADP-ferredoxin oxidoreductase. Another possibility not included in the scheme would be the reduction of ferredoxin by the pyruvate-clastic reaction (Neuer and Bothe 1982). Nitrogen fixation is always accompanied by production of hydrogen, which is reutilized by an uptake hydrogenase. ATP formation apparently is carried out by the respiratory electron-transport chain with NAD(P)H or in the light using either electron donor(s) for photosystem I or cyclic electron flow around photosystem I. This scheme is not meant to express intracellular localization or relative size of the reaction complexes

Hydrogen formed as an obligatory by-product during nitrogen fixation by nitrogenase is reutilized in heterocysts by an uptake hydrogenase (Kerfin and Böger 1982, Bothe 1982a,b). The occurrence of the latter enzyme is correlated with heterocyst formation (Peterson and Wolk 1978, Houchins and Burris 1982). This hydrogen-consuming enzyme, however, is absent in the filamentous non-heterocystous *Phormidium foveolarum,* which nevertheless is able to induce nitrogenase, fix atmospheric nitrogen and produce excess of hydrogen under anaerobic conditions in vegetative cells, after exhaustion of combined nitrogen in the medium (Weisshaar and Böger 1983, see also Phlips and Mitsui 1983). Electrons from hydrogenase can be transferred either to ferredoxin in a light-dependent reaction or to oxygen via a respiratory chain (Eisbrenner and Bothe 1979, Schrautemeier et al. 1984a). Both pathways are sensitive to DBMIB. Apparently, there is an interconnection of electron transfer from hydrogenase and NADH to photosystem I or to the respiratory chain. With isolated heterocysts, Böhme and Almon (1983) demonstrated that cytochromes f-556 and b-564 were oxidized either by PS I in the light or by oxygen in the dark. Thus, a common segment of electron transport including the plastoquinone pool, the cytochrome b-f complex, and most likely cytochrome c-553 or plastocyanin, can safely be assumed for heterocysts similarly to vegetative cells (Fig. 3, part I). Apparently, oxyhydrogen reaction and/or respiration can provide for a semi-anaerobic environment in the heterocysts to maintain nitrogenase activity.

ATP is provided for the nitrogenase reaction by two different reactions. Isolated heterocysts exhibited high rates of cyclic and noncyclic photophosphorylation (Almon and Böhme 1982). Furthermore, oxygen-dependent ATP formation with hydrogen as substrate was demonstrated by Ernst et al. (1983). The latter authors have shown that light-induced nitrogen fixation generally is limited by reductant supply, rather by energy change. Membrane potential or pH gradient are not decisive for nitrogenase activity (Schrautemeier et al. 1984a).

The possible interactions of heterocyst energy metabolism necessary to support the nitrogenase reaction are summarized in Fig. 4.

Acknowledgments. Studies on blue-green algae bioenergetics conducted in our research group were supported by the Deutsche Forschungsgemeinschaft.

Abbreviations

PS, II, I, photosystem II, I; Chl, chlorophyll (a); DCMU, 3-(3,4-dichlorophenyl)-1, 1-dimethylurea; DBMIB, 2,5-dibromo-3-methyl-6-isopropyl-p-benzoquinone; DNP-INT, 2,4-dinitrophenyl-2'-iodo-3'-methyl-4'-nitro-6'-isopropylphenyl ether

References

Almon H, Böhme H (1980) Biochim Biophys Acta 592:113–120
Almon H, Böhme H (1982) Biochim Biophys Acta 679:279–286
Alpes I, Stürzl E, Scherer S, Böger P (1984) Z Naturforsch 39c, in press
Aoki M, Katoh S (1982) Biochim Biophys Acta 682:307–314

Biggins J (1969) J Bacteriol 99:570–575
Binder A (1982) J Bioenerg Biomembr 14:271–286
Binder A, Bohler MC, Wolf M, Muster P (1981) In: Selman BR, Selman-Reimer S (eds) Energy coupling in photosynthesis. Elsevier, Amsterdam, pp 313–321
Bisalputra T, Brown DK, Weier TE (1969) J Ultrastruct Res 27:182–197
Böhme H, Almon H (1983) Biochim Biophys Acta 722:401–407
Bothe H (1982a) Experientia (Basel) 38:59–64
Bothe H (1982b) In: Carr HG, Whitton BA (eds) The biology of cyanobacteria. Blackwell Scientific Publ, Oxford, pp 87–104
Boyer PD, Chance B, Ernster L, Mitchell P, Racker E, Slater EC (1977) Annu Rev Biochem 46: 957–1026
Bradley S, Carr NG (1971) J Gen Microbiol 68:III
Brown AH, Webster GC (1953) Am J Bot 40:753–758
Cammack R, Tel-Or E, Stewart WDP (1976) FEBS Lett 70:241–244
Carr NG, Exell G, Flynn V, Hallaway M, Talukdar S (1967) Arch Biochem Biophys 120:505–507
Donze M, Haveman J, Schiereck P (1972) Biochim Biophys Acta 256:157–161
Eisbrenner G, Bothe H (1979) Arch Microbiol 123:37–45
Ernst A, Böhme H, Böger P (1983) Biochim Biophys Acta 723:83–90
Golecki JR, Drews G (1982) In: Carr NG, Whitton BA (eds) The biology of cyanobacteria. Blackwell Scientific Publ, Oxford, pp 125–141
Haselkorn T (1978) Annu Rev Plant Physiol 29:319–344
Hirano M, Satoh K, Katoh S (1980) Photosynth Res 1:149–162
Hoch G, Owens OH, Kok B (1963) Arch Biochem Biophys 101:171–180
Horton AA (1968) Biochem Biophys Res Commun 32:839–845
Houchins JP, Hind G (1982) Biochim Biophys Acta 682:86–96
Houchins JP, Burris Rh (1982) J Bacteriol 146:209–214
Imafuku H, Katoh T (1976) Plant Cell Physiol 17:515–524
Jones LW, Myers J (1963) Nature (Lond) 199:670–672
Kerfin W, Böger P (1982) Physiol Plant 54:93–98
Kienzl PF, Peschek GA (1982) Plant Physiol (Bethesda) 69:580–584
Krinner M, Hauska G, Hurt E, Lockau W (1982) Biochim Biophys Acta 681:110–117
Krogmann DW (1977) In: Trebst A, Avron M (eds) Encyclopedia of plant physiology, vol 5. Springer, Berlin Heidelberg New York, pp 625–636
Lang NJ (1968) Annu Rev Microbiol 22:15–46
Leach CK, Carr NG (1970) J Gen Microbiol 64:55–70
Lester RL, Crane FL (1959) J Biol Chem 234:2169–2175
Lightbody JJ, Krogmann DW (1966) Biochim Biophys Acta 120:57–64
Lockau W (1981) Arch Microbiol 128:336–340
Lockau W, Pfeffer S (1982) Z Naturforsch 37c:658–664
Lockau W, Pfeffer S (1983) Biochim Biophys Acta 733:124–132
Murata N, Sato N, Omata T, Kuwabara T (1981) Plant Cell Physiol 22:855–866
Neuer G, Bothe H (1982) Biochim Biophys Acta 716:358–365
Onishi T (1973) Biochim Biophys Acta 301:105–128
Paschinger H (1977) Arch Microbiol 113:285–291
Peterson RB, Wolk CP (1982) Plant Physiol (Bethesda) 61:688–691
Peschek GA (1980a) Arch Microbiol 125:123–131
Peschek GA (1980b) Biochem J 186:515–523
Peschek GA (1981a) Biochem Biophys Res Commun 98:72–79
Peschek GA (1981b) Biochim Biophys Acta 635:470–475
Peschek GA, Schmetterer G (1982) Biochem Biophys Res Commun 108:1188–1195
Peschek GA, Schmetterer G, Sleytr UB (1981) FEMS Microbiol Lett 11:121–124
Peschek GA, Muchl R, Kienzl PF, Schmetterer G (1982a) Biochim Biophys Acta 679:35–43
Peschek GA, Schmetterer G, Wagesreiter H (1982b) Arch Microbiol 133:222–224
Peschek GA, Schmetterer G, Laurtisch G, Muchl T, Kienzl PF, Nietschmann WH (1983) In: Papageorgiou GC, Packer L (eds) Photosynthetic prokaryotes: cell differentiation and function. Elsevier, Amsterdam, pp 147–162

Aoki M, Katoh S (1982) Biochim Biophys Acta 682:307–314
Peterson RB, Wolk CP (1978) Plant Physiol (Bethesda) 61:688–691
Phlips EJ, Mitsui A (1983) Appl Environ Microbiol 45:1212–1220
Raboy B, Padan E (1978) J Biol Chem 253:3287–3291
Ragan CI (1976) Biochim Biophys Acta 456:249–290
Reed RH, Rowell P, Stewart WDP (1981) Eur J Biochem 116:323–330
Sandmann G (1984) Photosynth Res, in press
Sandmann G, Böger P (1980) Plant Sci Lett 17:417–428
Sandmann G, Böger P (1981) In: Akoyunoglou G (ed) Photosynthesis, vol II. Electron transport and photophosphorylation. Balaban Int Science Services, Philadelphia, pp 623–632
Sandmann G, Malkin R (1983a) Biochim Biophys Acta, in press
Sandmann G, Malkin R (1983b) Arch Microbiol, in press
Sandmann G, Malkin R (1984) Arch Biochem Biophys, in press
Scherer S, Böger P (1982) Arch Microbiol 132:329–332
Scherer S, Böger P (1984) FEMS Microbiol Lett 22, in press
Scherer S, Stürzl E, Böger P (1981) Z Naturforsch 36:1036–1040
Scherer S, Stürzl E, Böger P (1982) Arch Microbiol 132:333–337
Scherer S, Stürzl E, Böger P (1984) J Bacteriol 158, in press
Scherer S, Stürzl E, Böger P (1983) In: Sybesma C (ed) Proc 6th Int Congr Photosynth, Brussels, in press
Scholes P, Mitchell P, Moyle J (1969) Eur J Biochem 8:450–454
Schrautemeier B, Böhme H, Böger P (1984a) Arch Microbiol 137:14–20
Schrautemeier B, Böhme H, Böger P (1984b) In: Sybesma C (ed) Proc 6th Int Congr Photosynth, Brussels, in press
Slater (1973) Biochim Biophys Acta 301:129–154
Smith AJ (1982) In: Carr NG, Whitton BA (eds) The biology of cyanobacteria. Blackwell Scientific Publ, Oxford, pp 47–85
Stewart WDP (1980) Annu Rev Microbiol 34:497–536
Stürzl E, Scherer S, Böger P (1982) Photosynth Res 3:191–201
Stürzl E, Scherer S, Böger P (1984) Physiol Plant, in press
Tel-Or E, Stewart WDP (1977) Proc R Soc Lond B Biol Sci 198:61–86
Van Liere L, Mur LR, Gibson CE, Herdman M (1979) Arch Microbiol 123:315–318
Weisshaar H, Böger P (1983) Arch Microbiol 136:270–274
Wolk CP (1982) In: Carr NG, Whitton BA (eds) The biology of cyanobacteria. Blackwell Scientific Publ, Oxford, pp 359–386

Recent Aspects of Heterocyst Biochemistry and Differentiation

H. BOTHE, H. NELLES, T. KENTEMICH, H. PAPEN and G. NEUER[1]

Among prokaryotes, cyanobacteria are the only clear-cut examples of intercellular compartmentation. Filamentous cyanobacteria can form the different cell types of vegetative cells, heterocysts and spore-forming akinetes. Heterocysts are often thought to be anaerobic compartments containing little else than nitrogenase. Experimental evidence of the past few years, however, indicates that the biochemistry of heterocysts is far more complex than believed hitherto. The process of the differentiation of vegetative cells to heterocysts is not fully understood at present, but aspects of this subject have become clear recently. As reviews on heterocyst biochemistry and differentiation are available [5, 8, 13, 24, 32, 44, 60, 69], only some newer findings, not discussed extensively in the preceding publications, will be summarized here.

Nitrogen Fixation in Cyanobacteria

All heterocystous cyanobacteria perform N_2-fixation. This capability is, however, not restricted to heterocystous species (Table 1). The best investigated example of a unicellular cyanobacterium performing N_2-fixation in air is *Gloeocapsa,* now designated *Gloeothece.* Some exceptional filamentous *Oscillatoria* species have been reported to express nitrogenase activity under aerobic conditions. More than 50% of the cyanobacterial strains tested express nitrogenase activity under reduced O_2-tensions. Some of them strictly require the exclusion of O_2, whereas others can tolerate moderate levels of it. For heterocystous species, there is now general agreement among the different experimentators that nitrogenase is confined to heterocysts under aerobic growth conditions. Several lines of evidence suggest that vegetative cells synthesize nitrogenase under microaerobic conditions. The experimental basis for these statements will not be repeated here (see [5, 69]).

Both nitrogenase subunit proteins are rapidly destroyed by exposure to O_2 with the iron-sulfur protein (protein II) being extremely sensitive to O_2. Nitrogenase in cyanobacteria has to cope with the approximately 21% O_2 of air and the O_2 evolved photosynthetically. The extra O_2 produced by photosynthesis obviously causes special problems to the cyanobacteria, and heterocystous and nonheterocystous species apparently solve the problem by different strategies. Nitrogenase is probably located on or near the surface of the thylakoid membranes and may therefore be directly ex-

1 Botanisches Institut, Universität Köln, Gyrhofstr. 15, D-5000 Köln 41, FRG

Compartments in Algal Cells and Their Interaction
ed. by W. Wiessner, D. Robinson, and R.C. Starr

Table 1. N_2-fixing cyanobacteria

a	Nonheterocystous species performing N_2-fixation in air	
	unicellular:	*Gloeothece* (= *Gloeocapsa*) [72]
		Aphanothece (= *Synechococcus* strain [58]
	filamentous:	*Microcoleus chtonoplastes* [50]
		Oscillatoria spp. [14]
b	Nonheterocystous species performing N_2-fixation under strictly anaerobic or under microaerobic conditions	
		Plectonema boryanum [61]
		more than 50% of the filamentous, nonheterocystous species [56]
c	Species possessing heterocysts	
		all species investigated perform N_2-fixation

Table 2. Mechanisms to protect nitrogenase from damage by O_2 in cyanobacteria

Negative evidence	
a	No conformational protection as in *Azotobacter*
b	No Fe-S protein (Shethna protein) which complexes with nitrogenase proteins as in *Azotobacter*
c	Slime does not protect N_2-fixation (a slimeless mutant of *Gloeothece* is a tolerant to O_2 as the wild strain)
Positive evidence	
a	A high turnover of nitrogenase proteins
b	Respiratory protection
c	Protection by the oxyhydrogen reaction
Possible devices	
a	The cell-wall of the heterocysts may be impermeable for the diffusion of gases. However, experimental evidence clearly indicates that O_2 reaches the inside of heterocysts in high concentrations.
b	The ascorbate-glutathione cycle of chloroplasts may occur, but has not yet been demonstrated for heterocysts

posed to the O_2 generated by the photosystem II dependent H_2O splitting reaction in nonheterocystous forms. The experiments of Gallon and coworkers (see [20]) showed that the unicellular *Gloeothece* can protect nitrogenase by separating photosynthesis and N_2-fixation temporarily. Upon illumination, batch cultures of *Gloeothece* show maximal nitrogenase activity 4-8 d and photosynthetic O_2-evolution 12-14 d after inoculation. When grown under alternating light-dark cycles, *Gloeothece* performs virtually all N_2-fixation during darkness and very little when O_2 is evolved in the light. Heterocystous species solve the problem by separating photosynthetic O_2-evolution and N_2-fixation by compartmentation. Vegetative cells perform CO_2-fixation and O_2-evolution, whereas heterocysts do not possess a Calvin cycle and a functionally active photosystem II.

All aerobically living cyanobacteria must protect their nitrogenase against the damage by O_2 of air. Table 2 summarizes positive and negative evidence for the occurrence of protective devices in cyanobacteria. They do not process the special iron-sulfur (Shethna) protein which protects nitrogenase in *Azotobacter* and have no conformational protection mechanism. Slime does not play a role because a slimeless mutant of *Gloeothece* is as tolerant to O_2 as the wild strain [34]. It is often stated that the thick cell wall of heterocysts forms a barrier to the diffusion of O_2 similar as the nodules of legumes or the vesicles of *Frankia* [42]. Indeed, the experiments of Walsby using the gas perfluorocyclobutane [64] gave indications that the envelope of heterocysts has a high resistance to the diffusion of O_2 and N_2. Walsby [65] suggests that the "strategy of the heterocyst envelope should be to reduce the rate of arrival of O_2 in the heterocysts by limiting the gas permeability to a minimum value necessary to satisfy the requirement of the nitrogenase system for N_2". Haury and Wolk [26] described mutants of *Anabaena variabilis* which were deficient in specific glycolipids of the heterocyst envelope and which had no or very low nitrogenase activity under aerobic growth conditions. They therefore suggested that glycolipids are part of the O_2-protective device in the envelope. On the other hand, large amounts of O_2 obviously reach the inside of heterocysts. Heterocysts respire with higher rates than the vegetative cells. When H_2 is added to heterocyst suspensions, H_2 is rapidly utilized in an oxyhydrogen reaction via the respiratory chain. Oxygen must not come through the heterocyst envelope, but may move along the microplasmodesmata which connect vegetative cells with heterocysts. Thus nitrogenase inside the heterocysts must be protected against the damage by O_2 by mechanisms other than an impermeable envelope.

The just mentioned respiratory and particularly the H_2-dependent O_2-uptake probably removes the gas from the nitrogenase site. H_2 is evolved during N_2-reduction catalyzed by nitrogenase in cell-free preparations. Intact cyanobacteria, however, evolve very little H_2 because they immediately reutilize the gas by the oxyhydrogen (= Knallgas) reaction [10]. The physiological function of this H_2-utilization catalyzed by a membrane-bound uptake hydrogenase may, indeed, consist in protecting nitrogenase. Nitrogenase (C_2H_2-reduction) activity was found to be less sensitive to exposure to O_2 when experiments with intact *Anabaena variabilis* [7], isolated heterocysts from the same organism [68] and heterocysts from the marine *Anabaena* CA [39] were performed in the presence of H_2. On the other hand, a mutant of *Gloeothece,* lacking uptake hydrogenase, expresses nitrogenase activity in air, indicating that H_2-utilization is not a prerequisite for nitrogenase to function in this cyanobacterium [35]. In addition to respiration and the oxyhydrogen reaction, a high turnover of the nitrogenase peptides may help to permit aerobic nitrogenase activity [3]. Mullineaux et al. [41] recently found that synthesis of nitrogenase is not repressed by O_2 in *Gloeothece* in contrast to the situation in *Azotobacter* and *Klebsiella.*

Other, still unresolved mechanisms presumably exist in heterocysts. Karni and Tel-Or [37] recently detected high levels of glutathione reductase in addition to superoxide dismutase in heterocysts of *Nostoc muscorum.* Reduced electron carriers (e.g., ferredoxin) in heterocysts may be reoxidized with O_2 yielding the superoxide radical which is converted to H_2O_2 by the superoxide dismutase reaction. The removal of H_2O_2 remains to be elucidated, as Karni and Tel-Or [37] could not demonstrate catalase or a glutathione peroxidase the latter of which occurring in animal cells and catalyzing the reaction 2 GSH + H_2O_2 → GSSG + 2 H_2O. Heterocysts possibly possess the ascorbate-glutathione cycle which is present in chloroplasts [23] and which employs superoxide dismutase and NADPH-dependent glutathione reductase and, in addition to these, an ascorbate peroxidase catalyzing the reaction ascorbate + H_2O_2 → 2 H_2O + dehydroascorbate and a dehydroascorbate reductase mediating the conversion dehydroascorbate + 2 GSH → ascorbate + GSSG. Such a cycle, if functioning in heterocysts, could scavenge toxic oxygen radicals and could also remove O_2 from the nitrogenase site.

The Biochemistry of Heterocysts

Nitrogenase in heterocysts does not show special features. The enzyme is synthesized inside these cells [33]. Heterocysts contain a 2 [Fe-2S] ferredoxin [62], and circumstantial evidence suggests that it functions as electron carrier in N_2-fixation. Under iron-deficiency, cyanobacteria synthesize the flavoprotein flavodoxin (= phytoflavin) which substitutes for ferredoxin with approximately the same efficiency in C_2H_2-reduction by cell-free nitrogenase [4]. Several pathways exist which provide the reducing equivalents for the reduction of ferredoxin in N_2-fixation (Fig. 1). The occurrence of glucose-6-P dehydrogenase and NADPH:ferredoxin oxidoreductase as well as C_2H_2-reduction dependent on glucose-6-P, the dehydrogenase, $NADP^+$, NADPH: ferredoxin oxidoreductase and ferredoxin have amply been demonstrated for heterocysts (see [5]). Because heterocysts do not perform photosynthetic CO_2-fixation and do not generate NADPH by the two photosystems at the expense of H_2O as the electron donor, NADPH:ferredoxin oxidoreductase probably catalyzes the reduction of ferredoxin under physiological conditions. The reaction is thermodynamically unfavourable and therefore requires a large excess of NADPH. Surprisingly, the activity of glucose-6-P dehydrogenase is negatively affected by NADPH and that of NADPH: ferredoxin oxidoreductase by $NADP^+$. Glucose-6-P dehydrogenase shows greatest activity at reduction charges < 0.3 and NADPH:ferredoxin oxidoreductase at those > 0.3 [1]. This indicates that other pathways probably exist for the generation of reducing equivalents in heterocysts. Isocitrate dehydrogenase is particularly active in heterocysts [36, 43, 49]. Reducing equivalents for N_2-fixation may, therefore, also come from isocitrate, the dehydrogenase, $NADP^+$, NADPH:ferredoxin oxidoreductase and ferredoxin. However, recent experiments showed that isocitrate dehydrogenase of cyanobacteria is negatively affected by high reduction charges as glucose-6-P dehydrogenase [49].

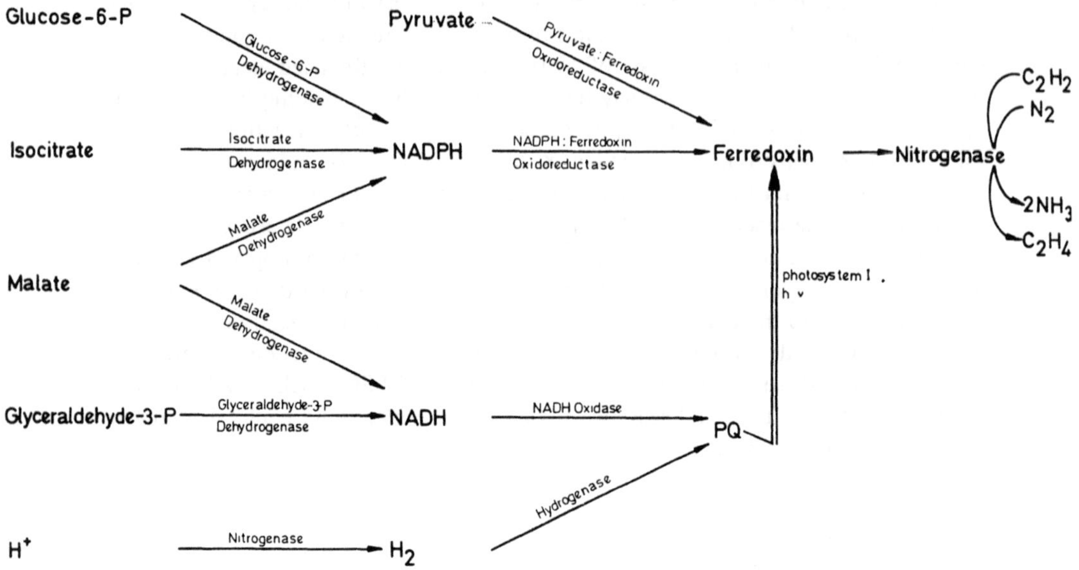

Fig. 1. The electron transport pathways to nitrogenase in cyanobacteria

A peculiar feature of heterocysts is the possession of the pyruvate: ferredoxin oxidoreductase catalyzing the conversion of pyruvate, coenzyme A and oxidized ferredoxin to acetylcoenzyme A, CO_2 and reduced ferredoxin [43]. The enzyme occurs in anaerobes and is very sensitive to exposure to O_2. Pyruvate and coenzyme A also support C_2H_2-reduction by isolated heterocysts. The activity is low in cell-free preparations due to technical difficulties which, however, must not reflect the situation in intact filaments. Another electron donor providing reducing equivalents for N_2-fixation may be glyoxylate which was shown to stimulate C_2H_2-reduction by intact cells [2].

Houchins and Hind [30] showed that NADH donates electrons both to the respiratory and photosynthetic electron transport chain in heterocyst preparations from *Anabaena* 7120. NADH also supported C_2H_2-reduction by isolated heterocysts in a light dependent reaction. We have recently confirmed that NADH oxidation by membrane preparations from heterocysts and vegetative cells is largely stimulated by light and methylviologen [8]. The light dependent NADH oxidation activity is about 1.5-fold higher in heterocysts than in vegetative cells of *Nostoc muscorum.* A NADH oxidase from a membrane preparation of *Nostoc muscorum* can be solubilized by treatment wth Triton X-100 and enriched by chromatography on DE-cellulose and Sephadex G-200. The enzyme contains flavin as demonstrated by fluorescence excitation and emission spectra [8]. The solubilized enzyme utilizes a variety of electron acceptors among which the quinones DBMIB, menadione and benzoquinone give the best activities (Table 3). In contrast, horse heart cytochrome c only poorly supports NADH oxidation. The effects of several inhibitors on NADH oxidation with menadione as the electron acceptor are given in Table 4. DBMIB and DNP-INT obviously

Table 3. Activities of the purified NADH:plastoquinone oxidoreductase in the presence of different electron acceptors

Electron acceptor	nmol NADH oxidized/min/mg protein
1. ---	0.9
2. 2×10^{-4} M menadione	120.5
3. 5×10^{-5} M DBMIB	158.6
4. 2×10^{-4} M benzoquinone	126.9
5. 1×10^{-3} M K-ferricyanide	50.7
6. 5×10^{-5} M cytochrome c (horse heart)	19.0
7. 1×10^{-5} M plastoquinone I	38.0
8. 5×10^{-5} M ubiquinone I	12.6
9. 2×10^{-4} M menadione, enzyme boiled	0.3

NADH oxidase was solubilized from membrane of *Nostoc muscorum* by treatment with Triton X-100 and enriched by $(NH_4)_2SO_4$ precipitation and chromatography on DE-52 cellulose and on Sephadex G-200. NADH-oxidation was determined spectrophotometrically at 340 nm, 25°C and in the dark. The assay contained in a final volume of 3 ml: enzyme (0.04 mg protein) and the following in mM: NADH 0.12, $MgCl_2$ 5, KH_2PO_4/KOH buffer pH 7.0, 50.

act as alternative electron acceptors better than menadione and therefore stimulate activity. NADH oxidation catalyzed by the solubilized enzyme is inhibited to some extent by 2×10^{-5} M rotenone; higher concentrations could not be tested due to the limited solubility of this substance. All the data suggest that heterocysts possess a membrane-bound NADH: plastoquinone oxidoreductase which had been described for the green alga *Chlamydomonas* [21] and the photosynthetic bacterium *Rhodopseudomonas palustris* [47] but which is absent in chloroplasts of higher plants.

Figure 1 also indicates that H_2 supports C_2H_2-reduction in a light dependent reaction. The different experimentators agree that H_2 is the most effective electron donor in supporting this reaction in cell-free preparations. H_2 can, however, not be the only electron donor, because aerobically living cyanobacteria can only reutilize the H_2 evolved during N_2-reduction catalyzed by nitrogenase and otherwise do not come into contact with H_2. The consumption of H_2 is catalyzed by a membrane-bound uptake hydrogenase and proceeds by two different pathways (for reviews see [6, 7]). In the one pathway, H_2 and uptake hydrogenase donate electrons to photosystem I and in the other to the respiratory chain. The electron acceptor for H_2 utilization is plastoquinone or a component close to plastoquinone similar as for NADH oxidation. Both pathways of H_2-utilization probably do not operate independently of each other. In cyanobacteria, photosynthesis and respiration appear to share a common segment of electron carriers consisting of the plastoquinone pool, cytochrome c_{553} and the cytochrome b/f complex on the thylakoid membranes [15, 30, 40, 53]. In addition, part of the respiratory activity may be located on the plasmalemma membrane which may be free of photosynthetic activity in some cyanobacteria [54].

Table 4. Effect of inhibitors on the activity of the purified NADH: plastoquinone oxidoreductase

Inhibitor	% Activity
1. ---	100.0
2. 5×10^{-5} M DBMIB	136.0
3. 2×10^{-5} M rotenone	84.6
4. 1×10^{-5} M DCCD	92.6
5. 1×10^{-4} M TTFA	61.0
6. 5×10^{-5} M DNP-INT	132.0

Experimental conditions as in Table 3. The assays contained 2×10^{-4} M menadione. 100% corresponded to 120.5 nmol NADH oxidized/min x mg protein. Abbreviations: DBMIB = 2,5-dibromo-3 methyl-6-isopropyl-p-benzoquinone = dibromothymoquinone; DCCD = N,N'-dicyclohexylcarbodiimide; TTFA = thenoyltrifluoracetone; DNP-INT = dinitro-phenylether of 2-iodo-4-nitrothymol. The quinones used were a kind gift of Dr. W. Oettmeier, Bochum, FRG.

The controversy about the number of hydrogenases in cyanobacteria does not appear to be settled at present. Houchins and Hind [27, 28] purified and characterized two different hydrogenases, referred to as uptake and reversible hydrogenase from *Anabaena* 7120. Uptake hydrogenase has a molecular weight of 56,000 and the two forms of the reversible hydrogenase had, surprisingly, almost the double and threefold molecular weight of 133,000 and 165,000, respectively. Both enzymes differed in several properties. Most important, reversible hydrogenase catalyzed the consumption of H_2 in the presence of various electron acceptors covering a wide range of midpoint potentials. Reversible hydrogenase also catalyzes the evolution of H_2 from methyl or benzyl viologen reduced by $Na_2S_2O_4$. Reversible hydrogenase, however, did not react with natural electron carriers, e.g., ferredoxin. In contrast, uptake hydrogenase, irrespectively of whether being particulate or solubilized by sonication, reacted rapidly only with electron acceptors of positive midpoint potential but not with methyl or benzylviologen. The activity of the reversible hydrogenase drastically increased by incubating *Anabaena* 7120 under anaerobic conditions [29]. Even under these conditions, where the enzyme was present in large quantities in the cells, H_2 and the reversible hydrogenase only poorly supported any of the reactions tested. H_2 did not support autotrophic growth which should be expected with such hydrogenase activities. Newer findings indicate that the reversible hydrogenase is reversibly activated upon removal of light and O_2 [59]. These authors suggested that the expression of the activity may be under redox control as in the case of the enzyme from *Escherichia coli*.

We observed that the relative activities of H_2-utilization and of methylviologen-dependent H_2-formation drastically change with the preparations of the extracts, and suggested that the reversible, soluble hydrogenase is an artifact of cell-free preparations [17]. Despite the convincinglooking experimental evidence of Houchins and Burris [27-29] we are not yet inclined to give up our suggestion. First of all, it is not true that the uptake hydrogenase is confined to heterocysts as claimed by Houchins and Burris [27-29]. It can readily be demonstrated in a number of cyanobacteria grown under anaerobic conditions and in the presence of H_2 [16,17]. The work of

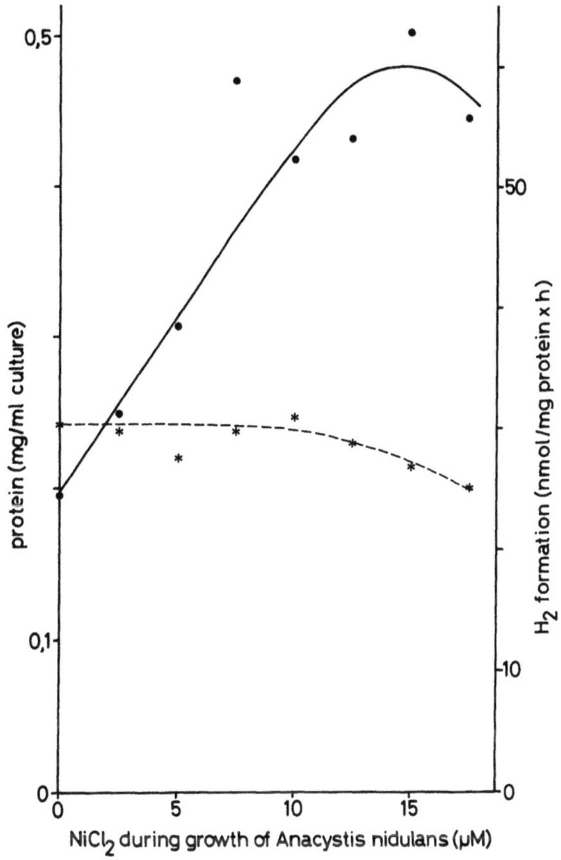

Fig. 2. Effect of $NiCl_2$ during growth on the development of the methylviologen-dependent H_2-evolution activity in *Anacystis nidulans. Anacystis nidulans* was grown for 2 d in the presence of 27 mM EDTA and of $NiCl_2$ the concentrations of which are given in the abscissa of the figure. The cyanobacteria were then concentrated 15 fold by centrifugation (4000 *g*, 5 min) and by suspending in 0.02 M Tris-HCl buffer pH 7.5. The suspension was incubated under a gas phase of 90% argon, 10% H_2 for 24 h and then directly-assayed (without breaking the cells) for methylviologen dependent H_2-evolution activity. The assay vessels contained in 100 ml flasks in a final volume of 15 ml: 8 ml cells with 24-30 mg protein and the following in mM: Tris-HCl buffer pH 7.5, 267; methylviologen, 6; $Na_2S_2O_4$, 13. Gas phase argon. The assay was performed for 60 min at 30°C in the dark and H_2-formation was determined by gaschromatography. A control culture grown without adding Ni and EDTA showed an activity of 48 nmol H_2 evolved/hx mg protein. • --- • H_2-evolution, * --- * protein content of the culture after growth for 2 d in the presence of EDTA and different amounts of Ni

Peschek clearly indicates that the two pathways of H_2-utilization and with them uptake hydrogenase are present in the unicellular *Anacystis nidulans* [51, 52]. The soluble hydrogenase was claimed to be assayed specifically by measuring H_2-evolution from reduced methylviologen [29]. Recent experiments from our laboratory indicate that the expression of this activity is dependent on the presence of Ni in the growth medium (Fig. 2). This can be shown by growing the cultures with EDTA which complexes the contaminant Ni in the medium. The control grown with EDTA but without Ni expressed only low H_2-evolution activity which was considerably enhanced when the cells were grown with Ni the optimum being at 10-15 mM (Fig. 2). The experiment indicates that Ni is either required for the biosynthesis of hydrogenase or is a constituent of the enzyme. The latter is the case in other organisms. Ni has, however, only been found in uptake hydrogenases, and we are not aware of any example (with the possible exception of *Desulfovibrio*) for a reversible hydrogenase containing this metal. Thus we wonder that the aerobically living cyanobacteria are so unique

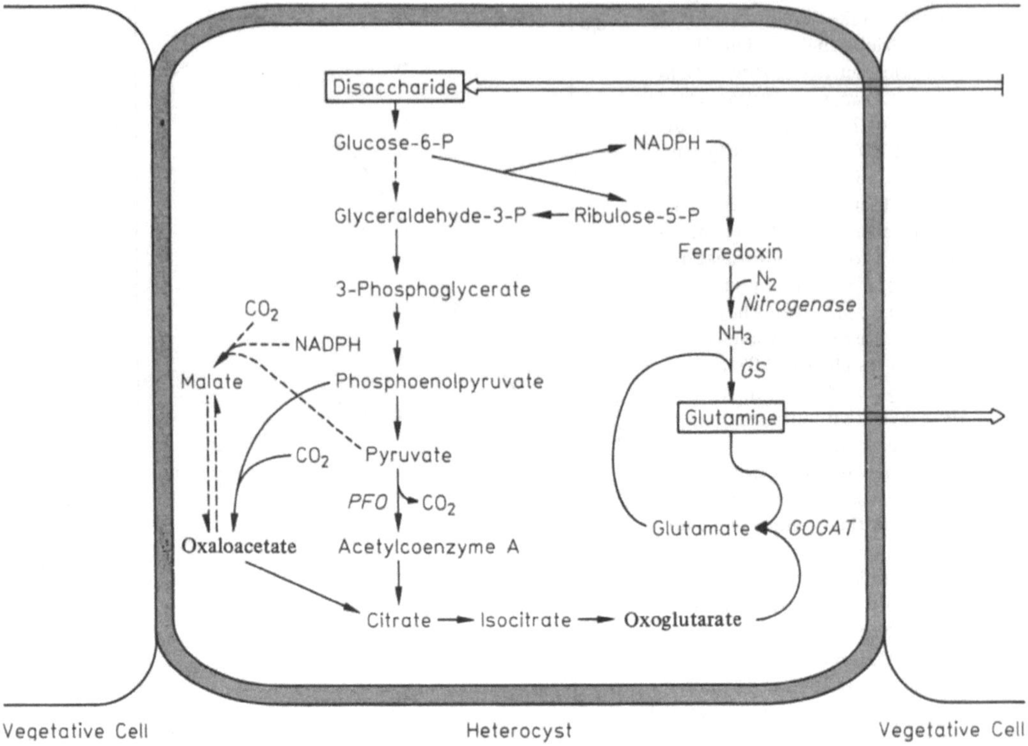

Fig. 3. The heterocyst-vegetative cell relationship [8]

in synthesizing a Ni-containing soluble, reversible hydrogenase with no obvious function.

As shown above, the electrons for N_2-fixation come from different sources (in the dark from NADPH generated from glucose-6-P and isocitrate or from pyruvate, in the light from NADH or H_2). Clearly, the transfer of electrons from different donors to nitrogenase needs to be regulated. Very little is known about the regulation of nitrogenase activity in heterocysts. Hawkesford et al. [31] found a positive correlation between the membrane potential components ($\Delta\psi$) of the proton motive force and nitrogenase activity in *Anabaena variabilis* and *Nostoc muscorum*. They suggested that the uphill electron transport from glucose-6-P to ferredoxin via glucose-6-P dehydrogenase is regulated by $\Delta\psi$. These authors cannot exclude the possibility that other reactions involved in N_2-fixation are controlled by $\Delta\psi$ (e.g., transfer of electrons from the Fe-protein to the Mo-Fe-protein of nitrogenase, binding of Mg^{++} to the membranes to avoid inhibition of $MgADP^-$, transport of metabolites from vegetative cells to heterocysts and vice versa). The transfer of reducing equivalents to nitrogenase may also be controlled by the energy charge or the ATP/ADP ratio in the cells. The experimental findings on this subject are controversial. Bottomley and Stewart [11, 12] and Privalle and Burris [55] did not find evidence for a regulation of nitrogenase activity by the supply with ATP. The opposite can be concluded from the experiments of Ohmori and Hattori [46], and Ernst et al. [19]. They recently found a positive correlation between the energy charge and nitrogenase activity in heterocysts. The latter authors now apparently come to opposite conclusions in the light of new experimental data [18].

Information on the occurrence of enzymes of carbon catabolism in heterocysts have recently been published from this laboratory ([43], for a review see [8]). The findings can be summarized schematically (Fig. 3). Heterocysts do not possess ribulose 1,5 bisphosphate carboxylase and therefore cannot fix CO_2 photosynthetically. They are supplied with fixed carbon from vegetative cells. The compound which is transported is probably a disaccharide which still remains to be identified. It is probably a disaccharide which still remains to be identified. It is probably converted to glucose-6-P which is then mainly degraded via the hexosemonophosphate shunt/oxidative pentosephosphate cycle to generate NADPH. In addition to this, heterocysts possess all the enzymes to convert glucose-6-P to glutamate via glycolysis and the tricarboxylic acid cycle. Among these, the occurrence of the pyruvate:ferredoxin oxidoreductase [43] and the ferredoxin-dependent glutamate synthase [22] is particularly noteworthy, because these enzymes are sensitive to O_2. Heterocysts obviously possess anaerobic compartments where these enzymes and nitrogenase are protected. They contain a NAD^+-dependent glyceraldehyde-3-P dehydrogenase and a rather active NADH-dependent malate dehydrogenase [44]. NADH formed by these enzyme reaction is reoxidized by the membrane-bound NADH: plastoquinone oxidoreductase (see above). Heterocysts do not possess an enzyme catalyzing the cleavage of oxoglutarate indicating that the tricarboxylic acid cycle is incomplete. Oxaloacetate and malate must therefore be formed by anaplerotic reactions. Neuer and Bothe [44] found phosphoenolpyruvate carboxylase and $NADP^+$-dependent malic enzyme in heterocysts of *Nostoc muscorum*. Thus heterocysts can perform CO_2-fixations which are, however, not active enough to meet all the requirements of heterocysts for fixed carbon but function as anaplerotic reactions. In exchange for the disaccharide, heterocysts supply vegetative cells with glutamine as convincingly shown by Wolk's group [63, 70]. These authors suggested that vegetative cells also supply heterocysts with glutamate. This latter statement may no longer be tenable, as all the enzymes necessary for the conversion of glucose-6-P to glutamate are found in heterocysts. At present it is uncertain whether the activities of these enzymes are high enough to catalyze the formation of all the glutamate required by heterocysts.

Heterocyst Differentiation

The structural changes taking place during heterocyst differentiation have been well investigated, but comparatively little is known about the accompanying physiological and biochemical events (see [45]). In almost all cyanobacteria, the availability of combined nitrogen (NO_3^-, NH_4^+ or organic nitrogen) in the medium suppresses nitrogenase biosynthesis and heterocyst formation. The experiments with methionine sulfoximine which specifically blocks glutamine synthetase showed that neither NO_3^- nor NH_4^+ itself represses heterocyst differentiation. In the presence of this inhibitor,

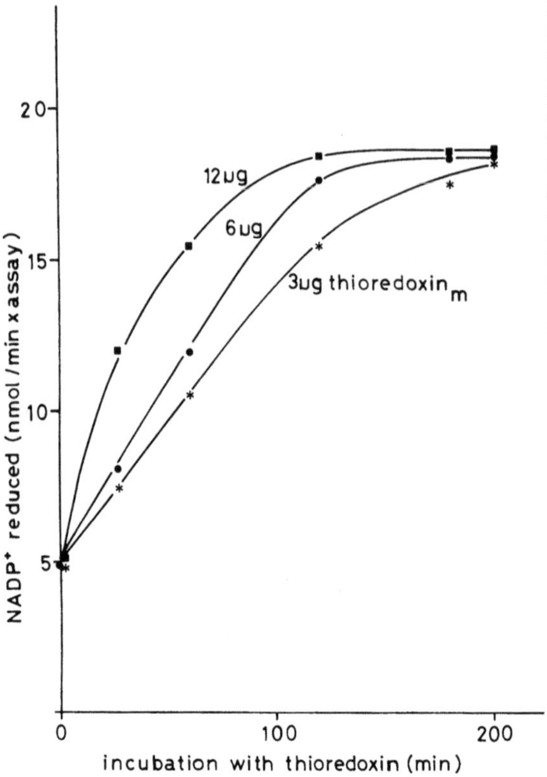

Fig. 4. Time dependence of the activation of isocitrate dehydrogenase by thioredoxin.
Extract (0.9 mg protein) from isolated heterocysts was preincubated with different amounts of thioredoxin$_m$ from *Anabaena cylindrica* as indicated in the figure in a total volume of 3 ml containing in μmol: KH_2PO_4/KOH-buffer pH 7.0, 20; DTT, 30. At the times indicated in the figure, 0.5 ml samples were taken out and injected into the reaction mixture containing in μmol: KH_2PO_4/KOH buffer pH 7.0, 99; $MgCl_2$ x 5 H_2O, 12; $NADP^+$, 0.6 and isocitrate, 5. $NADP^+$-reduction was measured spectrophotometrically at 340 nm

heterocysts are formed and nitrogenase activity is expressed even when the growth medium contains inorganic nitrogen compounds. It was concluded that glutamine synthetase itself or a product of the glutamine synthetase reaction acts as repressor of the genes that code for heterocyst differentiation. The addition of glutamine, glutamate, aspartate or alanine prevented the formation of heterocysts in methionine sulfoximine treated cultures of *Anabaena variabilis* and *Nostoc muscorum* [28, 48]. Ownby [48] did not observe a close correlation between the degree of heterocyst induction and the level of glutamine synthetase activity and suggested that glutamine, or a product formed from glutamine, rather than the enzyme itself, is involved in the regulation of heterocyst differentiation.

Does a low level of glutamine or a low N/C ratio [38] in a vegetative cell initiate the differentiation of a vegetative cell to a heterocyst? This is likely one of the factors contributing to differentiation, but the process is too complex to be explained just by this. It has been shown for *Anabaena catenula* [66, 67] that cell division is asymmetrical in the filament, and that heterocysts develop from a smaller daughter cell arising outside of an inhibitory zone of an existing heterocyst. Haselkorn and coworkers [24, 25] showed by protein analysis using polyacrylamide gel electrophoresis and autoradiography that the transfer of filaments into nitrogen-free medium is followed by the degradation of many existing proteins and the synthesis of many new ones. Two specific proteases, a phycocyanin degrading one and a Ca^{2+}-requiring en-

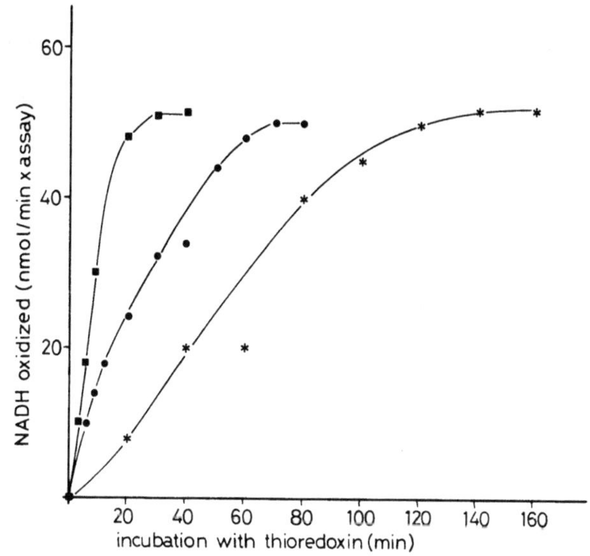

Fig. 5. Time dependence of the activation of NADPH dependent malate dehydrogenase by thioredoxin. Extract (0.9 mg) from isolated heterocysts of *Anabaena cylindrica* was preincubated with different amounts of thioredoxin$_m$ from *Anabaena cylindrica* as indicated in the figure in a final volume of 3 ml containing in μmol: Tris-HCl buffer pH 8.0, 50 and DTT, 30. At the times indicated in the figure, 0.5 ml samples were taken out and injected into the reaction mixture containing in a final volume of 3 ml in μmol: Tris-HCl-buffer pH 8.0, 100; NADPH, 0.9 and oxaloacetate, 10. NADPH oxidation was followed photometrically

zyme, become active within the first few hours of nitrogen depriviation. A vegetative cell too far away from a heterocyst may not be sufficiently supplied with glutamine from this cell, but amino acids are liberated probably in large quantities due to the activity of the proteases. Glutamine, though not sufficiently supplied from heterocysts, may be replenished by proteolysis. The "lack of glutamine" hypotheses can also not easily be reconciled with the fact that glutamine is required in large quantities for the synthesis of new proteins during the onset of heterocyst differentiation. Some investigators, therefore, assume the existence of an additional diffusible inhibitor produced by heterocysts which prevents neighboring cells from differentiating.

New data from our laboratory indicate that thioredoxin regulates the activity of enzymes involved in glutamine biosynthesis and may therefore play a role in the differentiation process. Papen et al. [49] reported that isocitrate dehydrogenase is weakly active in vegetative cells of *Anabaena cylindrica* and *Nostoc muscorum* and that the enzyme from heterocysts and vegetative cells can be activated by reduced thioredoxin. In the assays of that publication [49], the activation process was rapid, where the tests contained the isocitrate dehydrogenase preparation, substrates and dithiothreitol and were started by adding thioredoxin. The time required for maximal activation can better be resolved by preincubating the enzyme with dithiothreitol and thioredoxin and by assaying the enzyme activity upon the addition of the substrates (isocitrate, $NADP^+$) at the times given in the abscissa of the figure (Fig. 4). The activ-

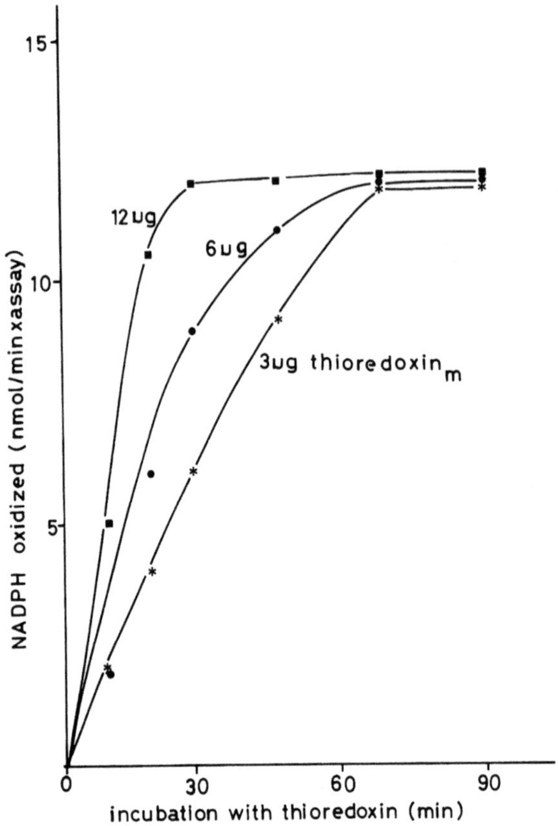

Fig. 6. Time dependence of the activation of glutamine synthetase by thioredoxin
Enzyme (0.15 mg) purified from crude extract of *Anabaena cylindrica* was preincubated with different concentrations of thioredoxin$_m$ from *Anabaena cylindrica* as indicated in the figure in a total volume of 3 ml containing in μmol: Tris-HCl buffer pH 7.0; 50 and DTT (=dithiothreitol), 30. At the times indicated in the figure, 0.5 ml samples were taken out and in injected into the reaction mixture containing lactate dehydrogenase (in glycerol, from Boehringer), 20 units; pyruvate kinase (in glycerol), 8 units and in μmol: Tris-HCl-buffer pH 7.0, 150; ATP, 3.0; NH_4Cl, 200, glutamate, 60; $MgCl_2 \times 5\ H_2O$, 150; KCl, 150; NADH, 0.45 and phosphoenolpyruvate, 0.5. NADH-oxidation was followed photometrically

ation process is dependent on the amount of thioredoxin added and requires about 100 min for maximimal activity at saturating concentrations of thioredoxin in the case of isocitrate dehydrogenase under these assay conditions (Fig. 4). The thioredoxin preparation from *Anabaena cylindrica* used in these experiments was prepared as described for thioredoxin$_m$ which activates NADPH-dependent malate dehydrogenase of chloroplasts [71]. This thioredoxin preparation also activates NADPH-dependent malate dehydrogenase of heterocysts (Fig. 5) and of vegetative cells (not documented) from *Anabaena cylindrica.* In the case of malate dehydrogenase, the activation is completed within about 20 min with saturating concentrations of thioredoxin. Figure 6 shows that thioredoxin also activates glutamine synthetase when the assays are performed at pH 7.0 and when the enzyme is tested by the biosynthetic assay in which the ADP formed is determined. This is achieved by converting ADP with phosphoenolpyruvate to pyruvate and ATP via the pyruvate kinase reaction and by determining the pyruvate formed by the NADH oxidation in the lactate dehydrogenase reaction. Other authors did not find an activation of glutamine synthetase by thioredoxin in cyanobacteria [24, 57]. The experimental discrepancies cannot fully by explained at present. However, the assay conditions, particularly the Mg^{2+}-concentration and the pH value, are critical. Currently we do not understand the full implications of our findings that the activity of enzymes of glutamine synthesis (glutamine synthetase, isocitrate dehydrogenase) is regulated by thioredoxin. However, if glutamine is, indeed, a key factor repressing the differentiation of vegetative cells to heterocysts, then a control of its biosynthesis by thioredoxin makes sense.

Acknowledgment. The authors are indebted to the Deutsche Forschungsgemeinschaft for financial support of their own experiments described or cited in the text.

References

1 Apte SK, Rowell P, Stewart WDP (1978) Proc R Soc Lond B Biol Sci 200:1–25
2 Bergman B (1981) Planta (Berl) 152:302–306
3 Bone DH (1972) Arch Microbiol 86:13–24
4 Bothe H (1969) In: Metzner H (ed and publ) Progress in Photosynthesis Research, vol III Tübingen, pp 1483–1491
5 Bothe H (1982) In: Carr NG, Whitton BA (eds) The biology of cyanobacteria. Blackwell, Oxford, pp 87–104
6 Bothe H, Eisbrenner G (1981) In: Bothe H, Trebst A (eds) The biology of inorganic nitrogen and sulfur. Springer, Berlin Heidelberg New York, pp 141–150
7 Bothe H, Distler E, Eisbrenner G (1978) Biochimie (Paris) 60:277–288
8 Bothe H, Nelles H, Häger KP, Papen H, Neuer G (1984) In: Veeger C, Newton WE (eds) Advances in nitrogen fixation reserach. Nyhoff Publ, The Hague-Pudoc, Wageningen, The Netherlands, pp 199–210
9 Bothe H, Neuer G, Kalbe I, Eisbrenner G (1980) In: Stewart WDP, Gallon JR (eds) Nitrogen Fixation. Academic, London, pp 83–112
10 Bothe H, Tennigkeit J, Eisbrenner G, Yates MG (1977) Planta (Berl) 133:237–242
11 Bottomley PJ, Stewart WDP (1976) Arch Microbiol 108:249–258
12 Bottomley PJ, Stewart WDP (1977) New Phytol 79:625–638
13 Carr NG (1979) In: Paris JH (ed) Developmental biology of prokaryotes. Blackwell, Oxford, pp 167–201
14 Carpenter EJ, Price CC IV (1976) Science N Y 191:1278–1280
15 Eisbrenner G, Bothe H (1979) Arch Microbiol 123:37–45
16 Eisbrenner G, Distler E, Floener L, Bothe H (1978) Arch Microbiol 118:177–184
17 Eisbrenner G, Roos P, Bothe H (1981) J Gen Microbiol 125:383–390
18 Ernst A, Almon H, Böhme H, Böger P (1983a) In: Abstract book of the 5th International Symposium on Nitrogen Fixation, Noordwijkerhout, The Netherlands, Abstract 4A–28
19 Ernst A, Böhme H, Böger P (1983b) Biochim Biophys Acta 723:83–90
20 Gallon JR (1981) TIBS 6:19–23
21 Godde D (1982) Arch Microbiol 131:197–202
22 Häger KP, Danneberg G, Bothe H (1983) FEMS Microbiol Lett 17:179–183
23 Halliwell B, Foyer CH, Charles SA (1981) In: Akoyunoglou G (ed) Proceedings of the V. International Photosynthesis Congress, vol II. Balaban Internat Sciences Service, Philadelphia, PA, pp 279–283
24 Haselkorn R (1978) Annu Rev Plant Physiol 29:319–344
25 Haselkorn R, Mazur B, Orr J, Rice D, Wood N, Rippka R (1980) In: Newton WE, Orme-Johnson WH (eds) Nitrogen Fixation, vol II. University Park Press, Baltimore, pp 259–278
26 Haury JF, Wolk CP (1978) J Bacteriol 136:688–692
27 Houchins JP, Burris RH (1981a) J Bacteriol 146:209–214
28 Houchins JP, Burris RH (1981b) J Bacteriol 146:215–221
29 Houchins JP, Burris RH (1981c) Plant Physiol 68:717–721
30 Houchins JP, Hind G (1982) Biochim Biophys Acta 682:86–96

31 Hawkesford MJ, Reed RH, Rowell P, Stewart WDP (1981) Eur J Biochem 115:519–523
32 Hawkesford MJ, Rowell P, Stewart WDP (1983) In: Papageorgiou GC, Packer L (eds) Photosynthetic prokaryotes; cell-differentiation and function. Elsevier Biomedical, New York, pp 195–218
33 Janaki S, Wolk CP (1982) Biochim Biophys Acta 696:187–192
34 Kallas T, Rippka R, Coursin T, Rebiere MC, Tandeau de Marsac N, Cohen-Bazire G (1983a) In: Papageorgiou GC, Packer L (eds) Photosynthetic prokaryotes: cell-differentiation and function. Elsevier Biomedical, New York, pp 281–302
35 Kallas T, Rippka R, Cohen-Bazire G (1983b) In: Abstract book of the 5th International Symposium on Nitrogen Fixation. Noordwijkerhout, The Netherlands, Abstract 4A–11
36 Karni L, Tel-Or E (1982) In: Abstract book of the IV. International Symposium on Photosynthetic Prokaryotes. Bombannes, France, Abstract B 55
37 Karni L, Tel-Or E (1983) In: Abstract book of the 5th International Symposium on Nitrogen Fixation. Noordwijkerhout, Abstract 4A–22
38 Kulasooriya SA, Lang NJ, Fay P (1972) Proc R Soc Lond B Biol Sci 181:199–209
39 Kumar A, Tabita FR, van Baalen C (1982) Arch Microbiol 133:103–109
40 Lockau W (1981) Arch Microbiol 128:336–340
41 Mullineaux PM, Chaplin AE, Gallon JR (1983) J Gen Microbiol 129:1689–1696
42 Murry MA, Fontaine MS, Torrey JG (1983) In: Abstract book of the 5th International Symposium on Nitrogen Fixation. Noordwijkerhout, The Netherlands, Abstract 4A–23
43 Neuer G, Bothe H (1982) Biochim Biophys Acta 716:358–365
44 Neuer G, Bothe H (1983) FEBS Lett 158:79–83
45 Neuer G, Papen H, Bothe H (1983) In: Papageorgiou GC, Packer L (eds) Cell-differentiation and function. Elsevier Biomedical, New York, pp 219–242
46 Ohmori M, Hattori A (1978) Arch Microbiol 117:17–20
47 Ohshima T, Drews G (1981) Z Naturforsch 36c:400–406
48 Ownby J (1977) Planta (Berl) 136:277–279
49 Papen H, Neuer G, Bothe H (1983) Arch Microbiol 134:73–79
50 Pearson HW, Howley R, Kjeldsen CK, Walsby AE (1979) FEMS Microbiol Lett 5:163–167
51 Peschek GA (1979a) Biochim Biophys Acta 548:187–202
52 Peschek GA (1979b) Biochim Biophys Acta 548:203–215
53 Peschek GA (1982) Naturwissenschaften 69:599–600
54 Peschek GA, Muchl R, Kienzl PF, Schmetterer G (1982) Biochim Biophys Acta 679:35–43
55 Privalle LS, Burris RH (1983) J Bacteriol 154:351–355
56 Rippka R, Stanier RY (1978) J Gen Microbiol 105:83–94
57 Rowell P, Cossar JD, Ip SM, Stewart WDP (1983) In: Abstract book of the 5th International Symposium on Nitrogen Fixation. Noordwijkerhout, The Netherlands, Abstract 4A–29
58 Singh PK (1973) Arch Microbiol 92:59–62
59 Spiller H, Brookjans G, Shanmugam KT (1983) J Bacteriol 155:129–137
60 Stewart WDP (1980) Annu Rev Microbiol 34:497–536
61 Stewart WDP, Lex M (1970) Arch Mikrobiol 73:250–260
62 Tel-Or E, Stewart WDP (1977) Proc R Soc Lond B Biol Sci 198:61–86
63 Thomas J, Meeks JC, Wolk CP, Shaffer PW, Austin SM, Chien WS (1977) J Bacteriol 129: 1545–1555
64 Walsby AE (1982a) J Gen Microbiol 128:1679–1684
65 Walsby AE (1982b) In: Abstract book of the IV. International Symposium on Photosynthetic Prokaryotes. Bombannes, France, Abstract C–33
66 Wilcox M, Mitchinson GJ, Smith RJ (1973a) J Cell Sci 12:707–723
67 Wilcox M, Mitchinson GJ, Smith RJ (1973b) J Cell Sci 13:637–649
68 Wolk CP (1979) In: Nichols JM (ed) Abstract book of the III. International Symposium on Photosynthetic Prokaryotes. The University of Liverpool. England, Abstract D1
69 Wolk CP (1982) In: Carr NG, Whitton BA (eds) The biology of cyanobacteria. Blackwell, Oxford, pp 359–385
70 Wolk CP, Thomas J, Shaffer PW, Austin SM, Chien WS (1976) J Biol Chem 251:5027–5034
71 Wolosiuk RA, Schürmann P, Buchanan BB (1980) Methods Enzym 69:382–391
72 Wyatt JT, Silvey JKG (1969) Science N Y 165:908–909

Taxonomic Index

Subject Index

The Prokaryotes

A Handbook on Habitats, Isolation, and Identification of Bacteria

Editors: **M. P. Starr, H. Stolp, H. G. Trüper, A. Balows, H. G. Schlegel**

1981. XLVIII, 2596 pages. (In two parts, not available separately). ISBN 3-540-08871-7

Contents: Introductory Essays. - The Phototrophic Prokaryotes. - The Gliding Bacteria. - The Sheathed Bacteria. Some Gas-Vacuolate Bacteria. - The Budding and/or Appendaged Bacteria. - The Spirochetes. - Some Helical and Curved Bacteria. - The Pseudomonads and Relatives. - Some Dinitrogen-Fixing Bacteria and Relatives. - Taxonomically Heterogeneous, Physiologically Defined Assemblages of Bacterial Genera. - The Obligately Chemolithotrophic Bacteria. - Small, Aerobic, Gram-Negative Bacteria of Medical Importance. - The Families Enterobacteriaceae and Vibrionaceae. Some Marine Bacteria. - Miscellaneous, Facultatively Anaerobic, Gram-Negative, Rod-Shaped Bacteria. - Anaerobic, Gram-Negative, Rod-Shaped Bacteria. - Gram-Negative Cocci and Related Rod-Shaped Bacteria. - Gram-Positive Cocci. - Gram-Positive, Asporogenous, Rod-Shaped Bacteria. - The Endospore-Forming Bacteria. - The Coryneform Bacteria. - The Actinomycetes. - Some Obligately Symbiotic Bacteria. - Wall-Deficient Bacteria. - Author Index. - Subject Index.

The Prokaryotes is destined to become the indispensable reference for microbiologists. Prepared under the supervision of a distinguished international editorial board, this comprehensive two volume Handbook will contain 169 contributions by acknowledged experts providing up-to-date, detailed information on the habitats, isolation, and identification of all the prokaryote groups.
This richly illustrated Handbook starts with six authoritative general essays covering diversity and habitats of prokaryotes, pathogenicity in man, animals, and plants, and principles of isolation, cultivation, conservation, characterization, and identification. Each of the following major sections on the prokaryote groups is also introduced by a general essay. Detailed procedures, many never before published, are presented in one convenient source.
The Prokaryotes contains a complete set of indices providing immediate access to the media, procedures, names of taxa, general topics covered, authors of cited literature, and contributions to the Handbook.
Every researcher working with these organisms will find The Prokaryotes an invaluable and essential source of information.

Springer-Verlag
Berlin
Heidelberg
New York
Tokyo

MIX
Papier aus verantwortungsvollen Quellen
Paper from responsible sources
FSC® C105338

If you have any concerns about our products,
you can contact us on
ProductSafety@springernature.com

In case Publisher is established outside the EU,
the EU authorized representative is:
Springer Nature Customer Service Center GmbH
Europaplatz 3, 69115 Heidelberg, Germany

Printed by Libri Plureos GmbH
in Hamburg, Germany